Mathematical Olympiads
1999–2000

Problems and Solutions
From Around the World

© 2002 by
The Mathematical Association of America (Incorporated)
Library of Congress Catalog Card Number 2001092456
ISBN 0-88385-805-3

Printed in the United States of America

Current Printing (last digit):
10 9 8 7 6 5 4 3 2 1

Mathematical Olympiads
1999–2000

Problems and Solutions
From Around the World

Edited by
Titu Andreescu
and
Zuming Feng

Published and distributed by
The Mathematical Association of America

PROBLEM BOOKS SERIES

Problem Books is a series of the Mathematical Association of America consisting of collections of problems and solutions from annual mathematical competitions; compilations of problems (including unsolved problems) specific to particular branches of mathematics; books on the art and practice of problem solving, etc.

Committee on Publications
Gerald L. Alexanderson, *Chair*

Problem Books Series Editorial Board
Roger Nelsen *Editor*

Irl Bivens	Clayton Dodge
Richard Gibbs	George Gilbert
Art Grainger	Gerald Heuer
Elgin Johnston	Kiran Kedlaya
Loren Larson	Margaret Robinson

Mathematical Olympiads 1998–1999: Problems and Solutions From Around the World, edited by Titu Andreescu and Zuming Feng

Mathematical Olympiads 1999–2000: Problems and Solutions From Around the World, edited by Titu Andreescu and Zuming Feng

USA and International Mathematical Olympiads 2000, edited by Titu Andreescu and Zuming Feng

The Inquisitive Problem Solver, Paul Vaderlind, Loren Larson, and Richard K. Guy

MAA Service Center
P. O. Box 91112
Washington, DC 20090-1112
1-800-331-1622   fax: 1-301-206-9789

Contents

Preface ix

Acknowledgments xi

1 1999 National Contests: Problems and Solutions   **1**
    1   Belarus   1
    2   Brazil   26
    3   Bulgaria   30
    4   Canada   40
    5   China   43
    6   Czech and Slovak Republics   49
    7   France   55
    8   Hong Kong (China)   61
    9   Hungary   64
   10   Iran   73
   11   Ireland   88
   12   Italy   95
   13   Japan   101
   14   Korea   106
   15   Poland   111
   16   Romania   116
   17   Russia   137
   18   Slovenia   170
   19   Taiwan   172
   20   Turkey   176
   21   Ukraine   183
   22   United Kingdom   185

23	United States of America	191
24	Vietnam	197

2 1999 Regional Contests: Problems and Solutions — 209

1	Asian Pacific Mathematical Olympiad	209
2	Austrian-Polish Mathematics Competition	214
3	Balkan Mathematical Olympiad	221
4	Czech and Slovak Match	225
5	Hungary-Israel Binational Mathematical Competition	230
6	Iberoamerican Mathematical Olympiad	239
7	Olimpiada Matemática del Cono Sur	243
8	St. Petersburg City Mathematical Olympiad (Russia)	246

3 2000 National Contests: Problems — 261

1	Belarus	261
2	Bulgaria	263
3	Canada	265
4	China	266
5	Czech and Slovak Republics	268
6	Estonia	269
7	Hungary	270
8	India	271
9	Iran	272
10	Israel	274
11	Italy	275
12	Japan	276
13	Korea	277
14	Mongolia	278
15	Poland	279
16	Romania	280
17	Russia	282
18	Taiwan	289
19	Turkey	290
20	United Kingdom	292
21	United States of America	293
22	Vietnam	294

4 2000 Regional Contests: Problems — 295

1	Asian Pacific Mathematical Olympiad	295
2	Austrian-Polish Mathematics Competition	297

Contents

3	Balkan Mathematical Olympiad	298
4	Mediterranean Mathematical Competition	299
5	St. Petersburg City Mathematical Olympiad (Russia) . . .	300

Glossary   **305**

Classification of Problems   **317**

Preface

This book is a continuation of *Mathematical Contests 1998–1999: Olympiad Problems and Solutions from around the World*, published by the American Mathematics Competitions. It contains solutions to the problems from 32 national and regional contests featured in the earlier book, together with selected problems (without solutions) from national and regional contests given during 2000. In many cases multiple solutions are provided in order to encourage students to compare different problem-solving strategies.

This collection is intended as practice for the serious student who wishes to improve his or her performance on the USAMO. Some of the problems are comparable to the USAMO in that they came from national contests. Others are harder, as some countries first have a national Olympiad, and later one or more exams to select a team for the IMO. And some problems come from regional international contests ("mini-IMOs").

Different nations have different mathematical cultures, so you will find some of these problems extremely hard and some rather easy. We have tried to present a wide variety of problems, especially from those countries that have often done well at the IMO.

Each contest has its own time limit. We have not furnished this information, because we have not always included complete exams. As a rule of thumb, most contests allow time ranging between one-half to one full hour per problem.

The problems themselves should provide much enjoyment for all those fascinated by solving challenging mathematics questions.

Acknowledgments

Thanks to George Lee and Kiran Kedlaya for the help they provided in ways too numerous to mention, and to the following students of the Mathematical Olympiad Summer Program who helped in preparing and proofreading solutions: Reid Barton, Gabriel Carroll, Luke Gustafson, Dani Kane, Ian Le, Zhihao Liu, Ricky Liu, Tiankai Liu, Po-Ru Loh, Alison Miller, Oaz Nir, David Shin, Tongke Xue, and Yan Zhang. In addition, thanks to Jerry Heuer and Elgin Johnston for proofreading the text.

Without their efforts this work would not have been possible.

Titu Andreescu   Zuming Feng

1999 National Contests: Problems and Solutions

1 Belarus

National Olympiad, Fourth Round

Problem 10.1 Determine all real numbers a such that the function $f(x) = \{ax + \sin x\}$ is periodic. Here, $\{y\}$ denotes the fractional part of y.

Solution. The solutions are $a = \frac{r}{\pi}$, $r \in \mathbb{Q}$.

First, suppose that $a = \frac{r}{\pi}$ for some $r \in \mathbb{Q}$. Write $r = \frac{s}{t}$ with $s, t \in \mathbb{Z}$ and $t > 0$. Then

$$f(x + 2t\pi) = \left\{ \frac{s}{t\pi}(x + 2t\pi) + \sin(x + 2t\pi) \right\}$$
$$= \left\{ \frac{s}{t\pi} x + 2s + \sin x \right\} = \left\{ \frac{s}{t\pi} x + \sin x \right\} = f(x).$$

Hence, f is periodic with period $2t\pi$.

On the other hand, suppose that f is periodic — i.e., there exists $p > 0$ such that $f(x) = f(x + p)$ for all $x \in \mathbb{R}$. Then for all $x \in \mathbb{R}$, $\{ax + \sin x\} = \{ax + ap + \sin(x+p)\}$ and hence $g(x) = ap + \sin(x+p) - \sin x$ is an integer. Because g is continuous, there exists a single integer k such that $g(x) = k$ for all $x \in \mathbb{R}$. For any $y \in \mathbb{R}$ and $n \in \mathbb{N}$, setting $x = y$, $y + p$, $y + 2p$, ..., $y + (n-1)p$ in the equation $g(x) = k$ and summing the resulting n equations yields

$$\sin(y + np) - \sin y = n(k - ap).$$

Because the absolute value of the left-hand side of this equation is bounded by 2, we conclude that $k = ap$ and $\sin(x + p) = \sin x$ for all $x \in \mathbb{R}$. In particular, $\sin\left(\frac{\pi}{2}+p\right) = \sin\left(\frac{\pi}{2}\right) = 1$. Hence, $p = 2m\pi$ for some $m \in \mathbb{N}$. Thus $a = \frac{k}{p} = \frac{k}{2m\pi} = \frac{r}{\pi}$ with $r = \frac{k}{2m} \in \mathbb{Q}$, as desired.

Problem 10.2 Prove that for any integer $n > 1$ the sum S of all divisors of n (including 1 and n) satisfies the inequalities
$$k\sqrt{n} < S < \sqrt{2kn},$$
where k is the number of divisors of n.

Solution. Let the divisors of n be $1 = d_1 < d_2 < \cdots < d_k = n$, so that $d_i d_{k+1-i} = n$ for each i. Then
$$S = \sum_{i=1}^{k} d_i = \sum_{i=1}^{k} \frac{d_i + d_{k+1-i}}{2} > \sum_{i=1}^{k} \sqrt{d_i d_{k+1-i}} = k\sqrt{n},$$
where the inequality is strict because equality does not hold for $\frac{d_1+d_k}{2} \geq \sqrt{d_1 d_k}$.

For the right inequality, let $S_2 = \sum_{i=1}^{k} d_i^2$. By the Power Mean Inequality,
$$\frac{S}{k} = \frac{\sum_{i=1}^{k} d_i}{k} \leq \sqrt{\frac{\sum_{i=1}^{k} d_i^2}{k}} = \sqrt{\frac{S_2}{k}}.$$
Hence, $S \leq \sqrt{kS_2}$. Now
$$\frac{S_2}{n^2} = \sum_{i=1}^{k} \frac{d_i^2}{n^2} = \sum_{i=1}^{k} \frac{1}{d_{k+1-i}^2} \leq \sum_{j=1}^{n} \frac{1}{j^2} < \frac{\pi^2}{6}$$
because $d_1, \ldots, d_k$ are distinct integers between 1 and n. Therefore,
$$S \leq \sqrt{kS_2} < \sqrt{\frac{kn^2\pi^2}{6}} < \sqrt{2kn}.$$

Problem 10.3 There is a 7×7 square board divided into 49 unit cells, and tiles of three types: 3×1 rectangles, 3-unit-square corners, and unit squares. Jerry has infinitely many rectangles and one corner, while Tom has only one square.

(a) Prove that Tom can put his square somewhere on the board (covering exactly one unit cell) in such a way that Jerry can not tile the rest of the board with his tiles.

(b) Now Jerry is given another corner. Prove that no matter where Tom puts his square (covering exactly one unit cell), Jerry can tile the rest of the board with his tiles.

Solution.

(a) Tom should place his square on the cell marked X in the boards below.

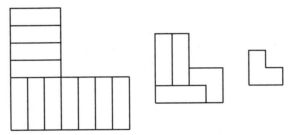

The grid on the left contains 17 1's, 15 2's and 16 3's. Because every 3×1 rectangle contains a 1, a 2 and a 3, Jerry's corner must cover a 3 and two 1's. Thus his corner must be oriented like a Γ. Every such corner covers a 1, a 2 and a 3 in the right grid, as does any 3×1 rectangle. Because the right grid also contains 17 1's, 15 2's and 16 3's, Jerry cannot cover the 48 remaining squares with his pieces.

(b) Consider the following figures.

The first figure can be rotated and placed on the 7×7 board so that Tom's square falls into its blank, untiled region. Similarly, the second figure can be rotated and placed within the remaining untiled 4×4 region so that Tom's square is still uncovered. Finally, the single corner can be rotated and placed without overlapping Tom's square.

Problem 10.4 A circle is inscribed in the isosceles trapezoid $ABCD$. Let the circle meet diagonal AC at K and L (with K between A and L). Find the value of
$$\frac{AL \cdot KC}{AK \cdot LC}.$$

First Solution.

Lemma. *Suppose that a (not necessarily isosceles) trapezoid $ABCD$ is circumscribed about a circle with radius r, where the circle touches* sides AB, BC, CD, DA at points P, Q, R, S, respectively. Let line AC intersect the circle at K and L, with K between A and L, and write $m = AP$ and $n = CR$. Then

$$AK \cdot LC = mn + 2r^2 - \sqrt{(mn + 2r^2)^2 - (mn)^2}$$

and

$$AL \cdot KC = mn + 2r^2 + \sqrt{(mn + 2r^2)^2 - (mn)^2}.$$

Proof. Assume, without loss of generality, that $AB \parallel CD$, and orient the trapezoid so that lines AB and CD are horizontal. Let $t = AK$, $u = KL$, and $v = LC$. Also let $\sigma = t + v$ and $\pi = tv$. By the power of a point theorem, $t(t + u) = m^2$ and $v(v + u) = n^2$. Multiplying these two equations yields $\pi(\pi + u\sigma + u^2) = m^2 n^2$. Also, A and C are separated by horizontal distance $m + n$ and vertical distance $2r$. Thus, $AC^2 = (m + n)^2 + (2r)^2$, and

$$(m+n)^2 + (2r)^2 = AC^2 = (t + u + v)^2$$
$$m^2 + 2mn + n^2 + 4r^2 = t(t + u) + v(v + u) + 2\pi + u\sigma + u^2$$
$$m^2 + 2mn + n^2 + 4r^2 = m^2 + n^2 + 2\pi + u\sigma + u^2$$
$$2mn + 4r^2 - \pi = \pi + u\sigma + u^2.$$

Multiplying both sides by π yields

$$\pi(2mn + 4r^2 - \pi) = \pi(\pi + u\sigma + u^2) = (mn)^2,$$

a quadratic in π with solutions

$$\pi = mn + 2r^2 \pm \sqrt{(mn + 2r^2)^2 - (mn)^2}.$$

Now, $m^2 n^2 = t(t + u)v(v + u) \geq t^2 v^2$, implying that $mn \geq \pi$. Therefore, $AK \cdot LC = \pi = mn + 2r^2 - \sqrt{(mn + 2r^2)^2 - (mn)^2}$. Also, $(AK \cdot AL) \cdot (CK \cdot CL) = m^2 \cdot n^2$. Hence, $AL \cdot KC = \frac{m^2 n^2}{\pi} = mn + 2r^2 + \sqrt{(mn + 2r^2)^2 - (mn)^2}$. ∎

As in the lemma, assume that $AB \parallel CD$ and let the given circle be tangent to sides AB, BC, CD, DA at points P, Q, R, S, respectively. Also define $m = AP = PB = AS = BQ$ and $n = DR = RC = DS = CQ$.

Drop perpendicular $\overline{AX}$ to line CD. Then $AD = m + n$, $DX = |m - n|$, and $AX = 2r$. Then by the Pythagorean Theorem on triangle ADX, we have $(m + n)^2 = (m - n)^2 + (2r)^2$, implying that $mn = r^2$.

Using the lemma, we find that $AK \cdot LC = (3-2\sqrt{2})r^2$ and $AL \cdot KC = (3+2\sqrt{2})r^2$. Thus, $\frac{AL \cdot KC}{AK \cdot LC} = 17 + 12\sqrt{2}$.

Second Solution. Suppose $A'B'C'D'$ is a square with side length s, and define K', L' analogously to K and L. Then $A'C' = s\sqrt{2}$, $K'L' = s$, $A'L' = K'C' = s\frac{\sqrt{2}+1}{2}$, and $A'K' = L'C' = s\frac{\sqrt{2}-1}{2}$. Thus,

$$\frac{A'L' \cdot K'C'}{A'K' \cdot L'C'} = \frac{(\sqrt{2}+1)^2}{(\sqrt{2}-1)^2} = (\sqrt{2}+1)^4 = 17 + 12\sqrt{2}.$$

Consider an arbitrary isosceles trapezoid $ABCD$ with inscribed circle ω, and assume that $AB \parallel CD$. Because no three of A, B, C, D are collinear, there is a projective transformation τ taking $ABCD$ to a parallelogram $A'B'C'D'$. This map takes ω to a conic ω' tangent to the four sides of $A'B'C'D'$. Let $P = BC \cap AD$, and let ℓ be the line through P parallel to line AB; then τ maps ℓ to the line at infinity. Because ω does not intersect ℓ, ω' is an ellipse. Thus, by composing τ with an affine transformation (which will map parallelograms to parallelograms) we may assume that ω' is a circle.

Let W, X, Y, Z be the tangency points of ω to sides AB, BC, CD, DA respectively, and W', X', Y', Z' their images under τ. By symmetry, line WY passes through the intersection of lines BC and AD, and line XZ is parallel to lines AB and CD. Thus, $W'Y' \parallel B'C' \parallel A'D'$ and $X'Z' \parallel A'B' \parallel C'D'$. Because ω' is tangent to the parallel lines $A'B'$ and $C'D'$ at W' and Y', $\overline{W'Y'}$ is a diameter of ω' and $W'Y' \perp A'B'$. Hence, $B'C' \perp A'B'$ and $A'B'C'D'$ is a rectangle. It follows that $A'B'C'D'$ must be a *square* because it has an inscribed circle.

Thus, we are in the case considered at the beginning of the problem. If K' and L' are the intersections of line $A'C'$ with ω', with K' between A' and L', then $\frac{A'L' \cdot K'C'}{A'K' \cdot L'C'} = 17 + 12\sqrt{2}$. Now τ maps $\{K, L\} = AC \cap \omega$ to $\{K', L'\} = A'C' \cap \omega'$ (but perhaps not in that order). If $\tau(K) = K'$ and $\tau(L) = L'$, then because projective transformations preserve cross-ratios, we would have

$$\frac{AL \cdot KC}{AK \cdot LC} = \frac{A'L' \cdot K'C'}{A'K' \cdot L'C'} = 17 + 12\sqrt{2}.$$

If instead $\tau(K) = L'$ and $\tau(L) = K'$, then we would obtain $\frac{AL \cdot KC}{AK \cdot LC} = \frac{1}{17+12\sqrt{2}} < 1$, which is impossible because $AL > AK$ and $KC > LC$. It follows that $\frac{AL \cdot KC}{AK \cdot LC} = 17 + 12\sqrt{2}$, as desired.

Problem 10.5 Let P and Q be points on the side AB of the triangle ABC (with P between A and Q) such that $\angle ACP = \angle PCQ = \angle QCB$, and let

$\overline{AD}$ be the angle bisector of $\angle BAC$. Line AD meets lines CP and CQ at M and N respectively. Given that $PN = CD$ and $3\angle BAC = 2\angle BCA$, prove that triangles CQD and QNB have the same area.

Solution. Because $3\angle BAC = 2\angle ACB$,

$$\angle PAN = \angle NAC = \angle ACP = \angle PCQ = \angle QCD.$$

Let θ equal this common angle measure. Thus $ACNP$ and $ACDQ$ are cyclic quadrilaterals, so

$$\theta = \angle ANP = \angle CQD = \angle CPN.$$

From angle-angle-side congruency we deduce that triangles NAP, CQD, and PCN are congruent. Hence $CP = CQ$, and by symmetry we have $AP = QB$. Thus, $[CQD] = [NAP] = [NQB]$.

Problem 10.6 Show that the equation

$$\{x^3\} + \{y^3\} = \{z^3\}$$

has infinitely many rational non-integer solutions. Here, $\{a\}$ denotes the fractional part of a.

Solution. Let $x = \frac{3}{5}(125k+1)$, $y = \frac{4}{5}(125k+1)$, $z = \frac{6}{5}(125k+1)$ for any integer k. These are never integers because 5 does not divide $125k+1$. Moreover,

$$125x^3 = 3^3(125k+1)^3 \equiv 3^3 \pmod{125}.$$

Hence, 125 divides $125x^3 - 3^3$ and $x^3 - \left(\frac{3}{5}\right)^3$ is an integer. Thus, $\{x^3\} = \frac{27}{125}$. Similarly, $\{y^3\} = \frac{64}{125}$ and $\{z^3\} = \frac{216}{125} - 1 = \frac{91}{125} = \frac{27}{125} + \frac{64}{125}$, and therefore $\{x^3\} + \{y^3\} = \{z^3\}$.

Problem 10.7 Find all integers n and real numbers m such that the squares of an $n \times n$ board can be labelled 1, 2, ..., n^2 with each number appearing exactly once in such a way that

$$(m-1)a_{ij} \le (i+j)^2 - (i+j) \le ma_{ij}$$

for all $1 \le i, j \le n$, where a_{ij} is the number placed in the intersection of the i-th row and j-th column.

Solution. Either $n = 1$ and $2 \le m \le 3$ or $n = 2$ and $m = 3$. It is easy to check that these work using the constructions below.

Now suppose we are given a labelling of the squares $\{a_{ij}\}$ which satisfies the given conditions. By assumption, $a_{11} \ge 1$, so

$$m - 1 \le (m-1)a_{11} \le (1+1)^2 - (1+1) = 2$$

and $m \le 3$. On the other hand, $a_{nn} \le n^2$, so

$$4n^2 - 2n = (n+n)^2 - (n+n) \le ma_{nn} \le mn^2$$

and

$$m \ge \frac{4n^2 - 2n}{n^2} = 4 - \frac{2}{n}.$$

Thus, $4 - \frac{2}{n} \le m \le 3$, which implies our initial claim.

Problem 11.1 Evaluate the product

$$\prod_{k=0}^{2^{1999}} \left(4 \sin^2 \frac{k\pi}{2^{2000}} - 3\right).$$

Solution. For all real x, let $f(x) = \sin\left(\frac{x\pi}{2^{2000}}\right)$.

When $k = 0$, the expression inside the parentheses equals -3. Using the triple-angle formula $\sin(3\theta) = 4\sin^3\theta - 3\sin\theta$, and noting that $f(k) \ne 0$ when $1 \le k \le 2^{1999}$, we can rewrite the given product as

$$-3 \prod_{k=1}^{2^{1999}} \frac{\sin\left(\frac{3k\pi}{2^{2000}}\right)}{\sin\left(\frac{k\pi}{2^{2000}}\right)} \quad \text{or} \quad -3 \prod_{k=1}^{2^{1999}} \frac{f(3k)}{f(k)}. \qquad (1)$$

Observe that

$$\prod_{k=1}^{2^{1999}} f(3k) = \prod_{k=1}^{\frac{2^{1999}-2}{3}} f(3k) \cdot \prod_{k=\frac{2^{1999}+1}{3}}^{\frac{2^{2000}-1}{3}} f(3k) \cdot \prod_{k=\frac{2^{2000}+2}{3}}^{2^{1999}} f(3k).$$

Because $\sin\theta = \sin(\pi - \theta) = -\sin(\pi + \theta)$, we have

$$f(x) = f(2^{2000} - x) = -f(x - 2^{2000}).$$

Hence, letting $S_i = \{k \mid 1 \le k \le 2^{1999}, k \equiv i \pmod 3\}$ for $i = 0, 1, 2$, the last expression equals

$$\prod_{k=1}^{\frac{2^{1999}-2}{3}} f(3k) \cdot \prod_{k=\frac{2^{1999}+1}{3}}^{\frac{2^{2000}-1}{3}} f(2^{2000} - 3k) \cdot \prod_{k=\frac{2^{2000}+2}{3}}^{2^{1999}} \left(-f(3k - 2^{2000})\right)$$

$$= \prod_{k \in S_0} f(k) \cdot \prod_{k \in S_1} f(k) \cdot \prod_{k \in S_2} \left(-f(k)\right)$$

$$= (-1)^{\frac{2^{1999}+1}{3}} \prod_{k=1}^{2^{1999}} f(k) = -\prod_{k=1}^{2^{1999}} f(k).$$

Combined with the expression in (1), this implies that the desired product is $(-3)(-1) = 3$.

Problem 11.2 Let m and n be positive integers. Starting with the list $1, 2, 3, \ldots$, we can form a new list of positive integers in two different ways.

(i) We first erase every m-th number in the list (always starting with the first); then, in the list obtained, we erase every n-th number. We call this *the first derived list*.

(ii) We first erase every n-th number in the list; then, in the list obtained, we erase every m-th number. We call this *the second derived list*.

Now, we call a pair (m, n) *good* if and only if the following statement is true: if some positive integer k appears in both derived lists, then it appears in the same position in each.

(a) Prove that $(2, n)$ is good for any positive integer n.

(b) Determine if there exists any good pair (m, n) such that $2 < m < n$.

Solution. Consider whether some positive integer j is in the first derived list. If it is congruent to $1 \pmod m$, then $j + mn$ is as well so they are both erased. If not, then suppose it is the t-th number remaining after we've erased all the integers congruent to $1 \pmod m$. There are n such numbers erased between j and $j + mn$, so $j + mn$ is the $(t + mn - n)$-th number remaining. Note that either t and $t + mn - n$ are both congruent to $1 \pmod n$ or both *not* congruent to $1 \pmod n$. Hence j is erased after our second pass if and only if $j + mn$ is as well.

A similar argument applies to the second derived list. Thus in either derived list, the locations of the erased numbers repeat with period mn. Also, among each mn consecutive numbers exactly $mn-(m+n-1)$ remain. (In the first list, $n+(\lfloor \frac{mn-n-1}{n} \rfloor +1) = n+(m-1+\lfloor \frac{-1}{n} \rfloor +1) = m+n-1$ of the first mn numbers are erased; similarly, $m+n-1$ of the first mn numbers are erased in the second list.)

These facts imply that the pair (m, n) is good if and only if when any $k \leq mn$ is in both lists, it appears at the same position.

(a) Given a pair $(2, n)$, the first derived list (up to $k = 2n$) is $4, 6, 8, \ldots, 2n$. If n is even, the second derived list is $3, 5, \ldots, n-1, n+2, n+4, \ldots, 2n$. If n is odd, the second derived list is $3, 5, \ldots, n-2, n, n+3, n+5, \ldots, 2n$. In either case, the first and second lists' common elements are the even numbers between $n+2$ and $2n$ inclusive. Each such $2n-i$ (with $i < \frac{n-1}{2}$) is the $(n-1-i)$-th number on both lists, showing that $(2, n)$ is good.

(b) Such a pair exists — in fact, the simplest possible pair $(m, n) = (3, 4)$ suffices. The first derived list (up to $k = 12$) is $3, 5, 6, 9, 11, 12$ and the second derived list is $3, 4, 7, 8, 11, 12$. The common elements are $3, 11, 12$, and these are all in the same positions.

Problem 11.3 Let $a_1, a_2, \ldots, a_{100}$ be an ordered set of numbers. At each move it is allowed to choose any two numbers a_n, a_m and change them to the numbers

$$\frac{a_n^2}{a_m} - \frac{n}{m}\left(\frac{a_m^2}{a_n} - a_m\right) \quad \text{and} \quad \frac{a_m^2}{a_n} - \frac{m}{n}\left(\frac{a_n^2}{a_m} - a_n\right)$$

respectively. Determine if it is possible, starting with the set with $a_i = \frac{1}{5}$ for $i = 20, 40, 60, 80, 100$ and $a_i = 1$ otherwise, to obtain a set consisting of integers only.

Solution. Suppose we change a_n to $a'_n = \frac{a_n^2}{a_m} - \frac{n}{m}\left(\frac{a_m^2}{a_n} - a_m\right)$ and a_m to $a'_m = \frac{a_m^2}{a_n} - \frac{m}{n}\left(\frac{a_n^2}{a_m} - a_n\right)$. Observe that $\frac{a'_n}{n} + \frac{a'_m}{m}$ equals

$$\left[\left(\frac{1}{n} \cdot \frac{a_n^2}{a_m} - \frac{1}{m} \cdot \frac{a_m^2}{a_n}\right) + \frac{a_m}{m}\right] + \left[\left(\frac{1}{m} \cdot \frac{a_m^2}{a_n} - \frac{1}{n} \cdot \frac{a_n^2}{a_m}\right) + \frac{a_n}{n}\right],$$

which equals $\frac{a_n}{n} + \frac{a_m}{m}$. Thus, the quantity $\sum_{i=1}^{100} \frac{a_i}{i}$ is invariant under the given operation. Originally, this sum equals

$$I_1 = \sum_{i=1}^{99} \frac{a_i}{i} + \frac{1}{500}.$$

When each of the numbers $\frac{a_1}{1}, \frac{a_2}{2}, \ldots, \frac{a_{99}}{99}$ is written as a fraction in lowest terms, none of their denominators are divisible by 125. On the other hand, 125 *does* divide the denominator of $\frac{1}{500}$. Thus when written as a fraction in lowest terms, I_1 must have a denominator divisible by 125.

Now suppose, by way of contradiction, that we could make all the numbers equal to integers $b_1, b_2, \ldots, b_{100}$ in that order. Then in $I_2 = \sum_{i=1}^{100} \frac{b_i}{i}$, the denominator of each of the fractions $\frac{b_i}{i}$ is not divisible by 125. Thus, when I_2 is written as a fraction in lowest terms, its denominator is not divisible by 125 either. Then I_2 cannot possibly equal I_1, a contradiction. Therefore, we can never obtain a set consisting of integers only.

Problem 11.4 A circle is inscribed in the trapezoid $ABCD$. Let K, L, M, N be the points of intersections of the circle with diagonals AC and BD respectively (K is between A and L and M is between B and N). Given that $AK \cdot LC = 16$ and $BM \cdot ND = \frac{9}{4}$, find the radius of the circle.

Solution. Let the circle touch sides AB, BC, CD, DA at P, Q, R, S, respectively, and let r be the radius of the circle. Let $w = AS = AP$, $x = BP = BQ$, $y = CQ = CR$, and $z = DR = DS$. As in problem 10.4, we have $wz = xy = r^2$ and thus $wxyz = r^4$. Also observe that from the lemma in problem 10.4, $AK \cdot LC$ depends only on r and $AP \cdot CR$, and $BM \cdot ND$ depends only on r and $BP \cdot DR$.

Now, draw a parallelogram $A'B'C'D'$ circumscribed about the same circle, with points P', Q', R', S' defined analogously to P, Q, R, S, such that $A'P' = C'R' = \sqrt{wy}$. Draw points K', L', M', N' analogously to K, L, M, N. Because $A'P' \cdot C'R' = wy$, by the observation in the first paragraph we must have $A'K' \cdot L'C' = AK \cdot LC = 16$. Therefore, $A'K' = L'C' = 4$. Also, as with quadrilateral $ABCD$, we have $A'P' \cdot B'P' \cdot C'R' \cdot D'R' = r^4 = wxyz$. Thus, $B'P' \cdot D'R' = xz$, and again by the observation we must have $B'M' \cdot N'D' = BM \cdot ND = \frac{9}{4}$. Therefore, $B'M' = N'D' = \frac{3}{2}$.

Letting O be the center of the circle, we have $A'O = 4 + r$ and $S'O = r$. By the Pythagorean Theorem, $A'S' = \sqrt{8r + 16}$. Similarly, $S'D' = \sqrt{3r + \frac{9}{4}}$. Because $A'S' \cdot S'D' = r^2$, we have

$$(8r + 16)\left(3r + \frac{9}{4}\right) = r^4,$$

which has positive solution $r = 6$ and, by Descartes' rule of signs, no other positive solutions.

Problem 11.5 Find the greatest real number k such that for any triple of positive real numbers a, b, c such that

$$kabc > a^3 + b^3 + c^3,$$

there exists a triangle with side lengths a, b, c.

Solution. Equivalently, we want the greatest real number k such that for any $a, b, c > 0$ with $a + b \leq c$, we have

$$kabc \leq a^3 + b^3 + c^3.$$

First pick $b = a$ and $c = 2a$. Then we must have

$$2ka^3 \leq 10a^3 \implies k \leq 5.$$

On the other hand, suppose $k = 5$. Then writing $c = a + b + x$, expanding $a^3 + b^3 + c^3 - 5abc$ gives

$$2a^3 + 2b^3 - 2a^2b - 2ab^2 + abx + 3(a^2 + b^2)x + 3(a + b)x^2 + x^3.$$

Either by rearrangement, by the AM-GM inequality, or from the inequality $(a + b)(a - b)^2 \geq 0$, we know that $2a^3 + 2b^3 - 2a^2b - 2ab^2 \geq 0$. The other terms in the expression above are also nonnegative, implying that $a^3 + b^3 + c^3 - 5abc \geq 0$, as desired.

Problem 11.6 Find all integers x and y such that

$$x^6 + x^3 y = y^3 + 2y^2.$$

Solution. The only solutions are (x, y) equals $(0, 0)$, $(0, -2)$, and $(2, 4)$.

If $x = 0$ then $y = 0$ or -2; if $y = 0$ then $x = 0$. Now assume that both x and y are nonzero, and rewrite the given equation as $x^3(x^3 + y) = y^2(y + 2)$.

We first show that $(x, y) = (ab, 2b^3)$, (ab, b^3), or $(ab, \frac{b^3}{2})$ for some integers a, b. Suppose some prime p divides y exactly $m > 0$ times (that is, y is divisible by p^m but not p^{m+1}). Because $x^6 = y^3 + 2y^2 - x^3 y$, p must divide x as well — say, $n > 0$ times.

First suppose $p > 2$. Then p divides the right-hand side $y^2(y + 2)$ exactly $2m$ times. If $3n < m$, then p divides the left-hand side $x^3(x^3 + y)$ exactly $6n$ times so that $6n = 2m$, a contradiction. If $3n > m$, then p

divides the left-hand side exactly $3n + m$ times so that $3n + m = 2m$ and $3n = m$, a contradiction. Therefore, $3n = m$.

Now suppose $p = 2$. If $m > 1$, then 2 divides the right-hand side exactly $2m + 1$ times. If $3n < m$, then 2 divides the left-hand side $6n$ times so that $6n = 2m + 1 > 2m$, a contradiction. If $3n > m$, then 2 divides the left-hand side $3n + m$ times so that $3n + m = 2m + 1$ and $3n = m + 1$. Alternatively, we could have $3n = m$.

We show that $(x, y) = (ab, 2b^3)$, (ab, b^3), or $(ab, \frac{b^3}{2})$ for some integers a and b. If 2 divides y only once, then from before (because $3n = m$ when $p > 2, m > 0$) we have $y = 2b^3$ and $x = ab$ for some a, b. If 2 divides y more than once, then (because $3n = m$ when $p > 2, m > 0$ and $3n = m$ or $m + 1$ when $p = 2, m > 1$) we either have $(x, y) = (ab, b^3)$ or $(x, y) = (ab, \frac{b^3}{2})$.

Now, we simply plug these possibilities into the equation. We then either have $a^6 + 2a^3 = 8b^3 + 8$, $a^6 + a^3 = b^3 + 2$, or $8a^6 + 4a^3 = b^3 + 4$.

In the first case, if $a > 1$ then $(a^2 + 1)^3 > (2b)^3 > (a^2)^3$. If $a < -2$ then $(a^2)^3 > (2b)^3 > (a^2 - 1)^3$. Thus either $a = -2, -1, 0$, or 1, and these possibilities yield no solutions.

In the second case, if $a > 1$ then $b^3 = a^6 + a^3 - 2$ and some algebra verifies that $(a^2 + 1)^3 > b^3 > (a^2)^3$, a contradiction. If $a < -1$ then we have $(a^2)^3 > b^3 > (a^2 - 1)^3$. Thus either $a = -1, 0$, or 1. The first possibility yields $b^3 = -2$, the second yields $x = 0$, and the third yields $b = 0$. However, we've assumed that b is an integer and that $x, y \neq 0$, so this case yields no solutions as well.

Finally, in the third case when $a > 1$ then $(2a^2 + 1)^3 > b^3 > (2a^2)^3$. When $a < -1$ then $(2a^2)^3 > b^3 > (2a^2 - 1)^3$. Thus either $a = -1, 0$, or 1, yielding the two possibilities $(a, b) = (-1, 0)$ and $(a, b) = (1, 2)$. Only the latter gives a solution where $x, y \neq 0$ — namely, $(x, y) = (2, 4)$. This completes the proof.

Problem 11.7 Let O be the center of circle ω. Two equal chords AB and CD of ω intersect at L such that $AL > LB$ and $DL > LC$. Let M and N be points on AL and DL respectively such that $\angle ALC = 2\angle MON$. Prove that the chord of ω passing through M and N is congruent to $\overline{AB}$ and $\overline{CD}$.

Solution. We work backward. Suppose that P is on minor arc $\widehat{AC}$ and Q is on minor arc $\widehat{BD}$ such that $PQ = AB = CD$, where line PQ hits $\overline{AL}$ at M' and $\overline{DL}$ at N'. We prove that $\angle ALC = 2\angle M'ON'$.

Let the midpoints of $\overline{AB}, \overline{PQ}, \overline{CD}$ be $T_1, T_2,$ and T_3. $\overline{CD}$ is the image of $\overline{AB}$ under the rotation about O through angle $\angle T_1 OT_3$. This angle also equals the measure of $\widehat{AC}$, which equals $\angle ALC$. Also, by symmetry we have $\angle T_1 OM' = \angle M'OT_2$ and $\angle T_2 ON' = \angle N'OT_3$. Therefore

$$\angle ALC = \angle T_1 OT_3 = \angle T_1 OT_2 + \angle T_2 OT_3$$
$$= 2(\angle M'OT_2 + \angle T_2 ON') = 2\angle M'ON',$$

as claimed.

We now return to the original problem. Because $\angle T_1 OT_3 = \angle ALC$, $\angle T_1 OL = \frac{1}{2}T_1 OT_3 = \frac{1}{2}\angle ALC$. Because $\angle MON = \frac{1}{2}\angle ALC = \angle T_1 OL$, M must lie on $\overline{T_1 L}$. Then consider the rotation about O that sends T_1 to M. It sends A to some P on $\widehat{AC}$, and B to some point Q on $\widehat{BD}$. Then $\overline{PQ}$ is a chord with length AB, passing through M on $\overline{AL}$ and N' on $\overline{DL}$. From the previous work, we know that $\angle ALC = 2\angle MON'$. Because $\angle ALC = 2\angle MON$, we must have $N = N'$. Thus the length of the chord passing through M and N indeed equals AB and CD, as desired.

IMO Selection Tests

Problem 1 Find all functions $h : \mathbb{Z} \to \mathbb{Z}$ such that

$$h(x+y) + h(xy) = h(x)h(y) + 1$$

for all $x, y \in \mathbb{Z}$.

Solution. There are three possible functions:

$$h(n) = 1 \text{ for all integers } n;$$
$$h(2n) = 1, h(2n+1) = 0 \text{ for all integers } n;$$
$$h(n) = n+1 \text{ for all integers } n.$$

Plugging $(x,y) = (0,0)$ into the functional equation, we find that $h(0)^2 - 2h(0) + 1 = 0$ and hence $h(0) = 1$. Plugging in $(x,y) = (1,-1)$ then yields $h(0) + h(-1) = h(1)h(-1) + 1$, or $h(-1)(1 - h(1)) = 0$. Thus, either $h(-1) = 0$ or $h(1) = 1$.

First, suppose that $h(1) \neq 1$ and $h(-1) = 0$. Plugging in $(x,y) = (2,-1)$ and $(x,y) = (-2,1)$ yields $h(1) + h(-2) = 1$ and $h(-2) = h(-2)h(1) + 1$. Substituting $h(-2) = 1 - h(1)$ into the second equation,

we find that
$$1 - h(1) = (1 - h(1))h(1) + 1,$$
$$h(1)^2 - 2h(1) = 0, \text{ and } h(1)(h(1) - 2) = 0,$$
implying that $h(1) = 0$ or $h(1) = 2$.

Thus, $h(1) = 0$, 1, or 2. Plugging $y = 1$ into the equation for each of these cases shows that h must be one of the three functions presented. Each of these three functions indeed satisfies the given functional equation.

Problem 2 Let $a, b, c \in \mathbb{Q}$, $ac \neq 0$. Given that the equation $ax^2 + bxy + cy^2 = 0$ has a nonzero solution of the form
$$(x, y) = (a_0 + a_1 \sqrt[3]{2} + a_2 \sqrt[3]{4}, b_0 + b_1 \sqrt[3]{2} + b_2 \sqrt[3]{4})$$
with $a_i, b_i \in \mathbb{Q}$, $i = 0, 1, 2$, prove that it has also has a nonzero rational solution.

Solution. Let $(\alpha, \beta) = (a_0 + a_1 \sqrt[3]{2} + a_2 \sqrt[3]{4}, b_0 + b_1 \sqrt[3]{2} + b_2 \sqrt[3]{4})$ be the given solution. If $\beta = 0$ then $a\alpha^2 = 0$, which is impossible because a and α must both be nonzero. Therefore, $\beta \neq 0$, and $\frac{\alpha}{\beta}$ is a root to the polynomial
$$at^2 + bt + c = 0.$$

Also, $\frac{\alpha}{\beta}$ is of the form $c_0 + c_1 \sqrt[3]{2} + c_2 \sqrt[3]{4}$ for some rationals c_0, c_1, c_2. Because it is a root to a quadratic with rational coefficients, applying the quadratic formula shows that it must also be of the form $d + e\sqrt{f}$ for rationals d, e, f.

Thus, $(c_0 - d) + c_1 \sqrt[3]{2} + c_2 \sqrt[3]{4} = e\sqrt{f}$. Writing $c_0' = c_0 - d$, the quantity $\left(c_0' + c_1 \sqrt[3]{2} + c_2 \sqrt[3]{4}\right)^2$ must be an integer. After expanding this square, we obtain an expression of the form $\alpha + \beta \sqrt[3]{2} + \gamma \sqrt[3]{4}$ for some integers α, β, γ. This expression is a quadratic in $\sqrt[3]{2}$ with integer coefficients. Because $x^3 - 2$ is irreducible over $\mathbb{Z}[x]$, it follows that $\alpha = \beta = \gamma = 0$.

Thus, $0 = \beta = 2(c_2^2 + c_0'c_1)$ and $0 = \gamma = 2c_0'c_2 + c_1^2$. From the first of these equations, $(c_0'c_1)^2 = -2c_0'^3 c_2$; from the second, $(c_0'c_1)^2 = c_2^4$. Hence $-2c_0'^3 c_2 = c_2^4$. This implies that either $c_2 = 0$ or $c_2 = -\sqrt[3]{2}c_0'$. In the latter case, because c_2 is rational we still must have $c_2 = c_0' = 0$.

Then $c_1 = 0$ as well, and $\frac{\alpha}{\beta} = c_0$ is *rational*. Thus, $(x, y) = (\frac{\alpha}{\beta}, 1)$ is a nonzero rational solution to the given equation.

Problem 3 Suppose a and b are positive integers such that the product of all positive divisors of a (including 1 and a) is equal to the product of all positive divisors of b (including 1 and b). Does it follow that $a = b$?

Solution. Yes, it follows that $a = b$. Let $d(n)$ denote the number of divisors of a positive integer n. The product of all positive divisors of n is

$$\prod_{k|n} k = \sqrt{\prod_{k|n} k \cdot \prod_{k|n} \frac{n}{k}} = \sqrt{\prod_{k|n} n} = n^{\frac{d(n)}{2}}.$$

Thus, the given condition implies that $a^{d(a)}$ and $b^{d(b)}$ equal the same number N. Because N is both a perfect $d(a)$-th power and a perfect $d(b)$-th power, it follows that it is also a perfect ℓ-th power of some number t, where $\ell = \text{lcm}(d(a), d(b))$. Then $a = t^{\frac{\ell}{d(a)}}$ and $b = t^{\frac{\ell}{d(b)}}$ are both powers of the same number t as well.

Now, if a is a bigger power of t than b, then it must have more divisors than b. Then $a = t^{\frac{\ell}{d(a)}} < t^{\frac{\ell}{d(b)}} = b$, a contradiction. Similarly, a cannot be a smaller power of t than b. Therefore, $a = b$, as claimed.

Problem 4 Let a, b, c be positive real numbers such that $a^2 + b^2 + c^2 = 3$. Prove that

$$\frac{1}{1+ab} + \frac{1}{1+bc} + \frac{1}{1+ca} \geq \frac{3}{2}.$$

Solution. Using the AM-HM inequality or the Cauchy-Schwarz inequality, we have

$$\frac{1}{x} + \frac{1}{y} + \frac{1}{z} \geq \frac{9}{x+y+z}$$

for $x, y, z \geq 0$. Also, notice that $a^2 + b^2 + c^2 \geq ab + bc + ca$ because this inequality is equivalent to $\frac{1}{2}(a-b)^2 + \frac{1}{2}(b-c)^2 + \frac{1}{2}(c-a)^2 \geq 0$. Thus,

$$\frac{1}{1+ab} + \frac{1}{1+bc} + \frac{1}{1+ca} \geq \frac{9}{3+ab+bc+ca}$$

$$\geq \frac{9}{3+a^2+b^2+c^2} = \frac{3}{2},$$

as desired.

Problem 5 Suppose triangle T_1 is similar to triangle T_2, and the lengths of two sides and the angle between them of T_1 are proportional to the lengths of two sides and the angle between them of T_2 (but not necessarily the corresponding ones). Must T_1 be congruent to T_2?

Solution. The triangles are not necessarily congruent. Let the vertices of T_1 be A, B, C with $AB = 4$, $BC = 6$, and $CA = 9$, and suppose that $\angle BCA = k\angle ABC$.

Then let the vertices of T_2 be D, E, F where $DE = \frac{8k}{3}$, $EF = 4k$, and $FD = 6k$. Triangles ABC and DEF are similar in that order, so

$\angle EFD = \angle BCA = k\angle ABC$. Also, $EF = k \cdot AB$ and $FD = k \cdot BC$. Therefore, these triangles satisfy the given conditions.

Because $AB < AC$, we have $\angle BCA < \angle ABC$ and $k < 1$. Thus, $DE = \frac{8k}{3} < \frac{8}{3} < AB$, and triangles ABC and DEF are not congruent, as desired.

Problem 6 Two real sequences $x_1, x_2, \ldots$ and $y_1, y_2, \ldots$ are defined in the following way:

$$x_1 = y_1 = \sqrt{3}, \quad x_{n+1} = x_n + \sqrt{1 + x_n^2}, \quad y_{n+1} = \frac{y_n}{1 + \sqrt{1 + y_n^2}}$$

for all $n \geq 1$. Prove that $2 < x_n y_n < 3$ for all $n > 1$.

First Solution. Let $z_n = \frac{1}{y_n}$ and notice that the recursion for y_n is equivalent to

$$z_{n+1} = z_n + \sqrt{1 + z_n^2}.$$

Also note that $z_2 = \sqrt{3} = x_1$. Because the x_i and z_i satisfy the same recursion, this means that $z_n = x_{n-1}$ for all $n > 1$. Thus,

$$x_n y_n = \frac{x_n}{z_n} = \frac{x_n}{x_{n-1}}.$$

Because the x_i are increasing, for $n > 1$ we have $3x_{n-1}^2 \geq 3x_1^2 > 1$. Thus,

$$4x_{n-1}^2 > 1 + x_{n-1}^2, \quad 2x_{n-1} > \sqrt{1 + x_{n-1}^2},$$

and

$$3x_{n-1} > x_{n-1} + \sqrt{1 + x_{n-1}^2} = x_n.$$

Also, $\sqrt{1 + x_{n-1}^2} > x_{n-1} \implies x_n > 2x_{n-1}$. Therefore, $2 < x_n y_n = \frac{x_n}{x_{n-1}} < 3$, as desired.

Second Solution. Writing $x_n = \tan a_n$ for $0° < a_n < 90°$, we have

$$x_{n+1} = \tan a_n + \sqrt{1 + \tan^2 a_n} = \tan a_n + \sec a_n$$
$$= \frac{1 + \sin a_n}{\cos a_n} = \tan\left(\frac{90° + a_n}{2}\right).$$

Because $a_1 = 60°$, we have $a_2 = 75°$, $a_3 = 82.5°$, and in general $a_n = 90° - \frac{30°}{2^{n-1}}$. Thus

$$x_n = \tan\left(90° - \frac{30°}{2^{n-1}}\right) = \cot\left(\frac{30°}{2^{n-1}}\right) = \cot \theta_n,$$

where $\theta_n = \frac{30°}{2^{n-1}}$.

Similar calculation shows that
$$y_n = \tan 2\theta_n = \frac{2\tan\theta_n}{1-\tan^2\theta_n},$$
implying that
$$x_n y_n = \frac{2}{1-\tan^2\theta_n}.$$

Because $0° < \theta_n < 45°$, we have $0 < \tan^2\theta_n < 1$ and $x_n y_n > 2$. For $n > 1$, we have $\theta_n < 30°$, implying that $\tan^2\theta_n < \frac{1}{3}$ and $x_n y_n < 3$.

Note. From the closed forms for x_n and y_n in the second solution, we can see the relationship $y_n = \frac{1}{x_{n-1}}$ used in the first solution.

Problem 7 Let O be the center of the excircle of triangle ABC opposite A. Let M be the midpoint of $\overline{AC}$, and let P be the intersection of lines MO and BC. Prove that if $\angle BAC = 2\angle ACB$, then $AB = BP$.

First Solution. Because O is the excenter opposite A, we know that O is equidistant from lines AB, BC, and CA. We also know that line AO bisects angle BAC. Thus, $\angle BAO = \angle OAC = \angle ACB$. Letting D be the intersection of $\overline{AO}$ and $\overline{BC}$, we then have $\angle DAC = \angle ACD$ and hence $DC = AD$.

Consider triangles OAC and ODC. From above, their altitudes from O are equal, and their altitudes from C are also clearly equal. Thus, $OA/OD = [OAC]/[ODC] = AC/DC$.

Next, because M is the midpoint of $\overline{AC}$ we have $[OAM] = [OMC]$ and $[PAM] = [PMC]$, and hence $[OAP] = [OPC]$ as well. Then
$$\frac{OA}{OD} = \frac{[OAP]}{[ODP]} = \frac{[OPC]}{[ODP]} = \frac{PC}{DP}.$$

Thus, $\frac{AC}{DC} = \frac{OA}{OD} = \frac{PC}{DP}$, and $\frac{AC}{CP} = \frac{DC}{DP} = \frac{AD}{DP}$. By the Angle Bisector Theorem, $\overline{AP}$ bisects $\angle CAD$.

It follows that $\angle BAP = \angle BAD + \angle DAP = \angle ACP + \angle PAC = \angle APB$, and therefore $BA = BP$, as desired.

Second Solution. Let R be the midpoint of the arc BC (not containing A) of the circumcircle of triangle ABC, and let I be the incenter of triangle ABC. We have
$$\angle RBI = \frac{1}{2}(\angle CAB + \angle ABC) = \frac{1}{2}(180° - \angle BRI).$$

Thus $RB = RI$ and similarly $RC = RI$, and hence R is the circumcenter of triangle BIC. Furthermore, because $\angle IBO = 90° = \angle ICO$, quadrilateral $IBOC$ is cyclic and R is also the circumcenter of triangle BCO.

Let lines AO and BC intersect at Q. Because M, O, and P are collinear we may apply Menelaus' Theorem to triangle AQC to get

$$\frac{AM}{CM}\frac{CP}{QP}\frac{QO}{AO} = 1.$$

We know that $\frac{AM}{CM} = 1$, and hence $\frac{CP}{PQ} = \frac{AO}{QO}$.

Because R lies on $\overline{AO}$ and $\overline{QO}$, we have

$$\frac{AO}{QO} = \frac{AR+RO}{QR+RO} = \frac{AR+RC}{CR+RQ},$$

which in turn equals $\frac{AC}{CQ}$ because triangles ARC and CRQ are similar. Also, $\frac{AC}{CQ} = \frac{AC}{AQ}$ because we are given that $\angle BAC = 2\angle ACB$; i.e., $\angle QAC = \angle QCA$ and $CQ = AQ$. Thus we have shown that $\frac{CP}{PQ} = \frac{AC}{AQ}$. By the Angle Bisector Theorem, this implies that line AP bisects $\angle QAC$, from which it follows that $\angle BAP = \frac{3}{2}\angle ACB = \angle BPA$ and $AB = BP$.

Problem 8 Let O, O_1 be the centers of the incircle and the excircle opposite A of triangle ABC. The perpendicular bisector of $\overline{OO_1}$ meets lines AB and AC at L and N respectively. Given that the circumcircle of triangle ABC touches line LN, prove that triangle ABC is isosceles.

Solution. Let M be the midpoint of arc $\widehat{BC}$ not containing A. Angle-chasing gives $\angle OBM = \frac{1}{2}(\angle A + \angle B) = \angle BOM$ and hence $MB = MO$.

Because $\angle OBC = \frac{\angle B}{2}$ and $\angle CBO_1 = \frac{1}{2}(180° - \angle B)$, $\angle OBO_1$ is a right angle. Because we know both that M lies on line AOO_1 (the angle bisector of $\angle A$) and that $MB = MO$, it follows that $\overline{BM}$ is a median to the hypotenuse of right triangle OBO_1. Thus, M is the midpoint of $\overline{OO_1}$.

Therefore, the tangent to the circumcircle of ABC at M must be perpendicular to line AM. This tangent is also parallel to line BC, implying that line AM, the angle bisector of $\angle A$, is perpendicular to line BC. This can only happen if $AB = AC$, as desired.

Problem 9 Does there exist a bijection f of

(a) a plane with itself

(b) three-dimensional space with itself

such that for any distinct points A, B line AB and line $f(A)f(B)$ are perpendicular?

Solution. (a) Yes: simply rotate the plane $90°$ about some axis perpendicular to it. For example, in the xy-plane we could map each point (x, y) to the point $(y, -x)$.

(b) Suppose such a bijection existed. Label the three-dimensional space with x-, y-, and z-axes. Given any point $P = (x_0, y_0, z_0)$, we also view it as the vector p from $(0, 0, 0)$ to (x_0, y_0, z_0). The given condition says that
$$(a - b) \cdot (f(a) - f(b)) = 0$$
for any vectors a, b.

Assume without loss of generality that f maps the origin to itself — otherwise, $g(p) = f(p) - f(0)$ is still a bijection and still satisfies the above equation. Plugging $b = (0, 0, 0)$ into the equation above, we have $a \cdot f(a) = 0$ for all a. Then the above equation reduces to
$$a \cdot f(b) + b \cdot f(a) = 0.$$

Given any vectors a, b, c and any reals m, n, we then have
$$m(a \cdot f(b) + b \cdot f(a)) = 0$$
$$n(a \cdot f(c) + c \cdot f(a)) = 0$$
$$a \cdot f(mb + nc) + (mb + nc) \cdot f(a) = 0.$$

Adding the first two equations and subtracting the third gives
$$a \cdot (mf(b) + nf(c) - f(mb + nc)) = 0.$$

Because this must be true for any vector a, we must have
$$f(mb + nc) = mf(b) + nf(c).$$

Therefore f is linear, and it is determined by how it transforms the unit vectors $\mathbf{i} = (1, 0, 0), \mathbf{j} = (0, 1, 0)$, and $\mathbf{k} = (0, 0, 1)$. If
$$f(\mathbf{i}) = (a_1, a_2, a_3), \quad f(\mathbf{j}) = (b_1, b_2, b_3), \text{ and } f(\mathbf{k}) = (c_1, c_2, c_3),$$
then for a vector x we have
$$f(x) = \begin{bmatrix} a_1 & b_1 & c_1 \\ a_2 & b_2 & c_2 \\ a_3 & b_3 & c_3 \end{bmatrix} x.$$

Applying $f(a) \cdot a = 0$ with $a = \mathbf{i}, \mathbf{j}, \mathbf{k}$, we have $a_1 = b_2 = c_3 = 0$. Then applying $a \cdot f(b) + b \cdot f(a)$ with $(a, b) = (\mathbf{i}, \mathbf{j}), (\mathbf{j}, \mathbf{k}), (\mathbf{k}, \mathbf{i})$ we have $b_1 = -a_2, c_2 = -b_3, c_1 = -a_3$.

Setting $k_1 = c_2$, $k_2 = -c_1$, and $k_3 = b_1$, we find that

$$f(k_1\mathbf{i} + k_2\mathbf{j} + k_3\mathbf{k}) = k_1 f(\mathbf{i}) + k_2 f(\mathbf{j}) + k_3 f(\mathbf{k}) = 0.$$

Because f is injective and $f(0) = 0$, this implies that $k_1 = k_2 = k_3 = 0$. Then $f(x) = 0$ for all x, contradicting the assumption that f was surjective.

Therefore our original assumption was false, and no such bijection exists.

Problem 10 A word is a finite sequence of two symbols a and b. The number of the symbols in the word is said to be the length of the word. A word is called 6-*aperiodic* if it does not contain a subword of the form cccccc for any word c. Prove that $f(n) > \left(\frac{3}{2}\right)^n$, where $f(n)$ is the total number of 6-aperiodic words of length n.

First Solution. Call a word 6-*periodic* if it is not 6-aperiodic.

For $n = 1$, $f(n) = 2 > \frac{3}{2}$. To prove the claim for $n > 1$, it suffices to prove that $f(n+1) > \frac{3}{2}f(n)$ for all $n \geq 1$.

We prove this inequality by strong induction on n. $f(2) = 4 > \frac{3}{2}f(1)$. Now assume that $n \geq 3$ and that the inequality is true for all smaller n.

For each positive k less than $\frac{n}{6}$, we count the words of the form $wcccccc$, where w is a 6-aperiodic word of length $n - 6k$ and c is a word of length k. There are $f(n-6k)$ words w and 2^k words c meeting these criteria, yielding $2^k f(n-6k)$ words of this form. If $6 \mid n$, we also count the $2^{\frac{n}{6}}$ words of the form $cccccc$. Therefore, letting $m = \lceil \frac{n}{6} \rceil$, our total count T equals

$$\sum_{k=1}^{m-1} 2^k f(n-6k) \text{ if } 6 \nmid n \quad \text{and} \quad 2^m + \sum_{k=1}^{m-1} 2^k f(n-6k) \text{ if } 6 \mid n.$$

Let S be the set of the $2f(n-1)$ words of the form $w_0 s$, where w_0 is a 6-aperiodic word of length $n-1$ and s is a single symbol. By construction, if a word w in S is 6-periodic, any subword of the form $cccccc$ must appear at the end of w. Thus, each of the 6-periodic words in S is counted at least once by T. As there are $2f(n-1)$ words in S and at most T 6-periodic words in S, it follows that there are at least $2f(n-1) - T$ 6-aperiodic words in S. Therefore, it suffices to prove that $T < \frac{1}{2}f(n-1)$.

Using the inductive hypothesis, a straightforward proof by induction on j shows that $\left(\frac{2}{3}\right)^{j-1} f(n-j) < f(n-1)$ for all $j \geq 1$. Then

$$\sum_{k=1}^{m-1} 2^k f(n-6k) < \sum_{k=1}^{m-1} 2^k \cdot \left(\frac{2}{3}\right)^{6k-1} f(n-1)$$

$$= \frac{3}{2} f(n-1) \sum_{k=1}^{m-1} \left(\frac{128}{729}\right)^k$$

$$< \frac{3}{2} \cdot \frac{\frac{128}{729}}{1 - \frac{128}{729}} f(n-1) = \frac{192}{601} f(n-1).$$

Thus, if $6 \nmid n$, $T < \frac{1}{2} f(n-1)$. Otherwise, suppose that $6 \mid n$.

If $n = 6$, observe that $f(n-1) = 32$ since all words of length 5 are 6-aperiodic. Hence $2^{\frac{n}{6}} < \frac{1}{6} f(n-1)$. Otherwise, $n \geq 12$. Then

$$\frac{1}{6} f(n-1) > \frac{1}{8} \cdot \left(\frac{3}{2}\right)^{n-1} > \frac{1}{8} \cdot 2^{\frac{n-1}{2}} = 2^{\frac{n-7}{2}} > 2^{\frac{n}{6}}.$$

In either case,

$$T < 2^{n/6} + \frac{192}{601} f(n-1) < \frac{1}{6} f(n-1) + \frac{1}{3} f(n-1) = \frac{1}{2} f(n-1),$$

as desired. This completes the induction and the proof.

Second Solution. Call a word $x_1 x_2 \ldots x_n$ *6-countable* if it satisfies the following conditions (each condition applies only for integers k such that the x_i in that condition are well-defined):

(i) $\{x_{5k+1} x_{5k+2} x_{5k+3}, x_{5k+6} x_{5k+7} x_{5k+8}\} \neq \{$ bab, aba $\}$;

(ii) $x_{5k-1} \neq x_k$;

(iii) $x_{5k} = x_k$.

The desired result follows from two lemmas.

Lemma 1. *Every 6-countable word is 6-aperiodic.*

Proof. Suppose, by way of contradiction, that there existed a 6-countable word $x = x_1 x_2 \ldots x_n$ that was not 6-aperiodic. Among all the words c such that $cccccc$ is a subword of x, let c' be a word of minimal length. Suppose that $x_m x_{m+1} \ldots x_{m+6\ell-1} = c'c'c'c'c'c'$.

If $5 \mid \ell$, let $m' = \lceil \frac{m}{5} \rceil$ and $\ell' = \frac{\ell}{5}$. By condition (iii), the subword $x_{m'} x_{m'+1} \ldots x_{m'+6\ell'-1}$ would be of the form $cccccc$, a contradiction.

Therefore, ℓ is not divisible by 5. For each integer k between 0 and $\ell-1$ inclusive, $m+k+r\ell$ is congruent to 4 modulo 5 for some $r \in \{0,1,2,3,4\}$.

Therefore, by condition (ii), $x_{m+k} = x_{m+k+r\ell} \neq x_{m+k+1+r\ell} = x_{m+k+1}$, and it follows that cccccc is of the form abab...ab or baba...ba. This, however, contradicts condition (i). Therefore our original assumption was false, and every 6-countable word is indeed 6-aperiodic. ∎

Lemma 2. *For each positive integer m, there are more than $\left(\frac{3}{2}\right)^m$ 6-countable words of length m.*

Proof. Let α_m be the number of 6-countable words of length m. Also, let β_m equal the number of 6-countable words of length m which end with aba (or equivalently, by symmetry, with bab), and let γ_m equal the number of 6-countable words of length m which *do not* end with aba (or equivalently, with bab). Observe that for $1 \leq m \leq 3$, $\alpha_m = 2^m > \left(\frac{3}{2}\right)^m$. Also, $\alpha_4 = 8 > \left(\frac{3}{2}\right)^4$. It now suffices to prove that $\alpha_m > \left(\frac{3}{2}\right)^m$ for $m \geq 5$.

Straightforward counting arguments show that for $t \geq 1$, $\beta_{5(t+1)} = \gamma_{5t}$, and $\gamma_{5(t+1)} = 6(\gamma_{5t} + \beta_{5t}) + \gamma_{5t}$. Manipulating these equations yields the recursive relations $\beta_{5(t+2)} = 7\beta_{5(t+1)} + 6\beta_{5t}$ and $\gamma_{5(t+2)} = 7\gamma_{5(t+1)} + 6\gamma_{5t}$. Because $\alpha_m = \beta_m + \gamma_m$, it follows that

$$\alpha_{5(t+2)} = 7\alpha_{5(t+1)} + 6\alpha_{5t}$$

for $t \geq 1$. After directly calculating $\alpha_5 = 8 > \left(\frac{3}{2}\right)^5$ and $\alpha_{10} = 62 > \left(\frac{3}{2}\right)^{10}$, with this equation it follows easily by induction that $\alpha_{5t} > \left(\frac{3}{2}\right)^{5t}$ for $t \geq 1$.

Observe that $(\alpha_{5t+1}, \alpha_{5t+2}, \alpha_{5t+3}, \alpha_{5t+4}) = (2\alpha_{5t}, 4\alpha_{5t}, 8\alpha_{5t}, 8\alpha_{5t})$ for $t \geq 1$. Combined with $\alpha_{5t} > \left(\frac{3}{2}\right)^{5t}$, this implies that $\alpha_m > \left(\frac{3}{2}\right)^m$ for all $m \geq 5$, as desired. ∎

Note. Another solution, similar to the second, uses the following conditions in the definition of *6-countable* rather than the above conditions (ii) and (iii):

(ii') $x_{5k-1} = $ a if and only if $t_{2k-2} = 0$;
(iii') $x_{5k} = $ b if and only if $t_{2k-1} = 0$,

where $t_0, t_1, \ldots$ is the Thue-Morse sequence defined by $t_0 = 0$ and the recursive relations $t_{2k} = t_k$, $t_{2k+1} = 1 - t_{2k}$ for $k \geq 1$.

Problem 11 Determine all positive integers n, $n \geq 2$, such that $\binom{n-k}{k}$ is even for $k = 1, 2, \ldots, \lfloor \frac{n}{2} \rfloor$.

Solution. Lucas's Theorem states: Let p be a prime and let a and b be positive integers. Express a and b in base p — that is, write $a = \sum_{i=0}^{r} a_i p^i$ and $b = \sum_{i=0}^{r} b_i p^i$, where a_i and b_i are integers between 0 and $p-1$ inclusive for $i = 1, 2, \ldots, r$. Then

$$\binom{a}{b} \equiv \binom{a_r}{b_r}\binom{a_{r-1}}{b_{r-1}} \cdots \binom{a_0}{b_0} \pmod{p}.$$

Suppose that $p = 2$, $a = 2^s - 1$, and $a_{s-1} = a_{s-2} = \cdots = a_0 = 1$. For any b with $0 \leq b \leq 2^s - 1$, each term $\binom{a_i}{b_i}$ in the above equation equals 1. Therefore, $\binom{a}{b} \equiv 1 \pmod{2}$.

This implies that $n+1$ is a power of two. Otherwise, let $s = \lfloor \log_2 n \rfloor$ and let $k = n - (2^s - 1) = n - \frac{2^{s+1}-2}{2} \leq n - \frac{n}{2} = \frac{n}{2}$. Then $\binom{n-k}{k} = \binom{2^s-1}{k}$ is odd, a contradiction.

Conversely, suppose that $n = 2^s - 1$ for some positive integer s. For $k = 1, 2, \ldots, \lfloor \frac{n}{2} \rfloor$, there is at least one 0 in the binary representation of $a = n - k$ (not counting leading zeros, of course). Whenever there is a 0 in the binary representation of $n-k$, there is a 1 in the corresponding digit of $b = k$. Then the corresponding $\binom{a_i}{b_i}$ equals 0, and by Lucas's Theorem, $\binom{n-k}{k}$ is even.

Therefore, $n = 2^s - 1$ for integers $s \geq 2$.

Problem 12 A set of n players took part in a chess tournament in which each pair of participants played against each other exactly once. After the tournament was over, it turned out that among any four players there was one who scored differently against the other three (i.e., he got a victory, a draw, and a loss). Prove that the largest possible n satisfies the inequality $6 \leq n \leq 9$.

Solution. Let $A_1 \Rightarrow A_2 \Rightarrow \cdots \Rightarrow A_n$ denote "A_1 beats A_2, A_2 beats A_3, $\ldots$, A_{n-1} beats A_n," and let $X \mid Y$ denote "X draws with Y."

First we show it is possible to have the desired results with $n = 6$: call the players A, B, C, D, E, F. Then let

$$A \Rightarrow B \Rightarrow C \Rightarrow D \Rightarrow E \Rightarrow A,$$

$$F \Rightarrow A, \ F \Rightarrow B, \ F \Rightarrow C, \ F \Rightarrow D, \ F \Rightarrow E,$$

and have all other games end in draws. Visually, we can view this arrangement as a regular pentagon $ABCDE$ with F at the center. There are three different types of groups of 4, represented by $ABCD$, $ABCF$, and $ABDF$. In these three respective cases, B (or C), A, and A are the players who score differently from the other three.

Alternatively, let

$$A \Rightarrow B \Rightarrow C \Rightarrow D \Rightarrow E \Rightarrow F \Rightarrow A,$$
$$B \Rightarrow D \Rightarrow F \Rightarrow B, \quad C \Rightarrow A \Rightarrow E \Rightarrow C,$$
$$A \mid D, \ B \mid E, \ C \mid F.$$

In this arrangement there are three different types of groups of four, represented by $\{A, B, C, D\}$, $\{A, B, D, E\}$, and $\{A, B, D, F\}$. (If the players are arranged in a regular hexagon, these correspond to a trapezoid-shaped group, a rectangle-shaped group, and a diamond-shaped group.) In these three cases, A, B (or D), and A (or D) are the players who score differently against the other three.

Now we show it is impossible to have the desired results with $n = 10$ and thus all $n \geq 10$. Suppose, by way of contradiction, it *were* possible. First we prove that all players draw exactly 4 times.

To do this, draw a graph with n vertices representing the players, and draw an edge between two vertices if they drew in their game. If V has degree 3 or less, then look at the remaining 6 or more vertices it is not adjacent to. By Ramsey's Theorem, either three of them (call them X, Y, Z) are all adjacent or all not adjacent. Then in the group $\{V, X, Y, Z\}$, none of the players draws exactly once with the other players, a contradiction.

Thus each vertex has degree at least 4; we now prove that every vertex has degree *exactly* 4. Suppose, by way of contradiction, that some vertex A were adjacent to at least 5 vertices B, C, D, E, F. None of these vertices can be adjacent to two others. For example, if B was adjacent to C and D, then in $\{A, B, C, D\}$ each vertex draws at least twice — but some player must draw exactly once in this group. Now in the group $\{B, C, D, E\}$ some pair must draw: without loss of generality, say B and C. In the group $\{C, D, E, F\}$ some pair must draw as well. Because C can't draw with D, E, or F from our previous observation, assume without loss of generality that E and F draw.

Now in $\{A, B, C, D\}$, D must beat one of B, C and lose to the other; without loss of generality, say D loses to B and beats C. Looking at $\{A, D, E, F\}$, we can similarly assume that D beats E and loses to F. Next, in $\{A, C, D, E\}$ players C and E can't draw; without loss of generality, say C beats E. Then in $\{A, C, E, F\}$, player C must lose to F. However, it follows that in $\{C, D, E, F\}$ no player scores differently against the other three players — a contradiction.

Now suppose A were adjacent to B, C, D, E, and without loss of generality assume $B \mid C$. Then ABC is a triangle. For each vertex K besides A, B, C, look at the group $\{A, B, C, K\}$: K must draw with one of A, B, C. By the Pigeonhole Principle, one of A, B, C draws with at least three of the K and thus has degree at least 5. From above, this is impossible.

It follows that it is *impossible* for n to be at least 10. Because n can be 6, the maximum n is between 6 and 9, as desired.

2 Brazil

Problem 1 Let $ABCDE$ be a regular pentagon such that the star region $ACEBD$ has area 1. Let $\overline{AC}$ and $\overline{BE}$ meet at P, and let $\overline{BD}$ and $\overline{CE}$ meet at Q. Determine $[APQD]$.

Solution. Let $R = \overline{AD} \cap \overline{BE}$, $S = \overline{AC} \cap \overline{BD}$, $T = \overline{CE} \cap \overline{AD}$. Now, $\triangle PQR \sim \triangle CAD$ because they are corresponding triangles in regular pentagons $QTRPS$ and $ABCDE$. Because $\triangle CAD \sim \triangle PAR$ as well, we have $\triangle PQR \cong \triangle PAR$. Thus,

$$[APQD] = \frac{[APQD]}{[ACEBD]} = \frac{2[APR] + [PQR] + [RQT]}{5[APR] + [PQR] + 2[RQT]}$$
$$= \frac{3[APR] + [RQT]}{6[APR] + 2[RQT]} = \frac{1}{2}.$$

Problem 2 Given a 10×10 board, we want to remove n of the 100 squares so that no 4 of the remaining squares form the corners of a rectangle with sides parallel to the sides of the board. Determine the minimum value of n.

Solution. The answer is 66. Consider the diagram below, in which a colored circle represents a square that has *not* been removed. The diagram demonstrates that n can be 66:

Now we proceed to show that n is at least 66. Suppose, for sake of contradiction, that it were possible with $n = 65$. Denote by a_i the number of squares left in row i ($i = 1, 2, \ldots, 10$). In row i, there are $\binom{a_i}{2}$ pairs of remaining squares. If no four remaining squares form the corners of a rectangle, then $N = \sum_{i=1}^{10} \binom{a_i}{2}$ must not exceed $\binom{10}{2} = 45$.

Note that, with a fixed $\sum_{i=1}^{10} a_i = 35$, the minimum of $\sum_{i=1}^{10} \binom{a_i}{2}$ is attained when and only when no two a_i's differ by more than 1. Thus, $45 = \sum_{i=1}^{10} \binom{a_i}{2} \geq 5 \cdot \binom{4}{2} + 5 \cdot \binom{3}{2} = 45$, i.e., this minimum *is* attained here, implying that five of the a_i's equal 4 and the rest equal 3. Then it is easy to see that aside from permutations of the row and columns, the first five rows of the board must be as follows:

We inspect this figure and notice that it is now impossible for another row to contain at least 3 remaining squares without forming the vertices of a rectangle with sides parallel to the sides of the board. This is a contradiction, because each of the remaining 5 rows is supposed to have 3 remaining squares. Thus, it is impossible for n to be less than 66, and we are done.

Problem 3 The planet Zork is spherical and has several cities. Given any city A on Zork, there exists an antipodal city A' (i.e., symmetric with respect to the center of the sphere). In Zork, there are roads joining pairs of cities. If there is a road joining cities P and Q, then there is a road joining P' and Q'. Roads don't cross each other, and any given pair of cities is connected by some sequence of roads. Each city is assigned a value, and the difference between the values of every pair of cities joined by a single road is at most 100. Prove that there exist two antipodal cities with values differing by at most 100.

Solution. Let $[A]$ denote the value assigned to city A. Name the pairs of cities
$$(Z_1, Z_1'), (Z_2, Z_2'), (Z_3, Z_3'), \ldots, (Z_n, Z_n')$$
with $0 \leq [Z_i] - [Z_i']$ for all i. Because any given pair of cities is connected by some sequence of roads, there must exist a, b such that Z_a and Z_b' are connected by a single road. Then Z_a' and Z_b are also connected by a single road. Thus, $[Z_a] - [Z_b'] \leq 100$ and $[Z_b] - [Z_a'] \leq 100$. Adding, we have
$$[Z_a] - [Z_a'] + [Z_b] - [Z_b'] \leq 200.$$
Hence, either $0 \leq [Z_a] - [Z_a'] \leq 100$ or $0 \leq [Z_b] - [Z_b'] \leq 100$. In either case, we are done.

Problem 4 In Tumbolia there are n soccer teams. We want to organize a championship such that each team plays exactly once with each other team. All games take place on Sundays, and a team can't play more than one game in the same day. Determine the smallest positive integer m for which it is possible to realize such a championship in m Sundays.

Solution. Let a_n be the smallest positive integer for which it is possible to realize a championship between n soccer teams in a_n Sundays. For $n > 1$, it is necessary that $a_n \geq 2\lceil \frac{n}{2} \rceil - 1$ — otherwise the total number of games played would not exceed

$$\left(2\left\lceil \frac{n}{2} \right\rceil - 2\right) \cdot \left\lfloor \frac{n}{2} \right\rfloor \leq \frac{(n-1)^2}{2} < \binom{n}{2},$$

a contradiction.

On the other hand, $2\lceil \frac{n}{2} \rceil - 1$ days suffice. Suppose that $n = 2t+1$ or $2t+2$. Number the teams from 1 to n and the Sundays from 1 to $2t+1$. On the i-th Sunday, let team i either sit out (if n is odd) or play against team $2t+2$ (if n is even), and have any other team j play against team $k \neq 2t+2$ where $j + k \equiv 2i \pmod{2t+1}$. Then each team indeed plays every other team, as desired.

Problem 5 Given a triangle ABC, show how to construct, with straightedge and compass, a triangle $A'B'C'$ with minimal area such that A', B', C' lie on AB, BC, CA, respectively, $\angle B'A'C' = \angle BAC$, and $\angle A'C'B' = \angle ACB$.

Solution. All angles are directed modulo $180°$.

For convenience, call any triangle $A'B'C'$ *proper* if A', B', C' lie on lines AB, BC, CA, respectively, and $\triangle ABC \sim \triangle A'B'C'$. The problem is, then, to construct the proper triangle with minimal area.

Suppose we have any proper triangle, and let P be the point (different from A') where the circumcircles of triangles $AA'C'$ and $BB'A'$ meet. Then

$$\angle B'PC' = 360° - \angle A'PB' - \angle C'PA'$$
$$= 360° - (180° - \angle CBA) - (180° - \angle BAC)$$
$$= 180° - \angle ACB,$$

so P also lies on the circumcircle of triangle $CC'B'$.

Next,

$$\angle PAB = \angle PC'A' = \angle B'C'A' - \angle B'C'P$$
$$= \angle B'CC' - \angle B'CP = \angle PCA,$$

and with similar reasoning we have

$$\angle PAB = \angle PC'A' = \angle PCA = \angle PB'C' = \angle PBC.$$

There is a unique point P (one of the Brocard points) satisfying $\angle PAB = \angle PBC = \angle PCA$, and thus P is fixed — independent of the choice of triangle $A'B'C'$. Because P is the corresponding point in similar triangles ABC and $A'B'C'$, we have

$$[A'B'C'] = [ABC]\left(\frac{PA'}{PA}\right)^2.$$

Thus $[A'B'C']$ is minimal when PA' is minimal, which occurs when $PA' \perp AB$ (and analogously, when $PB' \perp BC$ and $PC' \perp PA$). Thus, the proper triangle with minimal area is the pedal triangle $A'B'C'$ of P to triangle ABC. This triangle is indeed similar to triangle ABC: letting $\theta = \angle PAB$ be the Brocard angle, it is the image of triangle ABC under a rotation through $\theta - 90°$, followed by a homothety of ratio $|\sin\theta|$.

To construct this triangle, first draw the circles $\{X \mid \angle BXA = \angle BCA + \angle CAB\}$ and $\{Y \mid \angle CYB = \angle CAB + \angle ABC\}$ and let P' be their point of intersection (different from B). We have $\angle AP'C = \angle ABC + \angle BCA$, and

$$\angle P'AB = 180° - \angle ABP' - \angle BP'A$$
$$= 180° - (\angle ABC - \angle P'BC) - (\angle BCA + \angle CAB) = \angle P'BC,$$

and similarly $\angle P'BC = \angle P'CA$. Therefore $P = P'$. Finally, drop the perpendiculars from P' to the sides of triangle ABC to form A', B', C'. This completes the construction.

3 Bulgaria

National Olympiad, Third Round

Problem 1 Find all triples (x, y, z) of natural numbers such that y is a prime number, y and 3 do not divide z, and $x^3 - y^3 = z^2$.

Solution. We rewrite the equation in the form
$$(x - y)(x^2 + xy + y^2) = z^2.$$
Any common divisor of $x - y$ and $x^2 + xy + y^2$ also divides both z^2 and $(x^2 + xy + y^2) - (x + 2y)(x - y) = 3y^2$. Because z^2 and $3y^2$ are relatively prime by assumption, $x - y$ and $x^2 + xy + y^2$ must be relatively prime as well. Therefore, both $x - y$ and $x^2 + xy + y^2$ are perfect squares.

Writing $a = \sqrt{x - y}$, we have
$$x^2 + xy + y^2 = (a^2 + y)^2 + (a^2 + y)y + y^2 = a^4 + 3a^2 y + 3y^2$$
and
$$4(x^2 + xy + y^2) = (2a^2 + 3y)^2 + 3y^2.$$
Writing $m = 2\sqrt{x^2 + xy + y^2}$ and $n = 2a^2 + 3y$, we have
$$m^2 = n^2 + 3y^2$$
or
$$(m - n)(m + n) = 3y^2,$$
so $(m - n, m + n) = (1, 3y^2)$, $(y, 3y)$, or $(3, y^2)$.

In the first case, $2n = 3y^2 - 1$ and $4a^2 = 2n - 6y = 3y^2 - 6y - 1$. Hence, $a^2 \equiv 2 \pmod{3}$, which is impossible.

In the second case, $n = y < 2a^2 + 3y = n$, a contradiction.

In the third case, we have $4a^2 = 2n - 6y = y^2 - 6y - 3 < (y - 3)^2$. When $y \geq 10$ we have $y^2 - 6y - 3 > (y - 4)^2$. Hence, we must actually have $y = 2, 3, 5,$ or 7. In this case we have $a = \frac{\sqrt{y^2 - 6y - 3}}{2}$, which is real only when $y = 7$, $a = 1$, $x = y + a^2 = 8$, and $z = 13$. This yields the unique solution $(x, y, z) = (8, 7, 13)$.

Problem 2 A convex quadrilateral $ABCD$ is inscribed in a circle whose center O is inside the quadrilateral. Let $MNPQ$ be the quadrilateral whose vertices are the projections of the intersection point of the diagonals AC and BD onto the sides of $ABCD$. Prove that $2[MNPQ] \leq [ABCD]$.

Solution. The result actually holds even when $ABCD$ is not cyclic. We begin by proving the following result:

Lemma. *If $\overline{XW}$ is an altitude of triangle XYZ, then*
$$\frac{XW}{YZ} \leq \frac{1}{2}\tan\left(\frac{\angle Y + \angle Z}{2}\right).$$

Proof: X lies on an arc of a circle determined by $\angle YXZ = 180° - \angle Y - \angle Z$. Its distance from $\overline{YZ}$ is maximized when it is at the center of this arc, which occurs when $\angle Y = \angle Z$. At this point, $\frac{XW}{YZ} = \frac{1}{2}\tan\left(\frac{\angle Y+\angle Z}{2}\right)$. ∎

Suppose M, N, P, Q are on sides AB, BC, CD, DA, respectively. Also let T be the intersection of $\overline{AC}$ and $\overline{BD}$.

Let $\alpha = \angle ADB$, $\beta = \angle BAC$, $\gamma = \angle CAD$, $\delta = \angle DBA$. From the lemma, $MT \leq \frac{1}{2}AB \cdot \tan\left(\frac{\beta+\delta}{2}\right)$ and $QT \leq \frac{1}{2}AD \cdot \tan\left(\frac{\alpha+\gamma}{2}\right)$. Also, $\angle MTQ = 180° - \angle QAM = 180° - \angle DAB$. Thus $2[MTQ] = MT \cdot QT \sin\angle MTQ \leq \frac{1}{4}\tan\left(\frac{\alpha+\gamma}{2}\right)\tan\left(\frac{\beta+\delta}{2}\right)AB \cdot AD\sin\angle DAB$. Because $\frac{\alpha+\gamma}{2} + \frac{\beta+\delta}{2} = 90°$, this last expression exactly equals $\frac{1}{4}AB \cdot AD\sin\angle DAB = \frac{1}{2}[ABD]$. Thus, $2[MTQ] \leq \frac{1}{2}[ABD]$.

Likewise, $2[NTM] \leq \frac{1}{2}[BCA]$, $[PTN] \leq \frac{1}{2}[CDB]$, and $[QTP] \leq \frac{1}{2}[DAC]$. Adding these four inequalities shows that $2[MNPQ]$ is at most

$$\tfrac{1}{2}([ABD]+[CDB]) + \tfrac{1}{2}([BCA]+[DAC]) = [ABCD],$$

as desired.

Problem 3 In a competition 8 judges marked the contestants by *pass* or *fail*. It is known that for any two contestants, two judges marked both with *pass*; two judges marked the first contestant with *pass* and the second contestant with *fail*; two judges marked the first contestant with *fail* and the second contestant with *pass*; and finally, two judges marked both with *fail*. What is the largest possible number of contestants?

Solution. For a rating r (either *pass* or *fail*), let $\bar{r}$ denote the opposite rating. Also, whenever a pair of judges agree on the rating for some contestant, call this an *agreement*. We first prove that any two judges share at most three agreements. Suppose, by way of contradiction, this were false.

Assume, without loss of generality, that the judges (labelled with numbers) mark the first four contestants (labelled with letters) as follows in the left table:

	A	B	C	D
1	a	b	c	d
2	a	b	c	d
3	a	$\bar{b}$		
4	a	$\bar{b}$		
5	$\bar{a}$	$\bar{b}$		
6	$\bar{a}$	$\bar{b}$		
7	$\bar{a}$	b		
8	$\bar{a}$	b		

	A	B	C	D
1	a	b	c	d
2	a	b	c	d
3	a	$\bar{b}$	$\bar{c}$	$\bar{d}$
4	a	$\bar{b}$	$\bar{c}$	$\bar{d}$
5	$\bar{a}$	$\bar{b}$		
6	$\bar{a}$	$\bar{b}$		
7	$\bar{a}$	b		
8	$\bar{a}$	b		

	A	B	C	D
1	a	b	c	d
2	a	b	c	d
3	a	$\bar{b}$	$\bar{c}$	$\bar{d}$
4	a	$\bar{b}$	$\bar{c}$	$\bar{d}$
5	$\bar{a}$	$\bar{b}$	c	d
6	$\bar{a}$	$\bar{b}$	c	d
7	$\bar{a}$	b		
8	$\bar{a}$	b		

Applying the given condition to contestants A and C, judges 3 and 4 must both give C the rating $\bar{c}$; similarly, they must both give D the rating $\bar{d}$. Next, applying the condition to contestants B and C, judges 5 and 6 must both give C the rating c; similarly, they must both give D the rating d. The condition then fails for contestants C and D, a contradiction.

Thus, each pair of judges agrees on at most three ratings, as claimed. Hence, there are at most $3 \cdot \binom{8}{2} = 84$ agreements between all the judges. On the other hand, for each contestant exactly four judges mark him with *pass* and exactly four judges mark him with *fail*, implying that there are $\binom{4}{2} + \binom{4}{2} = 12$ agreements per contestant. It follows that there are at most $\frac{84}{12} = 7$ contestants. As the following table shows (with 1 representing *pass* and 0 representing *fail*), it is indeed possible to have exactly 7 contestants:

	A	B	C	D	E	F	G
1	1	1	1	1	1	1	1
2	1	1	1	0	0	0	0
3	1	0	0	1	1	0	0
4	1	0	0	0	0	1	1
5	0	1	0	1	0	0	1
6	0	1	0	0	1	1	0
7	0	0	1	1	0	1	0
8	0	0	1	0	1	0	1

Problem 4 Find all pairs (x, y) of integers such that

$$x^3 = y^3 + 2y^2 + 1.$$

Solution. When $y^2 + 3y > 0$, $(y+1)^3 > x^3 > y^3$. Thus we must have $y^2 + 3y \leq 0$, and $y = -3, -2, -1$, or 0 — yielding the solutions $(x, y) = (1, 0), (1, -2)$, and $(-2, -3)$.

Problem 5 Let B_1 and C_1 be points on the sides AC and AB of triangle ABC. Lines BB_1 and CC_1 intersect at point D. Prove that a circle can be inscribed inside quadrilateral AB_1DC_1 if and only if the incircles of the triangles ABD and ACD are tangent to each other.

Solution. Let the incircle of triangle ABD be tangent to $\overline{AD}$ at T_1 and let the incircle of triangle ACD be tangent to $\overline{AD}$ at T_2. Then $DT_1 = \frac{1}{2}(DA + DB - AB)$ and $DT_2 = \frac{1}{2}(DA + DC - AC)$.

First suppose a circle can be inscribed inside AB_1DC_1. Let it be tangent to sides AB_1, B_1D, DC_1, C_1A at points E, F, G, H, respectively. Using equal tangents, we have

$$AB - BD = (AH + HB) - (BF - DF)$$
$$= (AH + BF) - (BF - DF) = AH + DF$$

and similarly $AC - CD = AE + DG$. Also, $AH + DF = AE + DG$ by equal tangents, implying that $AB - BD = AC - CD$ and thus $DA + DB - AB = DA + DC - AC$. Therefore $DT_1 = DT_2$, $T_1 = T_2$, and the two given incircles are tangent to each other.

Next suppose the two incircles are tangent to each other. Then $DA + DB - AB = DA + DC - AC$. Let ω be the incircle of ABB_1, and let D' be the point on $\overline{BB_1}$ (different from B_1) such that line CD' is tangent to ω. Suppose, by way of contradiction, that $D \neq D'$. From the result in the last paragraph, we know that the incircles of triangles ABD' and ACD' are tangent and hence $D'A + D'B - AB = D'A + D'C - AC$. Because $DB - AB = DC - AC$ and $D'B - AB = D'C - AC$, we must have $DB - D'B = DC - D'C$ by subtraction. Thus $DD' = |DB - D'B| = |DC - D'C|$. The triangle inequality then fails in triangle $DD'C$, a contradiction. This completes the proof.

Problem 6 Each interior point of an equilateral triangle of side 1 lies in one of six congruent circles of radius r. Prove that

$$r \geq \frac{\sqrt{3}}{10}.$$

Solution. From the condition, we also know that every point inside or *on* the triangle lies inside or *on* one of the six circles.

Define $R = \frac{1}{1+\sqrt{3}}$. Orient the triangle so that B is directly to the left of C, and so that A is above $\overline{BC}$. Draw point W on $\overline{AB}$ such that $WA = R$, and then draw point X directly below W such that $WX = R$. In triangle WXB, $WB = 1 - R = \sqrt{3}R$ and $\angle BWX = 30°$, implying that $XB = R$ as well. Similarly draw Y on $\overline{AC}$ such that $YA = R$, and Z directly below Y such that $YZ = ZC = R$.

In triangle AWY, $\angle A = 60°$ and $AW = AY = R$, implying that $WY = R$. This in turn implies that $XZ = R$ and that $WX = YZ = R\sqrt{2}$.

Now if the triangle is covered by six congruent circles of radius r, each of the seven points A, B, C, W, X, Y, Z lies on or inside one of the circles, so some two of them are in the same circle. Any two of these points are at least $R \leq 2r$ apart, so $r \geq \sqrt{3}/10$.

National Olympiad, Fourth Round

Problem 1 A rectangular parallelepiped has integer dimensions. All of its faces of are painted green. The parallelepiped is partitioned into unit cubes by planes parallel to its faces. Find all possible measurements of the parallelepiped if the number of cubes without a green face is one third of the total number of cubes.

Solution. Let the parallelepiped's dimensions be a, b, c. These lengths must all be at least 3 or else every cube has a green face. The given condition is equivalent to

$$3(a-2)(b-2)(c-2) = abc,$$

or

$$3 = \frac{a}{a-2} \cdot \frac{b}{b-2} \cdot \frac{c}{c-2}.$$

If all the dimensions are at least 7, then $\frac{a}{a-2} \cdot \frac{b}{b-2} \cdot \frac{c}{c-2} \leq \left(\frac{7}{5}\right)^3 = \frac{343}{125} < 3$, a contradiction. Thus one of the dimensions — say, a — equals 3, 4, 5, or 6. Assume without loss of generality that $b \leq c$.

When $a = 3$ we have $bc = (b-2)(c-2)$, which is impossible.

When $a = 4$, rearranging the equation yields $(b-6)(c-6) = 24$. Thus $(b, c) = (7, 30), (8, 18), (9, 14)$, or $(10, 12)$.

When $a = 5$, rearranging the equation yields $(2b-9)(2c-9) = 45$. Thus $(b, c) = (5, 27), (6, 12)$, or $(7, 9)$.

Finally, when $a = 6$, rearranging the equation yields $(b-4)(c-4) = 8$. Thus $(b, c) = (5, 12)$ or $(6, 8)$.

Therefore, the parallelepiped may measure $4 \times 7 \times 30$, $4 \times 8 \times 18$, $4 \times 9 \times 14$, $4 \times 10 \times 12$, $5 \times 5 \times 27$, $5 \times 6 \times 12$, $5 \times 7 \times 9$, or $6 \times 6 \times 8$.

Problem 2 Let $\{a_n\}$ be a sequence of integers such that for $n \geq 1$

$$(n-1)a_{n+1} = (n+1)a_n - 2(n-1).$$

If 2000 divides a_{1999}, find the smallest $n \geq 2$ such that 2000 divides a_n.

Solution. First, we note that the sequence $a_n = 2n - 2$ works. Then writing $b_n = a_n - (2n - 2)$ gives the recursion

$$(n-1)b_{n+1} = (n+1)b_n.$$

For $n \geq 2$, observe that $b_n = b_2 \cdot \prod_{k=2}^{n-1} \frac{k+1}{k-1} = b_2 \cdot \frac{\prod_{k=3}^{n} k}{\prod_{k=1}^{n-2} k} = \frac{n(n-1)}{2} b_2$. Thus when $n \geq 2$, the solution to the original equation is of the form

$$a_n = 2(n-1) + \frac{n(n-1)}{2} c$$

for some constant c. Plugging in $n = 2$ shows that $c = a_2 - 2$ is an integer.

Now, because $2000 \mid a_{1999}$ we have $2(1999 - 1) + \frac{1999 \cdot 1998}{2} \cdot c \equiv 0 \Longrightarrow -4 + 1001c \equiv 0 \Longrightarrow c \equiv 4 \pmod{2000}$. Then $2000 \mid a_n$ exactly when

$$2(n-1) + 2n(n-1) \equiv 0 \pmod{2000}$$
$$\Longleftrightarrow (n-1)(n+1) \equiv 0 \pmod{1000}.$$

$(n-1)(n+1)$ is divisible by 8 exactly when n is odd, and it is divisible by 125 exactly when either $n-1$ or $n+1$ is divisible by 125. The smallest $n \geq 2$ satisfying these requirements is $n = 249$.

Problem 3 The vertices of a triangle have integer coordinates and one of its sides is of length $\sqrt{n}$, where n is a square-free natural number. Prove that the ratio of the circumradius to the inradius of the triangle is an irrational number.

Solution. Label the triangle ABC. Let r, R, K be the inradius, circumradius, and area of the triangle; let $a = BC$, $b = CA$, $c = AB$ and write $a = p_1\sqrt{q_1}$, $b = p_2\sqrt{q_2}$, $c = p_3\sqrt{q_3}$ for positive integers p_i, q_i with q_i square-free. By Pick's Theorem ($K = I + \frac{1}{2}B - 1$), K is rational. Also, $R = \frac{abc}{4K}$ and $r = \frac{2K}{a+b+c}$. Thus $\frac{R}{r} = \frac{abc(a+b+c)}{8K^2}$ is rational if and only if $abc(a + b + c) = a^2bc + ab^2c + abc^2$ is rational. Let this quantity equal m, and assume, by way of contradiction, that m is rational.

We have $a^2bc = m_1\sqrt{q_2q_3}$, $ab^2c = m_2\sqrt{q_3q_1}$, and $abc^2 = m_3\sqrt{q_1q_2}$ for positive integers m_1, m_2, m_3. Then $m_1\sqrt{q_2q_3} + m_2\sqrt{q_3q_1} = m - $

$m_3\sqrt{q_1q_2}$. Squaring both sides, we find that

$$m_1^2 q_2 q_3 + m_2^2 q_3 q_1 + 2m_1 m_2 q_3 \sqrt{q_1 q_2} = m^2 + m_3^2 - 2mm_3\sqrt{q_1 q_2}.$$

If $\sqrt{q_1 q_2}$ is not rational, then the coefficients of $\sqrt{q_1 q_2}$ must be the same on both sides. However, this is impossible because $2m_1 m_2 q_3$ is positive while $-2mm_3$ is not.

Hence, $\sqrt{q_1 q_2}$ is rational. Because q_1 and q_2 are square-free, this can only be true if $q_1 = q_2$. Similarly, $q_2 = q_3$.

Assume without loss of generality that $BC = \sqrt{n}$ so that $q_1 = q_2 = q_3 = n$ and $p_1 = 1$. Also assume that A is at $(0,0)$, B is at (w,x), and C is at (y, z). By the triangle inequality, we must have $p_2 = p_3$ and hence

$$w^2 + x^2 = y^2 + z^2 = p_2^2 n$$
$$(w - y)^2 + (x - z)^2 = n.$$

Notice that

$$n = (w-y)^2 + (x-z)^2 \equiv w^2 + x^2 + y^2 + z^2 = 2p_2^2 n \equiv 0 \pmod{2},$$

so n is even. Thus w and x have the same parity; and y and z have the same parity. Then w, x, y, z must *all* have the same parity because $w^2 + x^2 \equiv y^2 + z^2 \pmod{4}$. Then $n = (w-y)^2 + (x-z)^2 \equiv 0 \pmod{4}$, contradicting the assumption that n is square-free.

Therefore our original assumption was false, and the ratio of the circumradius to the inradius is indeed always irrational.

Note. Without the condition that n is square-free, the ratio *can* be rational. For example, the points $(i, 2j - i)$ form a grid of points $\sqrt{2}$ apart. In this grid, we can find a $3\sqrt{2}$-$4\sqrt{2}$-$5\sqrt{2}$ right triangle by choosing, say, the points $(0,0)$, $(3,3)$, and $(7,-1)$. Then $q_1 = q_2 = q_3$, and the ratio is indeed rational.

Problem 4 Find the number of all natural numbers n, $4 \leq n \leq 1023$, whose binary representations do not contain three consecutive equal digits.

Solution. A *binary string* is a finite string of digits, all either 0 or 1. Call such a string (perhaps starting with zeroes) *valid* if it does not contain three consecutive equal digits. Let a_n represent the number of valid n-digit strings; let s_n be the number of valid strings starting with two equal digits; and let d_n be the number of valid strings starting with two different digits. Observe that $a_n = s_n + d_n$ for all n.

An $(n+2)$-digit string starting with 00 is valid if and only if its last n digits form a valid string starting with 1. Similarly, an $(n+2)$-digit string starting with 11 is valid if and only if its last n digits form a valid string starting with 0. Thus, $s_{n+2} = a_n = s_n + d_n$.

An $(n+2)$-digit string starting with 01 is valid if and only if its last n digits form a valid string starting with 00, 01, or 10. Similarly, an $(n+2)$-digit string starting with 10 is valid if and only if its last n digits form a valid string starting with 11, 01, or 10. Thus, $d_{n+2} = s_n + 2d_n$.

Solving these recursions gives

$$s_{n+4} = 3s_{n+2} - s_n \quad \text{and} \quad d_{n+4} = 3d_{n+2} - d_n,$$

which when added together yield

$$a_{n+4} = 3a_{n+2} - a_n.$$

Thus we can calculate initial values of a_n and then use the recursion to find other values:

n	1	2	3	4	5	6	7	8	9	10
a_n	2	4	6	10	16	26	42	68	110	178

Now of the a_n valid n-digit strings, only half start with 1. Thus only half are binary representations of positive numbers, and exactly

$$\frac{1}{2}(a_1 + a_2 + \cdots + a_{10}) = 231$$

numbers between 1 and 1023 have the desired property. Ignoring 1, 2, and 3, we find that the answer is 228.

Problem 5 The vertices A, B and C of an acute-angled triangle ABC lie on the sides B_1C_1, C_1A_1 and A_1B_1 of triangle $A_1B_1C_1$ such that $\angle ABC = \angle A_1B_1C_1$, $\angle BCA = \angle B_1C_1A_1$, and $\angle CAB = \angle C_1A_1B_1$. Prove that the orthocenters of the triangle ABC and triangle $A_1B_1C_1$ are equidistant from the circumcenter of triangle ABC.

Solution. Let H and H_1 be the orthocenters of triangles ABC and $A_1B_1C_1$, respectively, and let O, O_A, O_B, O_C be the circumcenters of triangles ABC, A_1BC, AB_1C, and ABC_1, respectively.

First note that $\angle BA_1C = \angle C_1A_1B_1 = \angle CAB = 180° - \angle CHB$, showing that BA_1CH is cyclic. Moreover,

$$O_AA_1 = \frac{BC}{2\sin \angle BA_1C} = \frac{CB}{2\sin \angle CAB} = OA$$

so circles ABC and BA_1CH have the same radius. Similarly, CB_1AH and AC_1BH are cyclic with circumradius OA. Then $\angle HBC_1 = 180° - \angle C_1AH = \angle HAB_1 = 180° - \angle B_1CH = \angle HCA_1$, and hence angles $\angle HO_CC_1, \angle HO_AA_1, \angle HO_BB_1$ are equal as well.

Let $\angle(\vec{r}_1, \vec{r}_2)$ denote the angle between rays $\vec{r}_1$ and $\vec{r}_2$. Because $O_AC = O_AB = HB = HC$, quadrilateral BO_ACH is a rhombus and hence a parallelogram. Then

$$\angle(\overrightarrow{OA}, \overrightarrow{HO_A}) = \angle(\overrightarrow{OA}, \overrightarrow{OB}) + \angle(\overrightarrow{OB}, \overrightarrow{HO_A})$$
$$= 2\angle ACB + \angle(\overrightarrow{CO_A}, \overrightarrow{HO_A})$$
$$= 2\angle ACB + \angle CO_AH$$
$$= 2\angle ACB + 2\angle CBH$$
$$= 2\angle ACB + 2(90° - \angle ACB)$$
$$= 180°.$$

Similarly, $\angle(\overrightarrow{OB}, \overrightarrow{HO_B}) = \angle(\overrightarrow{OC}, \overrightarrow{HO_C}) = 180°$. Combining this result with $\angle HO_AA_1 = \angle HO_BB_1 = \angle HO_CC_1$ from above, we find that

$$\angle(\overrightarrow{OA}, \overrightarrow{O_AA_1}) = \angle(\overrightarrow{OB}, \overrightarrow{O_BB_1}) = \angle(\overrightarrow{OC}, \overrightarrow{O_CC_1}).$$

Let this common angle be θ.

We now use complex numbers with the origin at O, letting p denote the complex number representing point P. Because HBO_AC is a parallelogram we have $o_A = b + c$ and we can write $a_1 = b + c + xa$ where $x = \text{cis}\,\theta$. We also have $b_1 = c + a + xb$ and $c_1 = a + b + xc$ for the same x. We can rewrite these relations as

$$a_1 = a + b + c + (x-1)a,$$
$$b_1 = a + b + c + (x-1)b,$$
$$c_1 = a + b + c + (x-1)c.$$

Thus the map sending z to $a + b + c + (x-1)z = h + (x-1)z$ is a spiral similarity taking triangle ABC into triangle $A'B'C'$. It follows that this map also takes H to H_1, so

$$h_1 = h + (x-1)h = xh$$

and $OH_1 = |h_1| = |x||h| = |h| = OH$, as desired.

Problem 6 Prove that the equation
$$x^3 + y^3 + z^3 + t^3 = 1999$$
has infinitely many integral solutions.

Solution. Observe that $(m - n)^3 + (m + n)^3 = 2m^3 + 6mn^2$. Now suppose we want a general solution of the form
$$(x, y, z, t) = \left(a - b, a + b, \frac{c}{2} - \frac{d}{2}, \frac{c}{2} + \frac{d}{2}\right)$$
for integers a, b and odd integers c, d. One simple solution to the given equation is $(x, y, z, t) = (10, 10, -1, 0)$, so we try setting $a = 10$ and $c = -1$. Then
$$(x, y, z, t) = \left(10 - b, 10 + b, -\frac{1}{2} - \frac{d}{2}, -\frac{1}{2} + \frac{d}{2}\right)$$
is a solution exactly when
$$(2000 + 60b^2) - \frac{1 + 3d^2}{4} = 1999 \iff d^2 - 80b^2 = 1.$$
The second equation is a Pell's equation with solution $(d_1, b_1) = (9, 1)$. We can generate infinitely many more solutions by setting
$$(d_{n+1}, b_{n+1}) = (9d_n + 80b_n, 9b_n + d_n) \quad \text{for } n = 1, 2, 3, \ldots.$$
This can be proven by induction, and it follows from a general recursion
$$(p_{n+1}, q_{n+1}) = (p_1 p_n + q_1 q_n D, p_1 q_n + q_1 p_n)$$
for generating solutions to $p^2 - Dq^2 = 1$ given a nontrivial solution (p_1, q_1).

A quick check also shows that each d_n is odd. Thus because there are infinitely many solutions (b_n, d_n) to the Pell's equation (and with each d_n odd), there are infinitely many integral solutions
$$(x_n, y_n, z_n, t_n) = \left(10 - b_n, 10 + b_n, -\frac{1}{2} - \frac{d_n}{2}, -\frac{1}{2} + \frac{d_n}{2}\right)$$
to the original equation.

4 Canada

Problem 1 Find all real solutions to the equation $4x^2 - 40\lfloor x \rfloor + 51 = 0$, where $\lfloor x \rfloor$ denotes the greatest integer less than or equal to x.

Solution. Note that

$$(2x - 3)(2x - 17) = 4x^2 - 40x + 51 \leq 4x^2 - 40\lfloor x \rfloor + 51 = 0,$$

which gives $1.5 \leq x \leq 8.5$ and $1 \leq \lfloor x \rfloor \leq 8$. Then

$$x = \frac{\sqrt{40\lfloor x \rfloor - 51}}{2},$$

so it is necessary that

$$\lfloor x \rfloor = \left\lfloor \frac{\sqrt{40\lfloor x \rfloor - 51}}{2} \right\rfloor.$$

Testing $\lfloor x \rfloor \in \{1, 2, 3, \ldots, 8\}$ into this equation, we find that $\lfloor x \rfloor$ can only equal 2, 6, 7, or 8. Thus the only solutions for x are $\sqrt{29}/2$, $\sqrt{189}/2$, $\sqrt{229}/2$, and $\sqrt{269}/2$. A quick check confirms that these values work.

Problem 2 Let ABC be an equilateral triangle of altitude 1. A circle, with radius 1 and center on the same side of AB as C, rolls along the segment AB; as it rolls, it always intersects both $\overline{AC}$ and $\overline{BC}$. Prove that the length of the arc of the circle that is inside the triangle remains constant.

Solution. Let ω be "the circle." Let O be the center of ω. Let ω intersect segments $\overline{AC}$ and $\overline{BC}$ at M and N, respectively. Let the circle through O, C, and M intersect $\overline{BC}$ again at P. Now $\angle PMO = 180° - \angle OCP = 60° = \angle MCO = \angle MPO$, so $OP = OM = 1$, and P coincides with N. Thus, $\angle MON = \angle MOP = \angle MCP = 60°$. Therefore, the angle of the arc of ω that is inside the triangle ABC is constant, and hence the length of the arc must be constant as well.

Problem 3 Determine all positive integers n such that $n = d(n)^2$, where $d(n)$ denotes the number of positive divisors of n (including 1 and n).

Solution. Label the prime numbers $p_1 = 2, p_2 = 3, \ldots$. Because n is a perfect square, we have

$$n = \prod_{i=1}^{\infty} p_i^{2a_i}, \quad d(n) = \prod_{i=1}^{\infty} (2a_i + 1)$$

for some nonnegative integers a_i. Then $d(n)$ is odd and so is n, whence $a_1 = 0$. Because $\frac{d(n)}{\sqrt{n}} = 1$, we have

$$\prod_{i=1}^{\infty} \frac{2a_i + 1}{p_i^{a_i}} = 1.$$

By Bernoulli's inequality, we have $p_i^{a_i} \geq (p_i - 1)a_i + 1 > 2a_i + 1$ for all primes $p_i \geq 5$ that divide n. Also, $3^{a_2} \geq 2a_2 + 1$ with equality only when $a_2 \in \{0, 1\}$. Thus, for equality to hold above, we must have $a_1 = a_3 = a_4 = a_5 = \cdots = 0$ and $a_2 \in \{0, 1\}$. Therefore, $n = 1$ and $n = 9$ are the only solutions.

Problem 4 Suppose $a_1, a_2, \ldots, a_8$ are eight distinct integers from the set $S = \{1, 2, \ldots, 17\}$. Show that there exists an integer $k > 0$ such that the equation $a_i - a_j = k$ has at least three different solutions. Also, find a specific set of 7 distinct integers $\{b_1, b_2, \cdots, b_7\}$ from S such that the equation

$$b_i - b_j = k$$

does not have three distinct solutions for any $k > 0$.

Solution. For the first part of this problem, assume, without loss of generality, that $a_1 < a_2 < \cdots < a_8$. Also assume, for the purpose of contradiction, that there does *not* exist an integer $k > 0$ such that the equation $a_i - a_j = k$ has at least three different solutions. Let $\delta_i = a_{i+1} - a_i$ for $i = 1, 2, \ldots, 7$. Then

$$16 \geq a_8 - a_1 = \delta_1 + \ldots + \delta_7 \geq 1 + 1 + 2 + 2 + 3 + 3 + 4 = 16,$$

for otherwise three of the δ_i's would be equal, a contradiction. Because equality must hold, $\Pi = (\delta_1, \delta_2, \ldots, \delta_7)$ must be a permutation of $(1, 1, 2, 2, 3, 3, 4)$.

We say we have a *m-n pair* if some $(\delta_i, \delta_{i+1}) = (m, n)$ or (n, m). Note that we cannot have any 1-1 or 1-2 pairs (δ_i, δ_{i+1}) — otherwise we would have $a_{i+2} - a_i = 2$ or 3, giving at least three solutions to $a_i - a_j = 2$ or 3. Nor can we have two 1-3 pairs because then, along with $\delta_i = 4$, we'd have three solutions to $a_i - a_j = 4$. Then considering what entries each 1 is next to, we see that we must have $\Pi = (1, 4, \ldots, 3, 1)$, $(1, 3, \ldots, 4, 1)$, $(1, 4, 1, 3, \ldots)$, or $(\ldots, 3, 1, 4, 1)$.

Now we cannot have any 2-2 pairs—otherwise, along with the 1-3 pair and the $\delta_i = 4$, we'd have three solutions to $a_i - a_j = 4$. Thus we have $\Pi = (1, 4, 2, 3, 2, 3, 1)$, $(1, 3, 2, 3, 2, 4, 1)$, $(1, 4, 1, 3, 2, 3, 2)$, or

$(2, 3, 2, 3, 1, 4, 1)$. In any case, there are at least four solutions to $a_i - a_j = 5$, a contradiction.

Thus, regardless of the $\{a_1, a_2, \ldots, a_8\}$ that we choose, for some integer $k \in \{1, 2, 3, 4, 5\}$ the equation $a_i - a_j = k$ has at least three different solutions.

For the second part of the problem, let $(b_1, b_2, \ldots, b_7) = (1, 2, 4, 9, 14, 16, 17)$. Each of 1, 2, 3, 5, 7, 8, 12, 13, and 15 is the difference of exactly two pairs of the b_i, and each of 10, 14, and 16 is the difference of exactly one pair of the b_i. No number is the difference of more than two such pairs, and hence the set $\{b_1, b_2, \ldots, b_7\}$ suffices.

Problem 5 Let x, y, z be nonnegative real numbers such that $x + y + z = 1$. Prove that
$$x^2 y + y^2 z + z^2 x \leq \frac{4}{27},$$
and determine when equality occurs.

Solution. Assume without loss of generality that $x = \max\{x, y, z\}$.

If $x \geq y \geq z$, then
$$x^2 y + y^2 z + z^2 x \leq x^2 y + y^2 z + z^2 x + z(xy + (x-y)(y-z))$$
$$= (x+z)^2 y = 4\left(\frac{1}{2} - \frac{1}{2}y\right)\left(\frac{1}{2} - \frac{1}{2}y\right) y \leq \frac{4}{27},$$
where the last inequality follows from the AM-GM inequality. Equality occurs if and only if $z = 0$ (from the first inequality) and $y = \frac{1}{3}$, in which case $(x, y, z) = \left(\frac{2}{3}, \frac{1}{3}, 0\right)$.

If $x \geq z \geq y$, then
$$x^2 y + y^2 z + z^2 x = x^2 z + z^2 y + y^2 x - (x-z)(z-y)(x-y)$$
$$\leq x^2 z + z^2 y + y^2 x \leq \frac{4}{27},$$
where the second inequality is true from the result we proved for $x \geq y \geq z$ (except with y and z reversed). Equality holds in the first inequality only when two of x, y, z are equal, and in the second inequality only when $(x, z, y) = \left(\frac{2}{3}, \frac{1}{3}, 0\right)$. Because these conditions can't both be true, the inequality is actually strict in this case.

Therefore the inequality is indeed true, and equality holds when (x, y, z) equals $\left(\frac{2}{3}, \frac{1}{3}, 0\right)$, $\left(\frac{1}{3}, 0, \frac{2}{3}\right)$, or $\left(0, \frac{2}{3}, \frac{1}{3}\right)$.

5 China

Problem 1 Let ABC be an acute triangle with $\angle C > \angle B$. Let D be a point on side BC such that $\angle ADB$ is obtuse, and let H be the orthocenter of triangle ABD. Suppose that F is a point inside triangle ABC and is on the circumcircle of triangle ABD. Prove that F is the orthocenter of triangle ABC if and only if both of the following are true: $HD \parallel CF$, and H is on the circumcircle of triangle ABC.

Solution. All angles are directed modulo $180°$. First observe that if P is the orthocenter of triangle UVW, then

$$\angle VPW = (90° - \angle PWV) + (90° - \angle WVP)$$
$$= \angle WV + \angle UV = 180° - \angle VOW.$$

Suppose that F is the orthocenter of triangle ABC. Then

$$\angle ACB = 180° - \angle AFB = 180° - \angle ADB = \angle AHB,$$

so $ACHB$ is cyclic. As for the other condition, lines CF and HD are both perpendicular to side AB, so they are parallel.

Conversely, suppose that $HD \parallel CF$ and that H is on the circumcircle of triangle ABC. Because $AFDB$ and $AHCB$ are cyclic,

$$\angle AFB = \angle ADB = 180° - \angle AHB = 180° - \angle ACB.$$

Thus, F is an intersection point of the circle defined by $\angle AFB = 180° - \angle ACB$ and the line defined by $CF \perp AB$. There are only two such points: the orthocenter of triangle ABC and the reflection of C across line AB. The latter point lies outside of triangle ABC, and hence F must indeed be the orthocenter of triangle ABC.

Problem 2 Let a be a real number. Let $\{f_n(x)\}$ be a sequence of polynomials such that $f_0(x) = 1$ and $f_{n+1}(x) = xf_n(x) + f_n(ax)$ for $n = 0, 1, 2, \ldots$.

(a) Prove that

$$f_n(x) = x^n f_n\left(\frac{1}{x}\right)$$

for $n = 0, 1, 2, \ldots$.

(b) Find an explicit expression for $f_n(x)$.

Solution. When $a = 1$, we have $f_n(x) = (x+1)^n$ for all n, and part (a) is easily checked. Now assume that $a \neq 1$.

Observe that f_n has degree n and always has constant term 1. Write $f_n(x) = c_0 + c_1 x + \cdots + c_n x^n$. We prove by induction on n that
$$(a^i - 1)c_i = (a^{n+1-i} - 1)c_{i-1}$$
for $0 \leq i \leq n$ (where we let $c_{-1} = 0$).

The base case $n = 0$ is clear. Now suppose that
$$f_{n-1}(x) = b_0 + b_1 x + \cdots + b_{n-1} x^{n-1}$$
satisfies the claim: specifically, we know $(a^i - 1)b_i = (a^{n-i} - 1)b_{i-1}$ and $(a^{n+1-i} - 1)b_{i-2} = (a^{i-1} - 1)b_{i-1}$ for $i \geq 1$.

For $i = 0$, the claim states $0 = 0$. For $i \geq 1$, the given recursion gives $c_i = b_{i-1} + a^i b_i$ and $c_{i-1} = b_{i-2} + a^{i-1} b_{i-1}$. Then the claim is equivalent to

$$(a^i - 1)c_i = (a^{n+1-i} - 1)c_{i-1} \iff (a^i - 1)(b_{i-1} + a^i b_i)$$
$$= (a^{n+1-i} - 1)(b_{i-2} + a^{i-1} b_{i-1})$$
$$\iff (a^i - 1)b_{i-1} + a^i(a^i - 1)b_i$$
$$= (a^{n+1-i} - 1)b_{i-2} + (a^n - a^{i-1})b_{i-1}$$
$$\iff (a^i - 1)b_{i-1} + a^i(a^{n-i} - 1)b_{i-1}$$
$$= (a^{i-1} - 1)b_{i-1} + (a^n - a^{i-1})b_{i-1} \iff (a^n - 1)b_{i-1}$$
$$= (a^n - 1)b_{i-1},$$

so it is true.

Now by telescoping products, we have
$$c_i = \frac{c_i}{c_0} = \prod_{k=1}^{i} \frac{c_k}{c_{k-1}}$$
$$= \prod_{k=1}^{i} \frac{a^{n+1-k} - 1}{a^k - 1} = \frac{\prod_{k=n+1-i}^{n}(a^k - 1)}{\prod_{k=1}^{i}(a^k - 1)}$$
$$= \frac{\prod_{k=i+1}^{n}(a^k - 1)}{\prod_{k=1}^{n-i}(a^k - 1)} = \prod_{k=1}^{n-i} \frac{a^{n+1-k} - 1}{a^k - 1}$$
$$= \prod_{k=1}^{n-i} \frac{c_k}{c_{k-1}} = \frac{c_{n-i}}{c_0} = c_{n-i},$$

giving our explicit form. Also, $f_n(x) = x^n f_n\left(\frac{1}{x}\right)$ if and only if $c_i = c_{n-i}$ for $i = 0, 1, \ldots, n$, and from above this is indeed the case. This completes the proof.

Problem 3 There are 99 space stations. Each pair of space stations is connected by a tunnel. There are 99 two-way main tunnels, and all the other tunnels are strictly one-way tunnels. A group of 4 space stations is called *connected* if one can reach each station in the group from every other station in the group without using any tunnels other than the 6 tunnels which connect them. Determine the maximum number of connected groups.

Solution. In this solution, let $f(x) = \frac{x(x-1)(x-2)}{6}$, an extension of the definition of $\binom{x}{3}$ to all real numbers x.

In a group of 4 space stations, call a station *troublesome* if three one-way tunnels lead toward it or three one-way tunnels lead out of it. In each group there is at most one troublesome station of each type for a count of at most two troublesome stations.

If a group of four stations contains a troublesome station, say A, then it is either impossible to reach A from the other stations or impossible to reach the other stations from A.

We claim that if a group of four stations A, B, C, D contains neither a troublesome station nor a two-way tunnel, then it is connected. We start at any station and travel along one-way tunnels until we reach our starting position. If we pass through all four stations, we are finished. Otherwise, we must pass through exactly three — say, from A to B to C. From each of these three stations we can reach any other of the three. Now, because D is not troublesome, we may assume without loss of generality that a one-way tunnel leads from A to D; thus, we can reach D from any station. Similarly, we can reach any station from D. Therefore, our group of four stations is connected.

Label the stations $1, 2, \ldots, 99$. For $i = 1, 2, \ldots, 99$, let a_i one-way tunnels point into station i and b_i one-way tunnels point out. Station i is troublesome in $\binom{a_i}{3} + \binom{b_i}{3}$ groups of four. Adding over all stations, we obtain a total count of $\sum_{i=1}^{198}\left(\binom{a_i}{3} + \binom{b_i}{3}\right)$. This equals $\sum_{i=1}^{198} f(x_i)$ for nonnegative integers $x_1, x_2, \ldots, x_{198}$ with $\sum_{i=1}^{198} x_i = 96 \cdot 99$. Without loss of generality, suppose that $x_1, x_2, \ldots, x_k$ are at least 1 and $x_{k+1}, x_{k+2}, \ldots, x_{198}$ are zero. Because $f(x)$ is convex as a function of x for $x \geq 1$, this is at least $k\binom{96 \cdot 99/k}{2}$. Also, $mf(x) \geq f(mx)$ when $m \leq 1$ and $mx \geq 2$. Letting $m = k/198$ and $mx = 96 \cdot 99/198 = 48$, we find that our total count is at least $198\binom{48}{2}$. Because each unconnected group of 4 stations has at most two troublesome stations, there are at least $99\binom{48}{3}$ unconnected groups of four and at most $\binom{99}{4} - 99\binom{48}{3}$ connected groups.

All that is left to show is that this maximum can be attained. Arrange the stations around a circle, and put a two-way tunnel between any two adjacent stations. Given two distinct stations A and B that are not adjacent, place a one-way tunnel running from station A to station B if and only if A is 3, 5, ..., or 97 stations away clockwise from B. In this arrangement, every station is troublesome $2\binom{48}{3}$ times. It is easy to check that under this arrangement, any group of four stations containing a two-way tunnel is connected. Now suppose that station A is troublesome in a group of four stations A, B, C, D with B closest and D furthest away clockwise from A. If one-way tunnels lead from A to the other tunnels, three one-way tunnels must lead to D from the other tunnels; if one-way tunnels lead to A from the other tunnels, three one-way tunnels must lead from B to the other tunnels. Thus every unconnected group of four stations has exactly two troublesome stations. Hence equality holds in the previous paragraph, and there are indeed exactly $\binom{99}{4} - 99\binom{48}{3}$ connected groups.

Problem 4 Let m be a positive integer. Prove that there are integers a, b, k, such that both a and b are odd, $k \geq 0$, and

$$2m = a^{19} + b^{99} + k \cdot 2^{1999}.$$

Solution. The key observation is that if $\{t_1, \cdots, t_n\}$ equals $\{1, 3, 5, \ldots, 2^n - 1\}$ modulo 2^n, then $\{t_1^s, \cdots, t_n^s\}$ does as well for any odd positive integer s. To show this, note that for $i \neq j$,

$$t_i^s - t_j^s = (t_i - t_j)(t_i^{s-1} + t_i^{s-2}t_j + \cdots + t_j^{s-1}).$$

Because $t_i^{s-1} + t_i^{s-2}t_j + \cdots + t_j^{s-1}$ is an odd number, $t_i \equiv t_j \iff t_i^s \equiv t_j^s$ (mod 2^n).

Therefore there exists an odd number a_0 such that $2m - 1 \equiv a_0^{19}$ (mod 2^{1999}). Hence if we pick $a \equiv a_0$ (mod 2^{1999}) sufficiently negative so that $2m - 1 - a^{19} > 0$, then

$$(a, b, k) = \left(a, 1, \frac{2m - 1 - a^{19}}{2^{1999}}\right)$$

is a solution to the equation.

Problem 5 Determine the maximum value of λ such that if $f(x) = x^3 + ax^2 + bx + c$ is a cubic polynomial with all its roots nonnegative, then
$$f(x) \geq \lambda(x-a)^3$$
for all $x \geq 0$. Find the equality condition.

Solution. Let α, β, γ be the three roots. Without loss of generality, suppose that $0 \leq \alpha \leq \beta \leq \gamma$. We have
$$x - a = x + \alpha + \beta + \gamma \geq 0 \quad \text{and} \quad f(x) = (x-\alpha)(x-\beta)(x-\gamma).$$

If $0 \leq x \leq \alpha$, then (applying the AM-GM inequality to obtain the first inequality below)
$$-f(x) = (\alpha - x)(\beta - x)(\gamma - x) \leq \frac{1}{27}(\alpha + \beta + \gamma - 3x)^3$$
$$\leq \frac{1}{27}(x + \alpha + \beta + \gamma)^3 = \frac{1}{27}(x-a)^3,$$
so that $f(x) \geq -\frac{1}{27}(x-a)^3$. Equality holds exactly when $\alpha - x = \beta - x = \gamma - x$ in the first inequality and $\alpha + \beta + \gamma - 3x = x + \alpha + \beta + \gamma$ in the second; that is, when $x = 0$ and $\alpha = \beta = \gamma$.

If $\beta \leq x \leq \gamma$, then (again applying the AM-GM inequality to obtain the first inequality below)
$$-f(x) = (x - \alpha)(x - \beta)(\gamma - x) \leq \frac{1}{27}(x + \gamma - \alpha - \beta)^3$$
$$\leq \frac{1}{27}(x + \alpha + \beta + \gamma)^3 = \frac{1}{27}(x-a)^3,$$
so that again $f(x) \geq -\frac{1}{27}(x-a)^3$. Equality holds exactly when $x - \alpha = x - \beta = \gamma - x$ in the first inequality and $x + \gamma - \alpha - \beta = x + \alpha + \beta + \gamma$; that is, when $\alpha = \beta = 0$ and $\gamma = 2x$.

Finally, when $\alpha < x < \beta$ or $x > \gamma$ then
$$f(x) > 0 \geq -\frac{1}{27}(x-a)^3.$$

Thus, $\lambda = -\frac{1}{27}$ works. From the above reasoning we can find that λ must be at most $-\frac{1}{27}$ or else the inequality fails for the polynomial $f(x) = x^2(x-1)$ at $x = \frac{1}{2}$. Equality occurs either when $\alpha = \beta = \gamma$ and $x = 0$; or when $\alpha = \beta = 0$, γ is any nonnegative real, and $x = \frac{\gamma}{2}$.

Problem 6 A $4 \times 4 \times 4$ cube is composed of 64 unit cubes. The faces of 16 unit cubes are to be colored red. A coloring is called *interesting* if there is exactly 1 red unit cube in every $1 \times 1 \times 4$ rectangular box composed of 4 unit cubes. Determine the number of interesting colorings. (Two colorings are different even if one can be transformed into another by a series of rotations.)

Solution. Pick one face of the cube as our bottom face. For each unit square A on this face, we consider the vertical $1 \times 1 \times 4$ box with A at its bottom. If the i-th unit cube up (counted from A) in the box is colored, then write the number i in A. Each interesting coloring is mapped one-to-one to a 4×4 *Latin square* on the bottom face. (In an $n \times n$ Latin square, each row and column contains each of n symbols $a_1, \ldots, a_n$ exactly once.) Conversely, given a Latin square we can reverse this construction. Therefore, to solve the problem, we only need to count the number of distinct 4×4 Latin squares.

Note that switching rows of a Latin square will generate another Latin square. Thus if our four symbols are a, b, c, d, then each of the $4! \cdot 3!$ arrangements of the first row and column correspond to the same number of Latin squares. Therefore there are $4! \cdot 3! \cdot x$ four-by-four Latin squares, where x is the number of Latin squares whose first row and column both contain the symbols a, b, c, d in that order. The entry in the second row and second column equals either a, c, or d, yielding the Latin squares

$$\begin{bmatrix} a & b & c & d \\ b & a & d & c \\ c & d & a & b \\ d & c & b & a \end{bmatrix}, \begin{bmatrix} a & b & c & d \\ b & a & d & c \\ c & d & b & a \\ d & c & a & b \end{bmatrix},$$

$$\begin{bmatrix} a & b & c & d \\ b & c & d & a \\ c & d & a & b \\ d & a & b & c \end{bmatrix}, \begin{bmatrix} a & b & c & d \\ b & d & a & c \\ c & a & d & b \\ d & c & b & a \end{bmatrix}.$$

Thus $x = 4$, and there are $4! \cdot 3! \cdot 4 = 576$ interesting colorings.

6 Czech and Slovak Republics

Problem 1 In the fraction
$$\frac{29 \div 28 \div 27 \div \cdots \div 16}{15 \div 14 \div 13 \div \cdots \div 2}$$
parentheses may be repeatedly placed anywhere on the numerator, granted they are also placed on the identical locations in the denominator.

(a) Find the least possible integral value of the resulting expression.

(b) Find all possible integral values of the resulting expression.

Solution. (a) The resulting expression can always be written (if we refrain from cancelling terms) as a ratio $\frac{A}{B}$ of two integers A and B satisfying

$$AB = (2)(3)\cdots(29) = 29! = 2^{25} \cdot 3^{13} \cdot 5^6 \cdot 7^4 \cdot 11^2 \cdot 13^2 \cdot 17 \cdot 19 \cdot 23 \cdot 29.$$

(To find these exponents, we could either count primes directly factor by factor, or use the rule that

$$\left\lfloor \frac{n}{p} \right\rfloor + \left\lfloor \frac{n}{p^2} \right\rfloor + \left\lfloor \frac{n}{p^3} \right\rfloor + \cdots \tag{1}$$

is the exponent of p in $n!$.)

The primes that have an odd exponent in the factorization of $29!$ cannot "vanish" from the ratio $\frac{A}{B}$ even after making any cancellations. For this reason no integer value of the result can be less than

$$H = 2 \cdot 3 \cdot 17 \cdot 19 \cdot 23 \cdot 29 = 1292646.$$

On the other hand,

$$\frac{29 \div ((\cdots((28 \div 27) \div 26) \div \cdots \div 17) \div 16)}{15 \div ((\cdots((14 \div 13) \div 12) \div \cdots \div 3) \div 2)}$$

$$= \frac{29 \cdot 14}{15 \cdot 28} \cdot \frac{(27)(26) \cdots (16)}{(13)(12) \cdots (2)}$$

$$= \frac{29 \cdot 14^2}{28} \cdot \frac{27!}{(15!)^2}$$

$$= 29 \cdot 7 \cdot \frac{2^{23} \cdot 3^{13} \cdot 5^6 \cdot 7^3 \cdot 11^2 \cdot 13^2 \cdot 17 \cdot 19 \cdot 23}{(2^{11} \cdot 3^6 \cdot 5^3 \cdot 7^2 \cdot 11 \cdot 13)^2} = H.$$

(Again it helps to count exponents in factorials using (1).) The number H is thus the desired least value.

(b) Let's examine the products A and B more closely. In each of the fourteen pairs of numbers

$$\{29,15\},\{28,14\},\ldots,\{16,2\},$$

one of the numbers is a factor in A and the other is a factor in B. The resulting value V can then be written as a product

$$\left(\frac{29}{15}\right)^{\epsilon_1}\left(\frac{28}{14}\right)^{\epsilon_2}\cdots\left(\frac{16}{2}\right)^{\epsilon_{14}},$$

where each ϵ_i equals ± 1, and where $\epsilon_1 = 1$ and $\epsilon_2 = -1$ no matter how the parentheses are placed. Because the fractions $\frac{27}{13}, \frac{26}{12}, \ldots, \frac{16}{2}$ are greater than 1, the resulting value V (whether an integer or not) has to satisfy the estimate

$$V \leq \frac{29}{15}\cdot\frac{14}{28}\cdot\frac{27}{13}\cdot\frac{26}{12}\cdots\cdots\frac{16}{2} = H,$$

where H is the number determined in part (a). It follows that H is the *only* possible integer value of V!

Problem 2 In a tetrahedron $ABCD$ we denote by E and F the midpoints of the medians from the vertices A and D, respectively. (The median from a vertex of a tetrahedron is the segment connecting the vertex and the centroid of the opposite face.) Determine the ratio of the volumes of tetrahedra $BCEF$ and $ABCD$.

Solution. Let K and L be the midpoints of the edges BC and AD, and let A_0, D_0 be the centroids of triangles BCD and ABC, respectively. Both medians AA_0 and DD_0 lie in the plane AKD, and their intersection T (the centroid of the tetrahedron) divides them in $3:1$ ratios. T is also the midpoint of $\overline{KL}$, because

$$\vec{T} = \frac{1}{4}(\vec{A}+\vec{B}+\vec{C}+\vec{D})$$
$$= \frac{1}{2}\left(\frac{1}{2}(\vec{A}+\vec{D}) + \frac{1}{2}(\vec{B}+\vec{C})\right) = \frac{1}{2}(\vec{K}+\vec{L}).$$

It follows that

$$\frac{ET}{AT} = \frac{FT}{DT} = \frac{1}{3},$$

and hence $\triangle ATD \sim \triangle ETF$ and $EF = \frac{1}{3}AD$. Because the plane BCL bisects both segments AD and EF into halves, it also divides both tetrahedra $ABCD$ and $BCEF$ into two parts of equal volume. Let G be

the midpoint of $\overline{EF}$. The corresponding volumes then satisfy
$$\frac{[BCEF]}{[ABCD]} = \frac{[BCGF]}{[BCLD]} = \frac{GF}{LD} \cdot \frac{[BCG]}{[BCL]} = \frac{1}{3}\frac{KG}{KL} = \frac{1}{3} \cdot \frac{2}{3} = \frac{2}{9}.$$

Problem 3 Show that there exists a triangle ABC for which, with the usual labelling of sides and medians, it is true that $a \neq b$ and $a + m_a = b + m_b$. Show further that there exists a number k such that for each such triangle $a + m_a = b + m_b = k(a+b)$. Finally, find all possible ratios $a:b$ of the sides of these triangles.

Solution. We know that
$$m_a^2 = \frac{1}{4}(2b^2 + 2c^2 - a^2), \quad m_b^2 = \frac{1}{4}(2a^2 + 2c^2 - b^2),$$
so
$$m_a^2 - m_b^2 = \frac{3}{4}(b^2 - a^2).$$
If the desired condition $m_a - m_b = b - a \neq 0$ is to be satisfied, then necessarily $m_a + m_b = \frac{3}{4}(b+a)$. From the system of equations
$$m_a - m_b = b - a$$
$$m_a + m_b = \frac{3}{4}(b+a)$$
we find that we would then have $m_a = \frac{1}{8}(7b - a)$, $m_b = \frac{1}{8}(7a - b)$, and
$$a + m_a = b + m_b = \frac{7}{8}(a+b).$$
Thus $k = \frac{7}{8}$.

Now we examine for what $a \neq b$ there exists a triangle ABC with sides a, b and medians $m_a = \frac{1}{8}(7b-a), m_b = \frac{1}{8}(7a-b)$. We can find all three side lengths of the triangle AB_1G, where G is the centroid of the triangle ABC and B_1 is the midpoint of the side AC:
$$AB_1 = \frac{b}{2}, \quad AG = \frac{2}{3}m_a = \frac{2}{3} \cdot \frac{1}{8}(7b-a) = \frac{1}{12}(7b-a),$$
$$B_1G = \frac{1}{3}m_b = \frac{1}{3} \cdot \frac{1}{8}(7a-b) = \frac{1}{24}(7a-b).$$
Examining the triangle inequalities for these three lengths, we get the condition
$$\frac{1}{3} < \frac{a}{b} < 3,$$
from which the value $\frac{a}{b} = 1$ has to be excluded by assumption. This condition is also sufficient: once the triangle AB_1G has been constructed,

it can always be completed to a triangle ABC with $b = AC, m_a = AA_1, m_b = BB_1$. Then from the equality $m_a^2 - m_b^2 = \frac{3}{4}(b^2 - a^2)$ we would also have $a = BC$.

Problem 4 In a certain language there are only two letters, A and B. The words of this language satisfy the following axioms:

(i) There are no words of length 1, and the only words of length 2 are AB and BB.

(ii) A sequence of letters of length $n > 2$ is a word if and only if it can be created from some word of length less than n by the following construction: all letters A in the existing word are left unchanged, while each letter B is replaced by some word. (While performing this operation, the B's do not all have to be replaced by the same word.)

Show that for any n the number of words of length n equals

$$\frac{2^n + 2 \cdot (-1)^n}{3}.$$

Solution. Let us call any finite sequence of letters A, B a *string*. From here on, we let $\cdots$ denote a (possibly empty) string, while $***$ will stand for a string consisting of identical letters. (For example, $\underbrace{B***B}_{k}$ is a string of k B's.)

We show that an arbitrary string is a word if and only if it satisfies the following conditions: (a) the string terminates with the letter B; and (b) it either starts with the letter A, or else starts (or even wholly consists of) an even number of B's.

It is clear that these conditions are necessary: they are satisfied for both words AB and BB of length 2, and they are likewise satisfied by any new word created by the construction described in (ii) if they are satisfied by the words in which the B's are replaced.

We now show by induction on n that, conversely, any string of length n satisfying the conditions is a word. This is clearly true for $n = 1$ and $n = 2$. If $n > 2$, then a string of length n satisfying the conditions must have one of the forms

$$AA\cdots B, \quad AB\cdots B, \quad \underbrace{B***B}_{2k}A\cdots B, \quad \underbrace{B***B}_{2k+2},$$

where $2 \leq 2k \leq n - 2$. We have to show that these four types of strings arise from the construction in (ii) in which the B's are replaced by strings (of lengths less than n) satisfying the condition — that is, by *words* in view of the induction hypothesis.

The word $AA\cdots B$ arises as $A(A\cdots B)$ from the word AB. The word $AB\cdots B$ arises either as $A(B\cdots B)$ from the word AB, or as $(AB)(\cdots B)$ from the word BB, depending on whether its initial letter A is followed by an even or an odd number of B's. The word

$$\underbrace{B***B}_{2k}A\cdots B$$

arises as $(B***B)(A\cdots B)$ from the word BB, and the word

$$\underbrace{B***B}_{2k+2} \quad \text{as} \quad (\underbrace{B***B}_{2k})(BB)$$

from the word BB. This completes the proof by induction.

Now we show that the number p_n of words of length n is indeed given by the formula

$$p_n = \frac{2^n + 2\cdot(-1)^n}{3}.$$

It is clearly true for $n = 1$ and 2 because $p_1 = 0$ and $p_2 = 2$, and the formula will then follow by induction if we can show that $p_{n+2} = 2^n + p_n$ for each n. This recursion is obvious because each word of length $n+2$ is either of the form $A\cdots B$ where $\cdots$ is any of the 2^n strings of length n, or of the form $BB\cdots$ where $\cdots$ is any of the p_n words of length n.

Problem 5 In the plane an acute angle APX is given. Show how to construct a square $ABCD$ such that P lies on side BC and P lies on the bisector of angle BAQ where Q is the intersection of ray PX with CD.

Solution. Consider the rotation by 90° around the point A that maps B to D, and the points P, C, D into some points P', C', D', respectively. Because $\angle PAP' = 90°$, it follows from the nature of exterior angle bisectors that AP' bisects $\angle QAD'$. Consequently, the point P' has the same distance from $\overline{AD'}$ and $\overline{AQ}$, equal to the side length s of square $ABCD$. This distance is also the length of the altitude $\overline{AD}$ in triangle AQP'. Because the altitudes from A and P' in this triangle are equal, we thus have $AQ = P'Q$. Because we can construct P', we can also construct Q as the intersection of line PX with the perpendicular bisector of the segment AP'. The rest of the construction is obvious, and it is likewise clear that the resulting square $ABCD$ has the required property.

Problem 6 Find all pairs of real numbers a and b such that the system of equations
$$\frac{x+y}{x^2+y^2} = a, \qquad \frac{x^3+y^3}{x^2+y^2} = b$$
has a solution in real numbers (x, y).

Solution. If the given system has a solution (x, y) for $a = A$, $b = B$, then it clearly also has a solution (kx, ky) for $a = \frac{1}{k}A$, $b = kB$, for any $k \neq 0$. It follows that the existence of a solution of the given system depends only on the value of the product ab.

We therefore begin by examining the values of the expression
$$P(u, v) = \frac{(u+v)(u^3+v^3)}{(u^2+v^2)^2}$$
where the numbers u and v are normalized by the condition $u^2 + v^2 = 1$. This condition implies that
$$P(u, v) = (u+v)(u^3+v^3) = (u+v)^2(u^2 - uv + v^2)$$
$$= (u^2 + 2uv + v^2)(1 - uv) = (1 + 2uv)(1 - uv).$$

Under the condition $u^2 + v^2 = 1$ the product uv can attain all values in the interval $[-\frac{1}{2}, \frac{1}{2}]$ (if $u = \cos\alpha$ and $v = \sin\alpha$, then $uv = \frac{1}{2}\sin 2\alpha$). Hence it suffices to find the range of values of the function $f(t) = (1+2t)(1-t)$ on the interval $t \in [-\frac{1}{2}, \frac{1}{2}]$. From the formula
$$f(t) = -2t^2 + t + 1 = -2\left(t - \frac{1}{4}\right)^2 + \frac{9}{8}$$
it follows that this range of values is the closed interval with endpoints $f\left(-\frac{1}{2}\right) = 0$ and $f\left(\frac{1}{4}\right) = \frac{9}{8}$.

This means that if the given system has a solution, its parameters a and b must satisfy $0 \leq ab \leq \frac{9}{8}$, where the equality $ab = 0$ is possible only if $x + y = 0$ (then, however, $a = b = 0$).

Conversely, if a and b satisfy $0 < ab \leq \frac{9}{8}$, by our proof there exist numbers u and v such that $u^2 + v^2 = 1$ and $(u+v)(u^3+v^3) = ab$. Denoting $a' = u + v$ and $b' = u^3 + v^3$, the equality $a'b' = ab \neq 0$ implies that both ratios $\frac{a'}{a}$ and $\frac{b}{b'}$ have the same value $k \neq 0$. Then $(x, y) = (ku, kv)$ is clearly a solution of the given system for the parameter values a and b.

7 France

Problem 1

(a) What is the maximum volume of a cylinder that is inside a given cone and has the same axis of revolution as the cone? Express your answer in terms of the radius R and height H of the cone.

(b) What is the maximum volume of a ball that is inside a given cone? Again, express your answer in terms of R and H.

(c) Given fixed values for R and H, which of the two maxima you found is bigger?

Solution. Let $\ell = \sqrt{R^2 + H^2}$ be the slant height of the given cone. Also, orient the cone so that its base is horizontal and its tip is pointing upward.

(a) Any cylinder satisfying the described conditions can be enlarged, if necessary, so that the circumferences of its two circular faces lie on the cone. The top face of the cylinder cuts off a smaller cone at the top of the original cone. If the cylinder has radius r, then the smaller cone has height $r \cdot \frac{H}{R}$ and the cylinder has height $h = H - r \cdot \frac{H}{R}$. Then the volume of the cylinder is

$$\pi r^2 h = \pi r^2 H \left(1 - \frac{r}{R}\right) = 4\pi R^2 H \left(\frac{r}{2R} \cdot \frac{r}{2R} \cdot \left(1 - \frac{r}{R}\right)\right).$$

By the AM-GM inequality on $\frac{r}{2R}$, $\frac{r}{2R}$, and $1 - \frac{r}{R}$, this is at most

$$4\pi R^2 H \cdot \frac{1}{27} \left(\frac{r}{2R} + \frac{r}{2R} + \left(1 - \frac{r}{R}\right)\right)^3 = \frac{4}{27}\pi R^2 H,$$

with equality when $r/2R = 1 - r/R \iff r = \frac{2}{3}R$.

(b) Any sphere satisfying the described conditions can be translated, if necessary, so that its center lies on the cone's axis. It can then be enlarged and translated vertically, if necessary, so that it is tangent to the base and lateral face of the cone.

Let the sphere have radius r. Take a planar cross-section of the cone slicing through its axis, cutting off a triangle from the cone and a circle from the sphere. The triangle's side lengths are ℓ, ℓ, and $2R$, and its height (from the side of length $2R$) is H. The circle has radius r and is the incircle of this triangle.

The area K of the triangle is $\frac{1}{2}(2R)(H) = RH$ and its semiperimeter is $s = R + \ell$. Because $K = rs$ we have $r = \frac{RH}{R+\ell}$, and thus the volume of

the sphere is
$$\frac{4}{3}\pi r^3 = \frac{4}{3}\pi \left(\frac{RH}{R+\ell}\right)^3.$$

(c) We claim that when $h/R = \sqrt{3}$ or $2\sqrt{6}$, the two volumes are equal; when $\sqrt{3} < h/R < 2$, the sphere has larger volume; and when $0 < h/R < \sqrt{3}$ or $2 < h/R$, the cylinder has larger volume.

We wish to compare
$$\frac{4}{27}\pi R^2 H \quad \text{and} \quad \frac{4}{3}\pi \left(\frac{RH}{R+\ell}\right)^3.$$

Equivalently, multiplying by
$$\frac{27}{4\pi R^2 H}(R+\ell)^3,$$
we wish to compare $(R+\ell)^3$ and $9RH^2 = 9R(\ell^2 - R^2)$. Writing $\phi = \ell/R$, this is equivalent to comparing $(1+\phi)^3$ and $9(\phi^2 - 1)$. Now,
$$(1+\phi)^3 - 9(\phi^2 - 1) = \phi^3 - 6\phi^2 + 3\phi + 10 = (\phi+1)(\phi-2)(\phi-5).$$
Thus when $\phi = 2$ or 5, the volumes are equal; when $2 < \phi < 5$, the sphere has larger volume; and when $1 < \phi < 2$ or $5 < \phi$, the cylinder has larger volume. Comparing R and H instead of R and ℓ yields the conditions stated before.

Problem 2 Find all positive integers n such that $(n+3)^n = \sum_{k=3}^{n+2} k^n$.

Solution. $n = 2$ and $n = 3$ are solutions to the equations; we claim they are the only ones.

First observe that the function
$$f(n) = \left(\frac{n+3}{n+2}\right)^n = \left(1 + \frac{1}{n+2}\right)^n$$
is an increasing function for $n > 0$. To see this, note that the derivative of $\ln f(n)$ with respect to n is
$$\ln\left(1 + \frac{1}{n+2}\right) - \frac{n}{(n+2)(n+3)}.$$

By the Taylor expansion,
$$\ln\left(1 + \frac{1}{n+2}\right) = \sum_{j=1}^{\infty} \frac{1}{(n+2)^{2j}}\left[\frac{1}{2j-1}(n+2) - \frac{1}{2j}\right]$$
$$> \frac{2(n+2) - 1}{2(n+2)^2}$$

and hence
$$\frac{d}{dn}\ln f(n) = \ln\left(\frac{n+3}{n+2}\right) - \frac{n}{(n+2)(n+3)}$$
$$> \frac{2(n+2)-1}{2(n+2)^2} - \frac{n}{(n+2)^2} = \frac{3}{2(n+2)^2} > 0.$$

Thus $\ln f(n)$ and therefore $f(n)$ are indeed increasing.

Now, notice that if $f(n) > 2$ then we have
$$\left(\frac{2}{1}\right)^n > \left(\frac{3}{2}\right)^n > \cdots > \left(\frac{n+3}{n+2}\right)^n > 2$$
so that
$$(n+3)^n > 2(n+2)^n > \cdots > 2^j(n+3-j)^n > \cdots > 2^n \cdot (3)^n.$$

Then
$$3^n + 4^n + \cdots + (n+2)^n < \left(\frac{1}{2^n} + \frac{1}{2^{n-1}} + \cdots + \frac{1}{2}\right)(n+3)^n$$
$$= \left(1 - \frac{1}{2^n}\right)(n+3)^n < (n+3)^n,$$

so the equality does not hold.

Because $2 < f(6) < f(7) < \cdots$, the equality must fail for all $n \geq 6$. Quick checks show that it also fails for $n = 1, 4, 5$ (in each case, one side of the equation is odd while the other is even). Therefore the only solutions are $n = 2$ and $n = 3$.

Problem 3 For which acute-angled triangle is the ratio of the shortest side to the inradius maximal?

Solution. Let the sides of the triangle have lengths $a \leq b \leq c$; let the angles opposite them be A, B, C; let the semiperimeter be $s = \frac{1}{2}(a+b+c)$; and let the inradius be r. Without loss of generality, assume that the triangle have circumradius $R = \frac{1}{2}$ and that $a = \sin A$, $b = \sin B$, $c = \sin C$.

The area of the triangle equals both $rs = \frac{1}{2}r(\sin A + \sin B + \sin C)$ and $abc/4R = \frac{1}{2}\sin A \sin B \sin C$. Thus
$$r = \frac{\sin A \sin B \sin C}{\sin A + \sin B + \sin C}$$
and
$$\frac{a}{r} = \frac{\sin A + \sin B + \sin C}{\sin B \sin C}.$$

Because $A = 180° - B - C$, $\sin A = \sin(B + C) = \sin B \cos C + \cos B \sin C$ and we also have

$$\frac{a}{r} = \cot B + \csc B + \cot C + \csc C.$$

Note that $f(x) = \cot x + \csc x$ is a decreasing function along the interval $0° < x < 90°$. Now there are two cases: $B \leq 60°$, or $B > 60°$.

If $B > 60°$, because $C > B > 60°$, the triangle with $A' = B' = C' = 60°$ has a larger ratio a'/r'. Therefore we may assume that $B \leq 60°$.

We may further assume that $A = B$ — otherwise, the triangle with angles $A' = B' = \frac{1}{2}(A+B) \leq B$ and $C' = C$ has a larger ratio a'/r'. Because $C < 90°$ we have $45° < A \leq 60°$. Now,

$$\frac{a}{r} = \frac{\sin A + \sin B + \sin C}{\sin B \sin C} = \frac{2\sin A + \sin(2A)}{\sin A \sin(2A)}$$
$$= 2\csc(2A) + \csc A.$$

Now $\csc x$ has second derivative $\csc x(\csc^2 x + \cot^2 x)$, which is strictly positive when $0° < x < 180°$. Thus, both $\csc x$ and $\csc(2x)$ are strictly convex along the interval $0° < x < 90°$. Therefore, $g(A) = 2\csc(2A) + \csc A$, a convex function in A, is maximized in the interval $45° \leq A \leq 60°$ at one of the endpoints. Because $g(45°) = 2 + \sqrt{2} < 2\sqrt{3} = g(60°)$, it is maximized when $A = B = C = 60°$.

Therefore the maximum ratio is $2\sqrt{3}$, attained with an equilateral triangle.

Problem 4 There are 1999 red candies and 6661 yellow candies on a table, made indistinguishable by their wrappers. A *gourmand* applies the following algorithm until the candies are gone:

(a) If there are candies left, he takes one at random, notes its color, eats it, and goes to (b).

(b) If there are candies left, he takes one at random, notes its color, and

   (i) if it matches the last one eaten, he eats it also and returns to (b).

   (ii) if it does not match the last one eaten, he wraps it up again, puts it back, and goes to (a).

Prove that all the candies will eventually be eaten. Find the probability that the last candy eaten is red.

Solution. If there are finitely many candies left at any point, then at the next instant the gourmand must perform either step (a), part (i) of step (b), or part (ii) of step (b). He eats a candy in the first two cases; in the third case, he returns to step (a) and eats a candy. Because there are only finitely many candies, the gourmand must eventually eat all the candies.

We now prove by induction on the total number of candies that if we start with $r > 0$ red candies and $y > 0$ yellow candies immediately before step (a), then the probability is $\frac{1}{2}$ that the last candy eaten is red.

Suppose that the claim is true for all smaller amounts of candy. After the gourmand has completed steps (a) and (b) and is first directed by the algorithm to return to (a) suppose there are r' red candies and y' yellow candies left. We must have $r' + y' < r + y$. The probability that $r' = 0$ is

$$\frac{r}{r+y} \cdot \frac{r-1}{r+y-1} \cdots \frac{1}{y+1} = \frac{1}{\binom{r+y}{r}}.$$

Similarly, the probability that $y' = 0$ is

$$\frac{1}{\binom{r+y}{y}} = \frac{1}{\binom{r+y}{r}}.$$

(In the case $r = y = 1$, this proves the claim.)

Otherwise, the probability is

$$1 - \frac{2}{\binom{r+y}{r}}$$

that both r' and y' are still positive. By the induction hypothesis, in this case the last candy is equally likely to be red as it is yellow. Thus the overall probability that the last candy eaten is red is

$$\underbrace{\frac{1}{\binom{r+y}{r}}}_{y'=0} + \underbrace{\frac{1}{2}\left(1 - \frac{2}{\binom{r+y}{r}}\right)}_{r',y'>0} = \frac{1}{2}.$$

This completes the inductive step, and the proof.

Problem 5 From a given triangle, form three new points by reflecting each vertex about the opposite side. Show that these three new points are collinear if and only if the distance between the orthocenter and the circumcenter of the triangle is equal to the diameter of the circumcircle of the triangle.

Solution. Let the given triangle be ABC and let the reflections of A, B, C across the corresponding sides be D, E, F. Let A', B', C' be the midpoints of $\overline{BC}, \overline{CA}, \overline{AB}$, and as usual let G, H, O denote the triangle's centroid, orthocenter, and circumcenter. Let triangle $A''B''C''$ be the triangle for which A, B, C are the midpoints of $B''C''', C'''A'', A''B''$, respectively. Then G is the centroid and H is the circumcenter of triangle $A''B''C''$. Let $D', E,' F'$ denote the projections of O on the lines $B''C''', C'''A'', A''B''$, respectively.

Consider the homothety h with center G and ratio $-1/2$. It maps A, B, C, A'', B'', C''' into A', B', C', A, B, C, respectively. Note that $A'D' \perp BC$ because O is the orthocenter of triangle $A'B'C'$. This implies that $AD : A'D' = 2 : 1 = GA : GA'$ and $\angle DAG = \angle D'A'G$. We conclude that $h(D) = D'$. Similarly, $h(E) = E'$ and $h(F) = F'$. Thus, D, E, F are collinear if and only if D', E', F' are collinear. Now D', E', F' are the projections of O on the sides $B''C''', C'''A'', A''B''$, respectively. By Simson's theorem, they are collinear if and only if O lies on the circumcircle of triangle $A''B''C'''$. Because the circumradius of triangle $A''B''C'''$ is $2R$, O lies on its circumcircle if and only if $OH = 2R$.

8 Hong Kong (China)

Problem 1 Let $PQRS$ be a cyclic quadrilateral with $\angle PSR = 90°$, and let H and K be the respective feet of perpendiculars from Q to lines PR and PS. Prove that line HK bisects $\overline{QS}$.

First Solution. $\overline{QK}$ and $\overline{RS}$ are both perpendicular to $\overline{PS}$. Hence $\overline{QK}$ is parallel to $\overline{RS}$ and thus $\angle KQS = \angle RSQ$. Because $PQRS$ is cyclic, $\angle RSQ = \angle RPQ$; quadrilateral $PHKQ$ is also cyclic because $\angle PKQ = \angle PHQ = 90°$, and it follows that $\angle RPQ = \angle HPQ = \angle HKQ$. Thus, $\angle KQS = \angle HKQ$. It follows that line HK bisects the hypotenuse $\overline{QS}$ of right triangle KQS, as desired.

Second Solution. The Simson line from Q with respect to $\triangle PRS$ goes through H, K, and the foot F of the perpendicular from Q to $\overleftrightarrow{RS}$. Thus, line HK is line FK, a diagonal in rectangle $SFQK$, so it bisects the other diagonal, $\overline{QS}$.

Problem 2 The base of a pyramid is a convex nonagon. Each base diagonal and each lateral edge is colored either black or white. Both colors are used at least once. (Note that the sides of the base are not colored.) Prove that there are three segments colored the same color which form a triangle.

Solution. Let us assume the contrary. By the pigeonhole principle, five of the lateral edges must be of the same color. Without loss of generality, assume that they are black segments from the vertex V to B_1, B_2, B_3, B_4, and B_5 where $B_1 B_2 B_3 B_4 B_5$ is a convex pentagon (and where the B_i's are not necessarily adjacent vertices of the nonagon). $\overline{B_1 B_2}$, $\overline{B_2 B_3}$, $\overline{B_3 B_4}$, $\overline{B_4 B_5}$, and $\overline{B_5 B_1}$ cannot all be sides of the nonagon, so without loss of generality suppose that $\overline{B_1 B_2}$ is colored. Because triangle $V B_i B_j$ cannot have three sides colored black, $\overline{B_1 B_2}$, $\overline{B_2 B_4}$, $\overline{B_4 B_1}$ must be white. Now triangle $B_1 B_2 B_4$ has three sides colored white, a contradiction.

Problem 3 Let s and t be nonzero integers, and let (x, y) be any ordered pair of integers. A move changes (x, y) to $(x - t, y - s)$. The pair (x, y) is *good* if after some (possibly zero) number of moves it becomes a pair of integers that are not relatively prime.

(a) Determine whether (s, t) is a good pair;

(b) Prove that for any s and t, there exists a pair (x, y) which is not good.

Solution. (a) Let us assume that (s,t) is not good. Then, after one move, we obtain $(s-t, t-s)$, so we may assume without loss of generality that $s-t=1$ and $t-s=-1$ because these numbers must be relatively prime. Then $s+t$ cannot equal 0 because it is odd. Also, $s+t = (s-t)+2t \neq (s-t)+0 = 1$, and $s+t = (t-s)+2s \neq (t-s)+0 = -1$. Hence, some prime p divides $s+t$. After $p-1$ moves, (s,t) becomes $(s-(p-1)t, t-(p-1)s) \equiv (s+t, t+s) \equiv (0,0) \pmod{p}$, a contradiction. Thus, (s,t) is good.

(b) Let x and y be integers which satisfy $sx - ty = g$, where $g = \gcd(s,t)$. Writing $s = gs'$ and $t = gt'$, we have $s'x - t'y = 1$, so $\gcd(x,y) = 1$. Now suppose, by way of contradiction, that after k moves some prime p divides both $x - kt$ and $y - ks$. We then have

$$0 \equiv x - kt \equiv y - ks \Longrightarrow 0$$
$$\equiv s(x-kt) \equiv t(y-ks) \Longrightarrow\Longrightarrow 0$$
$$\equiv sx - ty = g \pmod{p}.$$

Thus, p divides g, which divides s and t, so the first equation above becomes $0 \equiv x \equiv y \pmod{p}$. However, x and y are coprime, a contradiction. Therefore, (x,y) is not good.

Problem 4 Let f be a function defined on the positive reals with the following properties:

(i) $f(1) = 1$;
(ii) $f(x+1) = xf(x)$;
(iii) $f(x) = 10^{g(x)}$,

where $g(x)$ is a function defined on the reals satisfying

$$g(ty + (1-t)z) \leq tg(y) + (1-t)g(z)$$

for all real y, z, and any $0 \leq t \leq 1$.

(a) Prove that

$$t[g(n) - g(n-1)] \leq g(n+t) - g(n) \leq t[g(n+1) - g(n)]$$

where n is any integer and $0 \leq t \leq 1$.

(b) Prove that

$$\frac{4}{3} \leq f\left(\frac{1}{2}\right) \leq \frac{4\sqrt{2}}{3}.$$

Solution. (a) Setting $t = \frac{1}{2}$ in the given inequality, we find that
$$g\left(\frac{1}{2}(y+z)\right) \leq \frac{1}{2}(g(y) + g(z)).$$
Now fix t (perhaps not equal to $\frac{1}{2}$) constant. Letting $y = n - t$ and $z = n + t$ in
$$g\left(\frac{1}{2}(y+z)\right) \leq \frac{1}{2}(g(y) + g(z))$$
gives $g(n) \leq \frac{1}{2}\bigl(g(n-t) + g(n+t)\bigr)$ or
$$g(n) - g(n-t) \leq g(n+t) - g(n). \tag{1}$$
Plugging in $z = n$, $y = n - 1$ into the given inequality gives
$$g\bigl(t(n-1) + (1-t)n\bigr) \leq tg(n-1) + (1-t)g(n),$$
or
$$t[g(n) - g(n-1)] \leq g(n) - g(n-t).$$
Combining this with (1) proves the inequality on the left side. The inequality on the right side follows from the given inequality with $z = n$, $y = n + 1$.

(b) From (ii), $f(\frac{3}{2}) = \frac{1}{2}f(\frac{1}{2})$ and $f(\frac{5}{2}) = \frac{3}{2}f(\frac{3}{2}) = \frac{3}{4}f(\frac{1}{2})$. Also $f(2) = 1 \cdot f(1) = 1$ and $f(3) = 2f(2) = 2$. Now, if we let $n = 2$ and $t = \frac{1}{2}$ in the inequality in part (a), we find
$$\frac{1}{2}[g(2) - g(1)] \leq g\left(\frac{5}{2}\right) - g(2) \leq \frac{1}{2}[g(3) - g(2)].$$
Exponentiating with base 10 yields
$$\sqrt{\frac{f(2)}{f(1)}} \leq \frac{f\left(\frac{5}{2}\right)}{f(2)} \leq \sqrt{\frac{f(3)}{f(2)}} \quad \text{or} \quad 1 \leq f\left(\frac{5}{2}\right) \leq \sqrt{2}.$$
Plugging in $f\left(\frac{5}{2}\right) = \frac{3}{4}f(\frac{1}{2})$ yields the desired result.

9 Hungary

Problem 1 I have $n \geq 5$ real numbers with the following properties:

(i) They are nonzero, but at least one of them is 1999.

(ii) Any four of them can be rearranged to form a geometric progression.

What are my numbers?

Solution. First suppose that the numbers are all nonnegative. If $x \leq y \leq z \leq w \leq v$ are any five of the numbers, then x, y, z, w; x, y, z, v; x, y, w, v; x, z, w, v; and y, z, w, v must all be geometric progressions. Comparing each two successive progressions in this list, we find that $x = y = z = w = v$. Thus, all our numbers are equal.

If some numbers are negative in our original list, replace each number x by $|x|$. Property (ii) is preserved, and thus from above all the values $|x|$ are equal. Hence, each original number was 1999 or -1999. We are also given $n \geq 5$, implying that some three numbers are equal. However, no geometric progression can be formed from three -1999s and a 1999, or from three 1999s and a -1999. Therefore, all the numbers must be equal — to 1999.

Problem 2 Let ABC be a right triangle with $\angle C = 90°$. Two squares S_1 and S_2 are inscribed in triangle ABC such that S_1 and ABC share a common vertex C, and S_2 has one of its sides on AB. Suppose that $[S_1] = 441$ and $[S_2] = 440$. Calculate $AC + BC$.

Solution. Let $S_1 = CDEF$ and $S_2 = KLMN$ with D and K on $\overline{AC}$ and N on $\overline{BC}$. Let $s_1 = 21, s_2 = \sqrt{440}$ and $a = BC, b = CA, c = AB$. Using ratios between similar triangles AED, ABC, EBF, we find that $c = AB = AE + EB = c(s_1/a + s_1/b)$ or $s_1(1/a + 1/b) = 1$. Because triangles ABC, AKL, NBM are similar, we have $c = AB = AL + LM + MB = s_2(b/a + 1 + a/b)$ and $s_2 = abc/(ab + c^2)$. Then

$$\frac{1}{s_2^2} - \frac{1}{s_1^2} = \left(\frac{1}{c} + \frac{c}{ab}\right)^2 - \left(\frac{1}{a} + \frac{1}{b}\right)^2$$

$$= \left(\frac{1}{c^2} + \frac{c^2}{a^2b^2} + \frac{2}{ab}\right) - \left(\frac{1}{a^2} + \frac{1}{b^2} + \frac{2}{ab}\right) = \frac{1}{c^2}.$$

Thus, $c = 1/\sqrt{1/s_2^2 - 1/s_1^2} = 21\sqrt{440}$. Solving $s_2 = abc/(ab + c^2)$ for ab yields $ab = s_2c^2/(c - s_2) = 21^2 \cdot 22$. Finally, $AC + BC = a + b = ab/s_1 = 21 \cdot 22 = 462$.

Problem 3 We are given a pyramid $PABCD$ such that $ABCD$ is a 2 by 2 square and the altitude from P passes through the center of $ABCD$. Let O and K be the centers of the respective spheres tangent to the faces, and the edges, of pyramid $PABCD$. Determine the volume of the pyramid if O and K are equidistant from the base.

Solution. Let r, R be the spheres' respective radii, and let h be the pyramid's height. By symmetry, O and K lie on the altitude through P.

Take a cross section of the pyramid with a plane perpendicular to the base, cutting the base at a line through its center parallel to $\overline{AB}$. It cuts off an isosceles triangle from the pyramid with base 2 and legs $\sqrt{h^2+1}$. The triangle's incircle is the cross-section of the sphere centered at O and hence has radius r. On one hand, the area of this triangle is the product of its inradius and semiperimeter, or $\frac{1}{2}r(2+2\sqrt{h^2+1})$. On the other hand, it equals half of the product of its base and height, or $\frac{1}{2} \cdot 2 \cdot h$. Setting these quantities equal, we find that $r = (\sqrt{h^2+1}-1)/h$.

Next, by symmetry the second sphere is tangent to $\overline{AB}$ at its midpoint M. Because K must be distance r from plane $ABCD$, we have $R^2 = KM^2 = r^2+1$. Furthermore, if the second sphere is tangent to $\overline{AP}$ at N, then by equal tangents we have $AN = AM = 1$.

Then $PN = PA-1 = \sqrt{h^2+2}-1$. Also, $PK = h+r$ if K is on the opposite side of plane $ABCD$ as O, and it equals $h-r$ otherwise. Thus

$$PK^2 = PN^2 + NK^2$$
$$(h \pm r)^2 = (\sqrt{h^2+2}-1)^2 + (r^2+1)$$
$$\pm 2rh = 4 - 2\sqrt{h^2+2}.$$

Recalling that $r = (\sqrt{h^2+1}-1)/h$, this gives

$$\pm(\sqrt{h^2+1}-1) = 2 - \sqrt{h^2+2}.$$

This equation has the unique solution $h = \sqrt{7}/3$. Thus, the volume of the pyramid is $\frac{1}{3} \cdot 4 \cdot \frac{\sqrt{7}}{3} = 4\sqrt{7}/9$.

Problem 4 For any given positive integer n, determine (as a function of n) the number of ordered pairs (x, y) of positive integers such that

$$x^2 - y^2 = 10^2 \cdot 30^{2n}.$$

Prove further that the number of such pairs is never a perfect square.

Solution. Because $10^2 \cdot 30^{2n}$ is even, x and y must have the same parity. Then (x, y) is a valid solution if and only if $(u, v) = \left(\frac{x+y}{2}, \frac{x-y}{2}\right)$ is a pair of positive integers that satisfies $u > v$ and $uv = 5^2 \cdot 30^{2n}$. Now $5^2 \cdot 30^{2n} = 2^{2n} \cdot 3^{2n} \cdot 5^{2n+2}$ has exactly $(2n+1)^2(2n+3)$ factors. Thus without the condition $u > v$ there are exactly $(2n+1)^2(2n+3)$ such pairs (u, v). Exactly one pair has $u = v$, and by symmetry half of the remaining pairs have $u > v$. It follows that there are $\frac{1}{2}\left((2n+1)^2(2n+3) - 1\right) = (n+1)(4n^2 + 6n + 1)$ valid pairs.

Now suppose that $(n+1)(4n^2 + 6n + 1)$ were a square. Because $n+1$ and $4n^2 + 6n + 1 = (4n+2)(n+1) - 1$ are coprime, $4n^2 + 6n + 1$ must be a square as well. However, $(2n+1)^2 < 4n^2 + 6n + 1 < (2n+2)^2$, a contradiction.

Problem 5 For $0 \leq x, y, z \leq 1$, find all solutions to the equation

$$\frac{x}{1+y+zx} + \frac{y}{1+z+xy} + \frac{z}{1+x+yz} = \frac{3}{x+y+z}.$$

Solution. Assume $x + y + z > 0$, because otherwise the equation is meaningless. $(1 - z)(1 - x) \geq 0 \Rightarrow 1 + zx \geq x + z$, and hence $x/(1+y+zx) \leq x/(x+y+z)$. Doing this for the other two fractions yields that the left-hand side is at most $(x+y+z)/(x+y+z) \leq 3/(x+y+z)$. If equality holds, we must have in particular that $x + y + z = 3 \Rightarrow x = y = z = 1$. This is indeed a solution to the given equation.

Problem 6 The midpoints of the edges of a tetrahedron lie on a sphere. What is the maximum volume of the tetrahedron?

Solution. Let the sphere have center O and radius r. First let A, B, C be any points on its surface. Then $[OAB] = \frac{1}{2}OA \cdot OB \sin \angle AOB \leq \frac{1}{2}r^2$. Likewise, the height from C to plane OAB is at most $CO = r$, whence tetrahedron $OABC$ has maximum volume $r^3/6$. Now, if $\{A, A'\}, \{B, B'\}, \{C, C'\}$ are pairs of antipodal points on the sphere, the octahedron $ABCA'B'C'$ can be broken up into 8 such tetrahedra with vertex O and therefore has maximum volume $4r^3/3$. Equality holds for a regular octahedron.

In the situation of the problem, shrink the tetrahedron T (with volume V) by a factor of $1/2$ about each vertex to obtain four tetrahedra, each with volume $V/8$. Then the six midpoints form an octahedron with volume $V/2$. Moreover, the segment connecting two opposite vertices C and D of this octahedron has T's centroid P as its midpoint. If $O \neq P$ then line OP is a perpendicular bisector of each segment, and then all these segments must lie in the plane through P perpendicular to line OP. Then $V/2 = 0$. Otherwise, the midpoints form three pairs of antipodal points, which form a polyhedron of volume at most $4r^3/3$. Therefore, $V \leq 8r^3/3$, with equality for a regular tetrahedron.

Problem 7 A positive integer is written in each square of an n^2 by n^2 chess board. The difference between the numbers in any two adjacent squares (sharing an edge) is less than or equal to n. Prove that at least $\lfloor n/2 \rfloor + 1$ squares contain the same number.

Solution. Consider the smallest and largest numbers a and b on the board. They are separated by at most $n^2 - 1$ squares horizontally and $n^2 - 1$ vertically, so there is a path from one to the other with length at most $2(n^2 - 1)$. Because any two successive squares differ by at most n, we have $b - a \leq 2(n^2 - 1)n$. All the numbers on the board are integers lying between a and b, so only $2(n^2 - 1)n + 1$ distinct numbers can exist. Therefore, because $n^4 > (2(n^2 - 1)n + 1)(n/2)$, more than $n/2$ squares contain the same number.

Problem 8 One year in the 20th century, Alex noticed on his birthday that adding the four digits of the year of his birth gave his actual age. That same day, Bernath — who shared Alex's birthday but was not the same age as him — also noticed this about his own birth year and age. That day, both were under 99. By how many years do their ages differ?

Solution. Let c be the given year. Alex's year of birth was either $18\underline{uv}$ or $19\underline{uv}$ respectively (where u and v are digits), and thus either $c = 18\underline{uv} + (9 + u + v) = 1809 + 11u + 2v$ or $c = 19\underline{uv} + (10 + u + v) = 1910 + 11u + 2v$.

Similarly, let Bernath's year of birth end in the digits u', v'. Alex and Bernath could not have been born in the same century. Otherwise, we would have $11u + 2v = 11u' + 2v' \Rightarrow 2(v - v') = 11(u' - u)$. Thus either $(u, v) = (u', v')$ or else $|v - v'| \geq 11$, which are both impossible. Now without loss of generality assume Alex was born in the 1800s, and that $1809 + 11u + 2v = 1910 + 11u' + 2v' \Rightarrow 11(u - u') + 2(v - v') = 101 \Rightarrow u - u' = 9, v - v' = 1$. The difference between their ages then equals $19\underline{u'v'} - 18\underline{uv} = 100 + 10(u' - u) + (v' - v) = 9$.

Problem 9 Let ABC be a triangle and D a point on the side AB. The incircles of the triangles ACD and CDB touch each other on $\overline{CD}$. Prove that the incircle of ABC touches $\overline{AB}$ at D.

Solution. Suppose that the incircle of a triangle XYZ touches sides YZ, ZX, XY at U, V, W. Then (using equal tangents) $XY + YZ + ZX =$

$(YW + YU) + (XW + ZU) + XZ = (2YU) + (XZ) + XZ$, and $YU = \frac{1}{2}(XY + YZ - ZX)$.

Thus, if the incircles of triangles ACD and CDB touch each other at E, then $AD + DC - CA = 2DE = BD + DC - CB \Rightarrow AD - CA = (AB - AD) - BC \Rightarrow AD = \frac{1}{2}(CA + AB - BC)$. If the incircle of ABC is tangent to $\overline{AB}$ at D', then $AD' = \frac{1}{2}(CA + AB - BC)$ as well — so $D = D'$, as desired.

Problem 10 Let R be the circumradius of a right pyramid with a square base. Let r be the radius of the sphere touching the four lateral faces and the circumsphere. Suppose that $2R = (1 + \sqrt{2})r$. Determine the angle between adjacent faces of the pyramid.

Solution. Let P be the pyramid's vertex, $ABCD$ the base, and M, N the midpoints of sides AB, CD. By symmetry, both spheres are centered along the altitude from P. Plane PMN intersects the pyramid in triangle PMN and meets the spheres in great circles. Let the smaller circle have center O. It is tangent to $\overline{PM}, \overline{PN}$, and the large circle at some points U, V, and W, respectively. By symmetry, W lies on the altitude from P, implying that it is diametrically opposite P on the larger circle. Thus $OP = 2R - r = \sqrt{2}r$, triangle OUP is a $45°$-$45°$-$90°$ triangle, and $\angle OPU = \angle OPV = 45°$. Therefore, triangle NPM is isosceles right, and the distance from P to plane $ABCD$ equals $BC/2$.

Hence, one can construct a cube with P as its center and $ABCD$ as a face, and this cube can be decomposed into six pyramids congruent to $PABCD$. In particular, three such pyramids have a vertex at A. Thus three times the dihedral angle between faces PAB, PAD forms one revolution, and this angle is $120°$. Stated differently, suppose that the three pyramids are $PABD, PADE, PAEB$; let P' be the midpoint of $\overline{AP}$, and let B', D', E' be points on planes PAB, PAD, PAE such that lines $B'P'$, $D'P'$, $E'P'$ are all perpendicular to line AP. The desired angle is the angle between any two of these lines. Because these three lines all lie in one plane (perpendicular to line AP), this angle must be $120°$.

Problem 11 Is there a polynomial $P(x)$ with integer coefficients such that $P(10) = 400, P(14) = 440$, and $P(18) = 520$?

Solution. If P exists, then by taking its remainder when divided by $(x - 10)(x - 14)(x - 18)$, we may assume that P is quadratic. Writing $P(x) = ax^2 + bx + c$, direct computation reveals $P(x + 4) + P(x - 4) - 2P(x) = 32a$ for all x. Plugging in $x = 14$ gives $40 = 32a$, which is impossible because a must be an integer. Therefore, no such polynomial exists.

Problem 12 Let a, b, c be positive numbers and $n \geq 2$ be an integer such that $a^n + b^n = c^n$. For which k is it possible to construct an obtuse triangle with sides a^k, b^k, c^k?

Solution. First observe that $a < c$ and $b < c$. Thus, for $m > n$ we have $c^m = c^{m-n}(a^n + b^n) > a^{m-n}a^n + b^{m-n}b^n = a^m + b^m$, while for $m < n$ we have $c^m = c^{m-n}(a^n + b^n) < a^{m-n}a^n + b^{m-n}b^n = a^m + b^m$. Now, a triangle with sides a^k, b^k, c^k exists if and only if $a^k + b^k > c^k$, and it is then obtuse if and only if $(a^k)^2 + (b^k)^2 < (c^k)^2$, i.e., $a^{2k} + b^{2k} < c^{2k}$. From our first observation, these conditions are equivalent to the inequalities $k < n$ and $2k > n$, respectively. Hence, $n/2 < k < n$.

Problem 13 Let $n > 1$ be an arbitrary real number and k be the number of positive primes less than or equal to n. Select $k + 1$ positive integers such that none of them divides the product of all the others. Prove that there exists a number among the $k + 1$ chosen numbers which is bigger than n.

Solution. Suppose otherwise, for the sake of contradiction. Our chosen numbers $a_1, \ldots, a_{k+1}$ must then have a total of at most k distinct prime factors (the primes less than or equal to n). Let $o_p(a)$ denote the highest value of d such that $p^d \mid a$. Also let $q = a_1 a_2 \cdots a_{k+1}$. Then for each prime p, $o_p(q) = \sum_{i=1}^{k+1} o_p(a_i)$, and it follows that there can be at most one *hostile* value of i for which $o_p(a_i) > o_p(q)/2$. Because there are at most k primes which divide q, there is some i which is not hostile for any such prime. Then $2o_p(a_i) \leq o_p(q) \Rightarrow o_p(a_i) \leq o_p(q/a_i)$ for each prime p dividing q, implying that $a_i \mid q/a_i$, a contradiction.

Problem 14 The polynomial $x^4 - 2x^2 + ax + b$ has four distinct real roots. Show that the absolute value of each root is smaller than $\sqrt{3}$.

Solution. Let the roots be p, q, r, s. We have $p + q + r + s = 0$, $pq + pr + ps + qr + qs + rs = -2$, and hence $p^2 + q^2 + r^2 + s^2 = 0^2 - 2(-2) = 4$. By the Cauchy-Schwarz inequality, $(1+1+1)(q^2 + r^2 + s^2) \geq (q + r + s)^2$ for any real q, r, s. Furthermore, because q, r, s must be distinct, the inequality becomes strict. Thus, $4 = p^2 + q^2 + r^2 + s^2 > p^2 + (-p)^2/3 = 4p^2/3$ or $|p| < \sqrt{3}$. The same argument holds for q, r, and s.

Problem 15 Each side of a convex polygon has integral length and the perimeter is odd. Prove that the area of the polygon is at least $\sqrt{3}/4$.

Solution.

Lemma 1. *If $0 \leq x, y \leq 1$, then*
$$\sqrt{1-x^2} + \sqrt{1-y^2} \geq \sqrt{1-(x+y-1)^2}.$$

Proof: Squaring and subtracting $2 - x^2 - y^2$ from both sides gives the equivalent inequality $2\sqrt{(1-x^2)(1-y^2)} \geq -2(1-x)(1-y)$, which is true because the left side is nonnegative and the right is at most 0. ∎

Lemma 2. *If $x_1 + \cdots + x_n \leq n - 1/2$ and $0 \leq x_i \leq 1$ for each i, then $\sum_{i=1}^{n} \sqrt{1-x_i^2} \geq \sqrt{3}/2$.*

Proof: We use induction on n. In the case $n = 1$, the statement is clear. If $n > 1$, then either $\min(x_1, x_2) \leq 1/2$ or $x_1 + x_2 > 1$. In the first case, we immediately have $\max(\sqrt{1-x_1^2}, \sqrt{1-x_2^2}) \geq \sqrt{3}/2$. In the second case, we can replace x_1, x_2 by the single number $x_1 + x_2 - 1$ and use the induction hypothesis together with the previous lemma. ∎

Let P, Q be vertices of our polygon such that $\ell = PQ$ is maximal. The polygon consists of two paths from P to Q, each of integer length greater than or equal to ℓ; these lengths are distinct because the perimeter is odd. Then the greater of the two lengths, m, is at least $\ell + 1$. Position the polygon in the coordinate plane with $P = (0,0)$, $Q = (\ell, 0)$ and the longer path in the upper half-plane. Because each side of the polygon has integer length, we can divide this path into line segments of length 1. Let the endpoints of these segments, in order, be $P_0 = P, P_1 = (x_1, y_1), P_2 = (x_2, y_2), \ldots, P_m = Q$. There exists some r such that y_r is maximal. Then either $r \geq x_r + 1/2$ or $(m - r) \geq (\ell - x_r) + 1/2$. Assume the former (otherwise, just reverse the choices of P and Q). We already know that $y_1 \geq 0$, and by the maximal definition of ℓ we must have $x_1 \geq 0$ as well. Because the polygon is convex, we must have $y_1 \leq y_2 \leq \cdots \leq y_r$ and $x_1 \leq x_2 \leq \cdots \leq x_r$. Now $y_{i+1} - y_i = \sqrt{1 - (x_{i+1} - x_i)^2}$, so

$$y_r = \sum_{i=0}^{r-1}(y_{i+1} - y_i) = \sum_{i=0}^{r-1} \sqrt{1 - (x_{i+1} - x_i)^2} \geq \sqrt{3}/2$$

by the second lemma. Hence triangle PP_rQ has base $\overline{PQ}$ with length at least 1 and height $y_r \geq \sqrt{3}/2$, implying that its area is at least $\sqrt{3}/4$. Because our polygon is convex, it contains this triangle and hence the area of the whole polygon is also at least $\sqrt{3}/4$.

Problem 16 Determine if there exists an infinite sequence of positive integers such that

(i) no term divides any other term;

(ii) every pair of terms has a common divisor greater than 1, but no integer greater than 1 divides all of the terms.

Solution. The desired sequence exists. Let $p_0, p_1, \ldots$ be the primes greater than 5 in order, and let $q_{3i} = 6, q_{3i+1} = 10, q_{3i+2} = 15$ for each nonnegative integer i. Then let $s_i = p_i q_i$ for all $i \geq 0$. The sequence $s_0, s_1, s_2, \ldots$ clearly satisfies (i) because s_i is not even divisible by p_j for $i \neq j$. For the first part of (ii), any two terms have their indices both in $\{0, 1\}$, both in $\{0, 2\}$, or both in $\{1, 2\}$ (mod 3), so they have a common divisor of 2, 3, or 5, respectively. For the second part, we just need to check that no prime divides all the s_i. Indeed, $2 \nmid s_2, 3 \nmid s_1, 5 \nmid s_0$, and no prime greater than 5 divides more than one s_i.

Problem 17 Prove that, for every positive integer n, there exists a polynomial with integer coefficients whose values at $1, 2, \ldots, n$ are different powers of 2.

Solution. It suffices to prove the claim when $n \geq 4$, because the same polynomials which work for $n \geq 4$ work for $n \leq 3$. For each $i = 1, 2, \ldots, n$, consider the product $s_i = \prod_{j=1, j \neq i}^{n}(i-j)$. Because $n \geq 4$, one of the terms $i - j$ equals 2 and s_i is even. Thus, we can write $s_i = 2^{q_i} m_i$ for positive integers q_i, m_i with m_i odd. Let L be the least common multiple of all the q_i, and let $r_i = L/q_i$. For each i, there are infinitely many powers of 2 which are congruent to 1 modulo $|m_i^{r_i}|$. (Specifically, by Euler's theorem, $2^{\phi(|m_i^{r_i}|)j} \equiv 1 \pmod{|m_i^{r_i}|}$ for all $j \geq 0$.) Thus there are infinitely many integers c_i such that $c_i m_i^{r_i} + 1$ is a power of 2. Choose such c_i, and define

$$P(x) = \sum_{i=1}^{n} c_i \left(\prod_{\substack{j=1 \\ j \neq i}}^{n}(x-j) \right)^{r_i} + 2^L.$$

For each $k, 1 \leq k \leq n$, in the sum each term $\left(\prod_{j=1, j \neq i}^{n}(x-j)\right)^{r_i}$ vanishes for all $i \neq k$. Then

$$P(k) = c_k \left(\prod_{\substack{j=1 \\ j \neq k}}^{n}(k-j) \right)^{r_i} + 2^L = 2^L(c_k m_k^{r_k} + 1),$$

a power of 2. Moreover, by choosing the c_i appropriately, we can guarantee that these values are all distinct, as needed.

Problem 18 Find all integers $N \geq 3$ for which it is possible to choose N points in the plane (no three collinear) such that each triangle formed by three vertices on the convex hull of the points contains exactly one of the points in its interior.

Solution. First, if the convex hull is a k-gon, then it can be divided into $k - 2$ triangles each containing exactly one chosen point. Because no three of the points are collinear, the sides and diagonals of the convex hull contain no chosen points on their interiors, giving $N = 2k - 2$.

Now we construct, by induction on $k \geq 3$, a set S of $2k - 2$ points consisting of a convex k-gon S_1 and a set S_2 of $k - 2$ points inside, such that (i) no three of the $2k-2$ points are collinear; (ii) each triangle formed by vertices of the k-gon contains exactly one point of S_2 in its interior. The case $k = 3$ is easy. Now, assume we have a convex k-gon $P_1 P_2 \ldots P_k$ and a set S_2. Certainly we can choose Q such that $P_1 P_2 \ldots P_k Q$ is a convex $(k+1)$-gon. Let R move along the line segment from P_k to Q. Initially (at $R = P_k$), for any indices $1 \leq i < j < k$, the triangle $P_i P_j R$ internally contains a point of S by assumption; if R moves less than some sufficiently small distance d_{ij}, the interior of triangle $P_i P_j R$ still contains this point and no others. Furthermore, by choosing the d_{ij} small enough, we can guarantee that whenever $0 < P_k R < d_{ij}$, R is not collinear with any two points from S. Similarly, for any index $1 \leq i \leq k$, $\overline{P_i P_k}$ contains the points P_i and P_k but none of the other points in $S' = \{P_1, P_2, \ldots, P_k\} \cup S$. Because there are only finitely many points in S', if $P_k R$ is less than a sufficiently small value e_i, the interior of triangle $P_i P_k R$ will not contain any of the points in S'.

Now fix a position of R such that $P_k R$ is less than the minimum d_{ij} and the minimum e_i; $P_1 P_2 \ldots P_k R$ is a convex $(k + 1)$-gon. Let P be an interior point of the triangle bounded by lines $P_1 P_k, R P_{k-1}, P_k R$ such that P is not collinear with any two points from $S \cup \{R\}$. We claim the polygon $P_1 P_2 \ldots P_k R$ and the set $S \cup \{P\}$ satisfy our condition. If we choose three of the P_i, they form a triangle containing a point of S by hypothesis, and no others. Any triangle $P_i P_j R$ ($i, j < k$) contains only the same internal point as triangle $P_i P_j P_k$, and each triangle $P_i P_k R$ contains only P. And by construction, no three points from our set our collinear. This completes the induction step.

10 Iran

First Round

Problem 1 Suppose that $a_1 < a_2 < \cdots < a_n$ are real numbers for some integer $n \geq 2$. Prove that

$$a_1 a_2^4 + a_2 a_3^4 + \cdots + a_n a_1^4 \geq a_2 a_1^4 + a_3 a_2^4 + \cdots + a_1 a_n^4.$$

First Solution. We prove the claim by induction on n. For $n = 2$, the two sides are equal. Now suppose the claim is true for $n - 1$, i.e.,

$$a_1 a_2^4 + a_2 a_3^4 + \cdots + a_{n-1} a_1^4 \geq a_2 a_1^4 + a_3 a_2^4 + \cdots + a_1 a_{n-1}^4.$$

Then the claim for n will follow from the inequality

$$a_{n-1} a_n^4 + a_n a_1^4 - a_{n-1} a_1^4 \geq a_n a_{n-1}^4 + a_1 a_n^4 - a_1 a_{n-1}^4$$

(Notice that this is precisely the case for $n = 3$.) Without loss of generality, suppose $a_n - a_1 = 1$ — otherwise, we can divide each of a_1, a_{n-1}, a_n by $a_n - a_1 > 0$ without affecting the truth of the inequality. Then by Jensen's inequality for the convex function x^4, we have

$$a_1^4(a_n - a_{n-1}) + a_n^4(a_{n-1} - a_1) \geq \bigl(a_1(a_n - a_{n-1}) + a_n(a_{n-1} - a_1)\bigr)^4$$
$$= \bigl(a_{n-1}(a_n - a_1)\bigr)^4 = a_{n-1}^4(a_n - a_1),$$

which rearranges to yield our desired inequality.

Second Solution. We use an elementary method to prove the case $n = 3$. Define

$$p(x, y, z) = xy^4 + yz^4 + zx^4 - yx^4 - zy^4 - xz^4.$$

We wish to prove that $p(x, y, z) \geq 0$ when $x \leq y \leq z$. Because $p(x, x, z) = p(x, y, y) = p(z, y, z) = 0$, we know that $(y-x)(z-y)(z-x)$ divides $p(x, y, z)$. In fact,

$$p(x, y, z) = yz^4 - zy^4 + zx^4 - xz^4 + xy^4 - yx^4$$
$$= zy(z^3 - y^3) + xz(x^3 - z^3) + xy(y^3 - x^3)$$
$$= zy(z^3 - y^3) + xz(y^3 - z^3) + xz(x^3 - y^3) + xy(y^3 - x^3)$$
$$= z(y - x)(z^3 - y^3) + x(z - y)(x^3 - y^3)$$
$$= (y - x)(z - y)\bigl(z(z^2 + zy + y^2) - x(x^2 + xy + y^2)\bigr)$$

$$\begin{aligned}
&= (y-x)(z-y)\big((z^3-x^3)+y^2(z-x)+y(z^2-x^2)\big) \\
&= (y-x)(z-y)(z-x)\big(z^2+zx+x^2+y^2+yz+yx\big) \\
&= \frac{1}{2}(y-x)(z-y)(z-x)\big((x+y)^2+(y+z)^2+(z+x)^2\big) \\
&\geq 0,
\end{aligned}$$

as desired.

Problem 2 Suppose that n is a positive integer. The n-tuple $(a_1,\ldots,a_n)$ of positive integers is said to be *good* if $a_1+\cdots+a_n=2n$ and if for every k between 1 and n, no k of the n integers add up to n. Find all n-tuples that are good.

Solution. Suppose we have a good n-tuple $(a_1,\ldots,a_n)$, and consider the sums $a_1, a_1+a_2, \ldots, a_1+a_2+\cdots+a_{n-1}$. All of these sums are between 0 and $2n$ exclusive. Thus, if any of the sums is 0 (mod n), it equals n and we have a contradiction. Also, if any two are congruent modulo n, we can subtract these two sums to obtain another partial sum that equals n, a contradiction again. Therefore, the sums must all be nonzero and distinct modulo n.

Specifically, $a_2 \equiv a_1+\cdots+a_k$ (mod n) for some $k \geq 1$. If $k > 1$ then we can subtract a_2 from both sides to find a partial sum that equals n. Therefore $k=1$ and $a_1 \equiv a_2$ (mod n). Similarly, all the a_i are congruent modulo n. Because the a_i have sum $2n$, one of the a_i is at most 2. Hence, the a_i are either all congruent to 1 modulo n, or all congruent to 2 modulo n.

If the a_i are all congruent to 1 modulo n, then $(a_1,a_2,\ldots,a_n)$ must be a permutation of $1,1,\ldots,1,n+1$. For every k between 1 and n, no k of these n integers add up to n, so these n-tuples are indeed good.

If the a_i are all congruent to 2 modulo n, then $(a_1,a_2,\ldots,a_n) = (2,2,\ldots,2)$. If n is odd, then no k of the a_i have sum n and this n-tuple is good; but if n is even, then any $\frac{n}{2}$ of the a_i have sum n, and the n-tuple is not good.

Problem 3 Let I be the incenter of triangle ABC and let AI meet the circumcircle of ABC at D. Denote the feet of the perpendiculars from I to BD and CD by E and F, respectively. If $IE+IF=AD/2$, calculate $\angle BAC$.

Solution. A well-known fact we will use in this proof is that $DB = DI = DC$. In fact, $\angle BDI = \angle C$ gives $\angle DIB = (\angle A + \angle B)/2$ while $\angle IBD = (\angle A + \angle B)/2$. Thus $DB = DI$, and similarly $DC = DI$.

Let $\theta = \angle BAD$. Then
$$\frac{1}{4} ID \cdot AD = \frac{1}{2} ID \cdot (IE + IF)$$
$$= \frac{1}{2} BD \cdot IE + \frac{1}{2} CD \cdot IF = [BID] + [DIC]$$
$$= \frac{ID}{AD}([BAD] + [DAC]) = \frac{1}{2} ID \cdot (AB + AC) \cdot \sin\theta,$$

whence $\frac{AD}{AB+AC} = 2\sin\theta$.

Let X be the point on ray $\overrightarrow{AB}$ different from A such that $DX = DA$. Because $\angle XBD = \angle DCA$ and $\angle DXB = \angle XAD = \angle DAC$, we have $\triangle XBD \cong \triangle ACD$, and $BX = AC$. Then
$$2\sin\theta = \frac{AD}{AB+AC} = \frac{AD}{AB+BX} = \frac{AD}{AX} = \frac{1}{2\cos\theta},$$
so that $2\sin\theta\cos\theta = \frac{1}{2}$, and $\angle BAC = 2\theta = 30°$ or $150°$.

Problem 4 Let ABC be a triangle with $BC > CA > AB$. Choose points D on $\overline{BC}$ and E on ray $\overrightarrow{BA}$ such that
$$BD = BE = AC.$$
The circumcircle of triangle BED intersects $\overline{AC}$ at P and the line BP intersects the circumcircle of triangle ABC again at Q. Prove that $AQ + QC = BP$.

First Solution. Except where indicated, all angles are directed modulo $180°$.

Let Q' be the point on line BP such that $\angle BEQ' = \angle DEP$. Then
$$\angle Q'EP = \angle AED - \angle BEQ' + \angle DEP = \angle BED.$$
Because $BE = BD$, $\angle BED = \angle EDB$, and because $BEPD$ is cyclic, $\angle EDB = \angle EPB$. Therefore $\angle Q'EP = \angle EPB = \angle EPQ'$ and $Q'P = Q'E$.

Because $BEPD$ and $BAQC$ are cyclic, we have
$$\angle BEQ' = \angle DEP = \angle DBP = \angle CAQ,$$
$$\angle Q'BE = \angle QBA = \angle QCA.$$

Combining this with $BE = AC$ yields that triangles EBQ' and ACQ are congruent. Thus $BQ' = QC$ and $EQ' = AQ$. Therefore

$$AQ + QC = EQ' + BQ' = PQ' + BQ',$$

which equals BP if Q' is between B and P.

E is on $\overrightarrow{BA}$ and P is on $\overrightarrow{CA}$, so E and P are on the same side of $\overline{BC}$ and thus $\overline{BD}$. D is on $\overrightarrow{BC}$ and P is on $\overrightarrow{AC}$, so D and P are on the same side of $\overline{BA}$ and thus $\overline{BE}$. Thus, $BEPD$ is cyclic in that order and (using undirected angles) $\angle BEQ' = \angle DEP < \angle BEP$. It follows that Q' lies on segment BP, as desired.

Second Solution. Because $BEPD$ and $BAQC$ are cyclic, we have

$$\angle PED = \angle PBD = \angle QBC = \angle QAC$$

and

$$\angle EPD = 180° - \angle DBE = 180° - \angle CBA = \angle AQC,$$

which together imply $\triangle PED \sim \triangle QAC$. Then

$$\frac{AC \cdot EP}{DE} = AQ \quad \text{and} \quad \frac{AC \cdot PD}{DE} = QC.$$

As in the first solution, $BEPD$ is cyclic in that order, so Ptolemy's Theorem implies that

$$BD \cdot EP + BE \cdot PD = BP \cdot DE.$$

Substituting $BD = BE = AC$, we have

$$\frac{AC \cdot EP}{DE} + \frac{AC \cdot PD}{DE} = BP,$$

or $AQ + QC = BP$, as desired.

Problem 5 Suppose that n is a positive integer and let

$$d_1 < d_2 < d_3 < d_4$$

be the four smallest positive integer divisors of n. Find all integers n such that

$$n = d_1^2 + d_2^2 + d_3^2 + d_4^2.$$

Solution. The answer is $n = 130$. Note that $x^2 \equiv 0 \pmod 4$ when x is even and that $x^2 \equiv 1 \pmod 4$ when x is odd.

If n is odd, then all the d_i are odd and $n \equiv d_1^2 + d_2^2 + d_3^2 + d_4^2 \equiv 1 + 1 + 1 + 1 \equiv 0 \pmod 4$, a contradiction. Thus, $2 \mid n$.

If $4 \mid n$ then $d_1 = 1$ and $d_2 = 2$, and $n \equiv 1 + 0 + d_3^2 + d_4^2 \not\equiv 0 \pmod 4$, a contradiction. Thus, $4 \nmid n$.

Therefore $\{d_1, d_2, d_3, d_4\} = \{1, 2, p, q\}$ or $\{1, 2, p, 2p\}$ for some odd primes p, q. In the first case, $n \equiv 3 \pmod 4$, a contradiction. Thus $n = 5(1 + p^2)$ and $5 \mid n$, so $p = d_3 = 5$ and $n = 130$.

Problem 6 Suppose that $A = (a_1, a_2, \ldots, a_n)$ and $B = (b_1, b_2, \ldots, b_n)$ are two 0-1 sequences. The difference $d(A, B)$ between A and B is defined to be the number of i for which $a_i \neq b_i$ $(1 \leq i \leq n)$. Suppose that A, B, C are three 0-1 sequences and that $d(A, B) = d(A, C) = d(B, C) = d$.

(a) Prove that d is even.

(b) Prove that there exists a 0-1 sequence D such that
$$d(D, A) = d(D, B) = d(D, C) = \frac{d}{2}.$$

Solution. (a) Modulo 2, we have
$$d(A, B) = (a_1 - b_1) + (a_2 - b_2) + \cdots + (a_n - b_n)$$
$$\equiv (a_1 + a_2 + \cdots + a_n) + (b_1 + b_2 + \cdots + b_n).$$

Thus,
$$3d \equiv d(A, B) + d(B, C) + d(C, A) = 2\left(\sum_{i=1}^n a_i + \sum_{i=1}^n b_i + \sum_{i=1}^n c_i\right),$$
so d must be divisible by 2.

(b) Define D as follows: for each i, if $a_i = b_i = c_i$, then let $d_i = a_i = b_i = c_i$. Otherwise, two of a_i, b_i, c_i are equal; let d_i equal that value. We claim this sequence D satisfies the requirements.

Let α be the number of i for which $a_i \neq b_i$ and $a_i \neq c_i$ (that is, for which a_i is "unique"). Define β and γ similarly, and note that $d(A, D) = \alpha$, $d(B, D) = \beta$, and $d(C, D) = \gamma$. We also have
$$d = d(A, B) = \alpha + \beta$$
$$d = d(B, C) = \beta + \gamma$$
$$d = d(C, A) = \gamma + \alpha.$$

Thus, $\alpha = \beta = \gamma = \frac{d}{2}$, as desired.

Second Round

Problem 1 Define the sequence $\{x_n\}_{n \geq 0}$ by $x_0 = 0$ and

$$x_n = \begin{cases} x_{n-1} + \dfrac{3^{r+1} - 1}{2}, & \text{if } n = 3^r(3k+1), \\ x_{n-1} - \dfrac{3^{r+1} + 1}{2}, & \text{if } n = 3^r(3k+2), \end{cases}$$

where k and r are nonnegative integers. Prove that every integer appears exactly once in this sequence.

First Solution. We prove by induction on $t \geq 1$ that

(i) $\{x_0, x_1, \ldots, x_{3^t - 2}\} = \left\{-\dfrac{3^t - 3}{2}, -\dfrac{3^t - 5}{2}, \ldots, \dfrac{3^t - 1}{2}\right\}$.

(ii) $x_{3^t - 1} = -\dfrac{3^t - 1}{2}$.

These claims imply the desired result, and they are easily verified for $t = 1$. Now supposing they are true for t, we show they are true for $t + 1$.

For any positive integer m, write $m = 3^r(3k + s)$ for nonnegative integers r, k, s, with $s \in \{1, 2\}$, and define $r_m = r$ and $s_m = s$. Then for $m < 3^t$, observe that

$$r_m = r_{m + 3^t} = r_{m + 2 \cdot 3^t} \quad \text{and} \quad s_m = s_{m + 3^t} = r_{m + 2 \cdot 3^t},$$

so that

$$x_m - x_{m-1} = x_{3^t + m} - x_{3^t + m - 1} = x_{2 \cdot 3^t + m} - x_{2 \cdot 3^t + m - 1}.$$

Setting $m = 1, 2, \ldots, k < 3^t$ and adding the resulting equations, we have

$$x_k = x_{3^t + k} - x_{3^t}$$
$$x_k = x_{2 \cdot 3^t + k} - x_{2 \cdot 3^t}.$$

Now, setting $n = 3^t$ in the recursion and using (ii) from the induction hypothesis, we have $x_{3^t} = 3^t$ — and

$$\{x_{3^t}, \ldots, x_{2 \cdot 3^t - 2}\} = \left\{\dfrac{3^t + 3}{2}, \ldots, \dfrac{3^{t+1} - 1}{2}\right\}$$

$$x_{2 \cdot 3^t - 1} = \dfrac{3^t + 1}{2}.$$

Then setting $n = 2 \cdot 3^t$ in the recursion we have $x_{2 \cdot 3^t} = -3^t$ — giving

$$\{x_{2 \cdot 3^t}, \ldots, x_{3^{t+1} - 2}\} = \left\{-\dfrac{3^{t+1} - 3}{2}, \ldots, -\dfrac{3^t + 1}{2}\right\}$$

$$x_{2 \cdot 3^{t+1}-1} = -\frac{3^{t+1}-1}{2}.$$

Combining this with (i) and (ii) from the induction hypothesis proves the claims for $t+1$. This completes the proof.

Second Solution. For $n_i \in \{-1, 0, 1\}$, let the number

$$[n_m n_{m-1} \cdots n_0]$$

in *base* $\bar{3}$ equal $\sum_{i=0}^m n_i \cdot 3^i$. It is simple to prove by induction on k that the base $\bar{3}$ numbers with at most k digits equal

$$\left\{-\frac{3^k-1}{2}, -\frac{3^k-3}{2}, \ldots, \frac{3^k-1}{2}\right\},$$

which implies that every integer has a unique representation in base $\bar{3}$.

Now we prove by induction on n that if $n = a_m a_{m-1} \cdots a_0$ in base 3, then $x_n = [b_m b_{m-1} \cdots b_0]$ in base $\bar{3}$, where $b_i = -1$ if $a_i = 2$ and $b_i = a_i$ for all other cases.

For the base case, $x_0 = 0 = [0]$. Now assume the claim is true for $n-1$. If $n = a_m a_{m-1} \cdots a_{r+1} 1 \underbrace{0\,0\cdots\,0}_{r}$, then

$$x_n = x_{n-1} + \frac{3^{r+1}-1}{2}$$
$$= [b_m b_{m-1} \ldots b_i\, 0\, \underbrace{-1\,-1\,\ldots\,-1}_{r}] + [\underbrace{1\,1\ldots 1}_{r+1}]$$
$$= [b_m b_{m-1} \ldots b_i\, 1\, \underbrace{0\,0\,\ldots\,0}_{r}].$$

If instead $n = a_m a_{m-1} \cdots a_i\, 2\, \underbrace{0\,0\ldots 0}_{r}$, then

$$x_n = x_{n-1} + \left(-\frac{3^{r+1}+1}{2}\right)$$
$$= [b_m b_{m-1} \ldots b_i\, 1\, \underbrace{-1\,-1\,\ldots\,-1}_{r}] + [-1\, \underbrace{1\,1\ldots 1}_{r+1}]$$
$$= [b_m b_{m-1} \ldots b_i\, -1\, \underbrace{0\,0\,\ldots\,0}_{r}].$$

In either case, the claim is true for n, completing the induction.

To finish the proof, note that every integer appears exactly once in base $\bar{3}$. Thus each integer appears exactly once in $\{x_n\}_{n \geq 0}$, as desired.

Problem 2 Suppose that $n(r)$ denotes the number of points with integer coordinates on a circle of radius $r > 1$. Prove that

$$n(r) < 6\sqrt[3]{\pi r^2}.$$

Solution. Consider a circle of radius r containing n lattice points. We must prove that $n < 6\sqrt[3]{\pi r^2}$.

Because $r > 1$ and $6\sqrt[3]{\pi} > 8$, we may assume $n > 8$. Label the n lattice points on the circle $P_1, P_2, \ldots, P_n$ in counterclockwise order. Because the sum of the (counterclockwise) arcs $P_1P_3, P_2P_4, \ldots, P_nP_2$ is 4π, one of the arcs P_iP_{i+2} has measure at most $\frac{4\pi}{n}$ — say, without loss of generality, arc P_1P_3.

Consider a triangle ABC inscribed in an arc of angle $\frac{4\pi}{n}$. Its area is maximized by moving A and C to the endpoints of the arc and then moving B to the midpoint (where the distance to line AC is greatest). At that point $\angle CAB = \angle BCA = \frac{\pi}{n}$ and $\angle ABC = 180° - \frac{2\pi}{n}$, so

$$[ABC] = \frac{abc}{4r} = \frac{(2r\sin\frac{\pi}{n})(2r\sin\frac{2\pi}{n})(2r\sin\frac{\pi}{n})}{4r}$$

$$\leq \frac{(2r\frac{\pi}{n})(2r\frac{2\pi}{n})(2r\frac{\pi}{n})}{4r}$$

$$= \frac{4r^2\pi^3}{n^3}.$$

Because triangle $P_1P_2P_3$ is inscribed in an arc of measure $\frac{4\pi}{n}$, by the preceding argument, $[P_1P_2P_3] \leq \frac{4r^2\pi^3}{n^3}$. Because P_1, P_2, and P_3 are lattice points, the area $[P_1P_2P_3]$ is at least $\frac{1}{2}$ (this can be proven by either Pick's Formula $K = I + \frac{1}{2}B - I$ or the formula $K = \frac{1}{2}|x_1y_2 - x_2y_1 + x_2y_3 - x_3y_2 + x_3y_1 - x_1y_3|$). Therefore,

$$\frac{1}{2} \leq [P_1P_2P_3] \leq \frac{4r^2\pi^3}{n^3} \Longrightarrow n^3$$

$$\leq 8r^2\pi^3 \Longrightarrow n$$

$$\leq \sqrt[3]{8r^2\pi^3} = 2\pi\sqrt[3]{r^2} < 6\sqrt[3]{\pi r^2},$$

as desired.

Problem 3 Find all functions $f : \mathbb{R} \to \mathbb{R}$ satisfying

$$f(f(x) + y) = f(x^2 - y) + 4f(x)y$$

for all $x, y \in \mathbb{R}$.

Solution. Let $(x, y) = (x, x^2)$. Then

$$f(f(x) + x^2) = f(0) + 4x^2 f(x). \tag{1}$$

Let $(x, y) = (x, -f(x))$. Then

$$f(0) = f(x^2 + f(x)) - 4f(x)^2. \tag{2}$$

Adding (1) and (2) gives $4f(x)(f(x) - x^2) = 0$. This implies that for each individual x, either $f(x) = 0$ or $f(x) = x^2$. (Alternatively, plugging $y = \frac{x^2 - f(x)}{2}$ into the original equation also yields this result.) Clearly $f(x) = 0$ and $f(x) = x^2$ satisfy the given equation. We now show that f cannot equal some combination of the two functions.

Suppose that there is an $a \neq 0$ such that $f(a) = 0$. Plugging in $x = a$ into the original equation, we have

$$f(y) = f(a^2 - y).$$

If $y \neq \frac{a^2}{2}$, then $y^2 \neq (a^2 - y)^2$ so $f(y) = f(a^2 - y) = 0$. Thus $f(y) = 0$ for all $y \neq \frac{a^2}{2}$. As for $f(\frac{a^2}{2})$, by choosing $x = 2a$ or some other value in the original equation, we can similarly show that $f(\frac{a^2}{2}) = 0$.

Therefore $f(x) = 0$ for all x or $f(x) = x^2$ for all x, as claimed.

Problem 4 In triangle ABC, the angle bisector of $\angle BAC$ meets BC at D. Suppose that ω is the circle which is tangent to BC at D and passes through A. Let M be the second point of intersection of ω and AC. Let P be the second point of intersection of ω and BM. Prove that P lies on a median of triangle ABD.

Solution. Extend $\overline{AP}$ to meet $\overline{BD}$ at E. We claim that $BE = ED$ and thus $\overline{AP}$ is a median of triangle ABD, as desired. In fact,

$$BE = ED \iff BE^2 = ED^2 = EP \cdot EA$$
$$\iff \triangle BEP \sim \triangle AEB \iff \angle EBP = \angle BAE.$$

Let N be the second intersection of ω with AB. Using directed angles and arc measures, because $\overline{AD}$ bisects the angle between lines AN and AC, we have $\widehat{DM} = \widehat{ND}$ and

$$\angle BAE = \angle NAP = \frac{\widehat{ND} - \widehat{PD}}{2} = \frac{\widehat{DM} - \widehat{PD}}{2} = \angle DBM = \angle EBP,$$

as desired.

Problem 5 Let ABC be a triangle. If we paint the points of the plane in red and green, prove that either there exist two red points which are one unit apart or three green points forming a triangle congruent to ABC.

Solution. We call a polygon green (respectively, red) if its vertices are all green (respectively, red); we call a segment green (respectively, red) if its endpoints are both green (respectively, red).

Suppose, by way of contradiction, there were no such red or green points, and let the sides of triangle ABC be a, b, and c. Without loss of generality assume that $a \leq b, c$.

First we prove no red segment has length a. If XY were a red segment of length a, then the unit circles around X and Y must be completely green. Now draw Z so that $\triangle XYZ \cong \triangle ABC$. The unit circle around Z must be completely red, or else it would form an illegal triangle with the corresponding points around X and Y. On this unit circle we can find a red unit segment, a contradiction.

Now, the whole plane cannot be green so there must be some red point R. The circle ω around R with radius a must be completely green. Then pick two points D, E on ω with $DE = a$. Because $a \leq b, c$, we can construct F outside ω so that $\triangle DEF \cong \triangle ABC$; F must be red. Thus if we rotate DE around R, F forms a completely red circle of radius greater than a — and on this circle we can find two red points distance a apart, a contradiction.

Third Round

Problem 1 Suppose that $S = \{1, 2, \ldots, n\}$ and that $A_1, A_2, \ldots, A_k$ are subsets of S such that for every $1 \leq i_1, i_2, i_3, i_4 \leq k$, we have

$$|A_{i_1} \cup A_{i_2} \cup A_{i_3} \cup A_{i_4}| \leq n - 2.$$

Prove that $k \leq 2^{n-2}$.

Solution. For a set T, let $|T|$ denote the numbers of elements in T. We call a set $T \subseteq S$ 2-coverable if $T \subseteq A_i \cup A_j$ for some i and j (not necessarily distinct). Among the subsets of S that are not 2-coverable, let A be a subset with minimum $|A|$.

Consider the family of sets $S_1 = \{A \cap A_1, A \cap A_2, \ldots, A \cap A_k\}$. ($A \cap A_i$ might equal $A \cap A_j$ for some distinct i and j, but we ignore any duplicate sets.) Because A is not 2-coverable, if $X \in S_1$, then $A - X \notin S_1$. Thus, at most half the subsets of $|A|$ are in S_1, and $|S_1| \leq 2^{|A|-1}$.

On the other hand, let $B = S - A$ and consider the family of sets $S_2 = \{B \cap A_1, B \cap A_2, \ldots, B \cap A_k\}$. We claim that if $X \in S_2$, then $B - X \notin S_2$. Suppose on the contrary that both $X, B - X \in S_2$ for some $X = B \cap A_\ell$ and $B - X = B \cap A_{\ell'}$. By the minimal definition of A, there are A_i and A_j such that $A_i \cup A_j = A \setminus \{m\}$ for some i, j, and m. Then

$$|A_\ell \cup A_{\ell'} \cup A_i \cup A_j| = n - 1,$$

a contradiction. Thus our assumption is false and $|S_2| \leq 2^{|B|-1} = 2^{n-|A|-1}$.

Because every set A_i is uniquely determined by its intersection with sets A and $B = S - A$, it follows that $k \leq |S_1| \cdot |S_2| \leq 2^{n-2}$.

Problem 2 Let ABC be a triangle and let ω be a circle passing through A and C. Sides AB and BC meet ω again at D and E, respectively. Let γ be the incircle of the circular triangle EBD and let S be its center. Suppose that γ touches the arc DE at M. Prove that the angle bisector of $\angle AMC$ passes through the incenter of triangle ABC.

Solution. Let O and R be the center and radius of ω, r be the radius of γ, and I be the incenter of triangle ABC. Extend lines AI and CI to hit ω at M_A and M_C respectively. Also let line AD be tangent to γ at F, and let line CE be tangent to γ at G. Finally, let d be the length of the exterior tangent from M_A to ω. Notice that because line AM_A bisects $\angle DAC$, we have $DM_A = M_AC$. Similarly, $EM_C = M_CA$.

Applying Generalized Ptolemy's Theorem to the circles M_A, C, D, and γ (some of them degenerate) externally tangent to ω gives

$$CG \cdot DM_A = M_AC \cdot DF + d \cdot CD$$

$$d^2 = M_AC^2 \left(\frac{CG - DF}{CD}\right)^2.$$

Note that d^2 equals the power of M_A with respect to γ, so $d^2 = M_AS^2 - r^2$.

By Stewart's Theorem on cevian M_AM in triangle SOM_A, we also have

$$M_AS^2 \cdot OM + M_AO^2 \cdot MS = M_AM^2 \cdot SO + SM \cdot MO \cdot SO$$

$$M_AS^2 \cdot R + R^2 \cdot r = M_AM^2 \cdot (R+r) + r \cdot R \cdot (R+r)$$

$$M_AM^2(R+r) = (M_AS^2 - r^2)R = d^2R,$$

Combining the two equations involving d^2, we find

$$M_A C^2 \left(\frac{CG - DF}{CD}\right)^2 = \frac{M_A M^2 (R+r)}{R}$$

$$\left(\frac{M_A M}{M_A C}\right)^2 = \left(\frac{R}{R+r}\right) \left(\frac{CG - DF}{CD}\right)^2.$$

Similarly,

$$\left(\frac{M_C M}{M_C A}\right)^2 = \left(\frac{R}{R+r}\right) \left(\frac{AF - EG}{AE}\right)^2.$$

Now

$$CG - DF = (CG + GB) - (DF + FB) = CB - DB$$

and similarly

$$AF - EG = (AF + FB) - (EG + GB) = AB - BE.$$

Furthermore, because $ACDE$ is cyclic, some angle-chasing gives $\angle BDC = \angle AEC$ and $\angle DCB = \angle BAE$, so $\triangle CBD \sim \triangle ABE$ and

$$\frac{CG - DF}{CD} = \frac{CB - DB}{CD} = \frac{AB - BE}{EA} = \frac{AF - EG}{AE}.$$

Therefore we have

$$\frac{M_A M}{M_A C} = \frac{M_C M}{M_C A} \implies \frac{\sin \angle M A M_A}{\sin \angle M_A A C} = \frac{\sin \angle M_C C M}{\sin \angle A C M_C}.$$

By the trigonometric form of Ceva's theorem in triangle AMC applied to lines AM_A, CM_C, and MI, we have

$$\frac{\sin \angle M A M_A}{\sin \angle M_A A C} \cdot \frac{\sin \angle A C M_C}{\sin \angle M_C C M} \cdot \frac{\sin \angle C M I}{\sin \angle I M A} = 1$$

so that

$$\sin \angle CMI = \sin \angle IMA \implies \angle CMI = \angle IMA$$

because $\angle AMC < 180°$. Therefore, line MI bisects $\angle AMC$, so the angle bisector of $\angle AMC$ indeed passes through the incenter I of triangle ABC.

Problem 3 Suppose that $C_1, C_2, \ldots, C_n$ are circles of radius 1 in the plane such that no two of them are tangent and the subset of the plane formed by the union of these circles is connected (i.e., for any partition of $\{1, 2, \ldots, n\}$ into nonempty subsets A and B, $\bigcup_{a \in A} C_a$ and $\bigcup_{b \in B} C_b$ are

not disjoint). Prove that $|S| \geq n$, where
$$S = \bigcup_{1 \leq i < j \leq n} C_i \cap C_j,$$
the set of intersection points of the circles. (Each circle is viewed as the set of points on its circumference, not including its interior.)

Solution. Let $T = \{C_1, C_2, \ldots, C_n\}$. For every $s \in S$ and $C \in T$ define
$$f(s, C) = \begin{cases} 0, & \text{if } s \notin C, \\ \frac{1}{k}, & \text{if } s \in C, \end{cases}$$
where k is the number of circles passing through s (including C). Thus
$$\sum_{C \in T} f(s, C) = 1$$
for every $s \in S$.

On the other hand, for a fixed circle $C \in T$, let $s_0 \in S \cap C$ be a point such that
$$f(s_0, C) = \min\{f(s, C) \mid s \in S \cap C\}.$$
Suppose that $C, C_2, \ldots C_k$ are the circles which pass through s_0. Then C meets $C_2, \ldots, C_k$ again in distinct points $s_2, \ldots, s_k$. Therefore
$$\sum_{s \in C} f(s, C) \geq \frac{1}{k} + \frac{k-1}{k} = 1.$$
We have
$$|S| = \sum_{s \in S} \sum_{C \in T} f(s, C) = \sum_{C \in T} \sum_{s \in S} f(s, C) \geq n,$$
as desired.

Problem 4 Suppose that $-1 \leq x_1, x_2, \ldots, x_n \leq 1$ are real numbers such that $x_1 + x_2 + \ldots + x_n = 0$. Prove that there exists a permutation σ of $\{1, 2, \ldots, n\}$ such that, for every $1 \leq p \leq q \leq n$,
$$|x_{\sigma(p)} + \cdots + x_{\sigma(q)}| < 2 - \frac{1}{n}.$$
Also prove that the expression on the right hand side cannot be replaced by $2 - \frac{4}{n}$.

Solution. If $n = 1$ then $x_1 = 0$, and the permutation $\sigma(1) = 1$ suffices. If $n = 2$ then $|x_1|, |x_2| \leq 1$ and $|x_1 + x_2| = 0$, and the permutation $(\sigma(1), \sigma(2)) = (1, 2)$ suffices. Now assume $n \geq 3$.

View the x_i as vectors. The problem is equivalent to saying that if we start at a point on the number line, we can travel along the n vectors $x_1, x_2, \ldots, x_n$ in some order so that we stay within an interval $(m, m+2-\frac{1}{n}]$.

Call x_i *long* if $|x_i| \geq 1 - \frac{1}{n}$ and call it *short* otherwise. Also call x_i *positive* if $x_i \geq 0$ and *negative* if $x_i < 0$. Suppose without loss of generality that there are at least as many long positive vectors as long negative vectors — otherwise, we could replace each x_i by $-x_i$. We make our trip in two phases:

(i) First, travel alternating along long positive vectors and long negative vectors until no long negative vectors remain. Suppose at some time we are at a point P. Observe that during this phase of our trip, travelling along a pair of vectors changes our position by at most $\frac{1}{n}$ in either direction. Thus if we travel along $2t \leq n$ vectors after P, we stay within $\frac{t}{n} \leq \frac{1}{2}$ of P; if we travel along $2t+1$ vectors after P, we stay within $\frac{t}{n} + 1 \leq \frac{3}{2} < 2 - \frac{1}{n}$ of P. Therefore, during this phase, we indeed stay within an interval $I = (m, m+2-\frac{1}{n}]$ of length $2 - \frac{1}{n}$.

(ii) After phase (i), we claim that as long as vectors remain unused and we are inside I, there is an unused vector we can travel along while remaining in I. This fact implies that we can finish the trip while staying in I.

If there are no positive vectors, then we can travel along any negative vector, and vice versa. Thus, assume there are positive *and* negative vectors remaining. Because all the long negative vectors were used in phase (i), only short negative vectors remain.

Now if we are to the right of $m + 1 - \frac{1}{n}$, we can travel along a short negative vector without reaching or passing m. If we are instead on or to the left of $m + 1 - \frac{1}{n}$, we can travel along a positive vector (short or long) without passing $m + 2 - \frac{1}{n}$.

Therefore it is possible to complete our journey, and it follows that the desired permutation σ indeed exists.

However, suppose $\frac{1}{n}$ is changed to $\frac{4}{n}$. This bound is never attainable for $n = 1$, and it is not always attainable when $n = 2$ (when $x_1 = 1$, $x_2 = -1$, for example).

If $n = 2k+1 \geq 3$ or $2k+2 \geq 4$, suppose that $x_1 = x_2 = \cdots = x_k = 1$ and $x_{k+1} = x_{k+2} = \cdots = x_{2k+1} = -\frac{k}{k+1}$. If n is even, we can let $x_n = 0$ and ignore this term in the permutation.

If two adjacent numbers in the permutation are equal then their sum is either $2 \geq 2 - \frac{4}{n}$ or $-2 \cdot \frac{k}{k+1} \leq -2 + \frac{4}{n}$. Therefore in the permutation, the vectors must alternate between $-\frac{k}{k+1}$ and 1, starting and ending with $-\frac{k}{k+1}$.

Then the outer two vectors add up to $-2 \cdot \frac{k}{k+1}$, so the middle $2k - 1$ vectors add up to $2 \cdot \frac{k}{k+1} \geq 2 - \frac{4}{n}$, a contradiction. Therefore, $\frac{1}{n}$ cannot be replaced by $\frac{4}{n}$.

Problem 5 Suppose that $r_1, \ldots, r_n$ are real numbers. Prove that there exists $S \subseteq \{1, 2, \ldots, n\}$ such that

$$1 \leq |S \cap \{i, i+1, i+2\}| \leq 2,$$

for $1 \leq i \leq n - 2$, and

$$\left| \sum_{i \in S} r_i \right| \geq \frac{1}{6} \sum_{i=1}^{n} |r_i|.$$

Solution. Let $\sigma = \sum_{i=1}^{n} |r_i|$ and for $i = 0, 1, 2$, define

$$s_i = \sum_{r_j \geq 0, \, j \equiv i} r_j \quad \text{and} \quad t_i = \sum_{r_j < 0, \, j \equiv i} r_j,$$

where congruences are taken modulo 3. Then we have $\sigma = s_1 + s_2 + s_3 - t_1 - t_2 - t_3$, and 2σ equals

$$(s_1 + s_2) + (s_2 + s_3) + (s_3 + s_1) - (t_1 + t_2) - (t_2 + t_3) - (t_3 + t_1).$$

Therefore, there exist $i_1 \neq i_2$ such that either $s_{i_1} + s_{i_2} \geq s/3$ or $t_{i_1} + t_{i_2} \leq -s/3$ or both. Without loss of generality, we assume that $s_{i_1} + s_{i_2} \geq \sigma/3$ and $|s_{i_1} + s_{i_2}| \geq |t_{i_1} + t_{i_2}|$. Thus, $s_{i_1} + s_{i_2} + t_{i_1} + t_{i_2} \geq 0$. We have $[s_{i_1} + s_{i_2} + t_{i_1}] + [s_{i_1} + s_{i_2} + t_{i_2}] \geq s_{i_1} + s_{i_2} \geq \sigma/3$. Therefore at least one of $s_{i_1} + s_{i_2} + t_{i_1}$ and $s_{i_1} + s_{i_2} + t_{i_2}$ is bigger or equal to $\sigma/6$, and we are done.

11 Ireland

Problem 1 Find all nonzero reals $x > -1$ which satisfy
$$\frac{x^2}{(x+1-\sqrt{x+1})^2} < \frac{x^2+3x+18}{(x+1)^2}.$$

Solution. Make the substitution $y = \sqrt{x+1}$, so that $y \in (0,1) \cup (1, \infty)$ and $x = y^2 - 1$. Then the inequality is equivalent to
$$\frac{(y^2-1)^2}{(y^2-y)^2} < \frac{(y^2-1)^2 + 3(y^2-1) + 18}{y^4}$$
$$\iff \frac{(y+1)^2}{y^2} < \frac{y^4+y^2+16}{y^4}$$
$$\iff (y+1)^2 y^2 < y^4 + y^2 + 16$$
$$\iff 2y^3 < 16,$$

so the condition is satisfied exactly when $y < 2$; i.e., exactly when $y \in (0,1) \cup (1,2)$ and $x \in (-1, 0) \cup (0, 3)$.

Problem 2 Show that there is a positive number in the Fibonacci sequence which is divisible by 1000.

Solution. In fact, for any natural number n, there exist infinitely many positive Fibonacci numbers divisible by n.

The Fibonacci sequence is defined thus: $F_0 = 0$, $F_1 = 1$, and $F_{k+2} = F_{k+1} + F_k$ for all $k \geq 0$. Consider ordered pairs of consecutive Fibonacci numbers $(F_0, F_1), (F_1, F_2), \ldots$ taken modulo n. Because the Fibonacci sequence is infinite and there are only n^2 possible ordered pairs of integers modulo n, two such pairs (F_j, F_{j+1}) must be congruent: $F_i \equiv F_{i+m}$ and $F_{i+1} \equiv F_{i+m+1} \pmod{n}$ for some i and m.

If $i \geq 1$ then $F_{i-1} \equiv F_{i+1} - F_i \equiv F_{i+m+1} - F_{i+m} \equiv F_{i+m-1} \pmod{n}$. Likewise $F_{i+2} \equiv F_{i+1} + F_i \equiv F_{i+m+1} + F_{i+m} \equiv F_{i+2+m} \pmod{n}$. Continuing similarly, we have $F_j \equiv F_{j+m} \pmod{n}$ for all $j \geq 0$. In particular, $0 = F_0 \equiv F_m \equiv F_{2m} \equiv \cdots \pmod{n}$, so the numbers F_m, $F_{2m}, \ldots$ are all positive Fibonacci numbers divisible by n. Applying this to $n = 1000$, we are done.

Problem 3 Let D, E, F be points on the sides BC, CA, AB, respectively, of triangle ABC such that $AD \perp BC$, $AF = FB$, and BE is the angle bisector of $\angle B$. Prove that AD, BE, CF are concurrent if and only

if
$$a^2(a-c) = (b^2 - c^2)(a+c),$$
where $a = BC$, $b = CA$, $c = AB$.

Solution. By Ceva's Theorem, the cevians AD, BE, CF in $\triangle ABC$ are concurrent if and only if (using directed line segments)
$$\frac{AF}{FB} \cdot \frac{BD}{DC} \cdot \frac{CE}{EA} = 1.$$
In this problem, $\frac{AF}{FB} = 1$, and $\frac{CE}{EA} = \frac{a}{c}$. Thus AD, BE, CF are concurrent if and only if $\frac{BD}{DC} = \frac{c}{a}$.

This in turn is true if and only if $BD = \frac{ac}{a+c}$ and $DC = \frac{a^2}{a+c}$. Because $AB^2 - BD^2 = BD^2 = AC^2 - CD^2$, these last conditions hold exactly when the following equations are true:
$$AB^2 - \left(\frac{ac}{a+c}\right)^2 = AC^2 - \left(\frac{a^2}{a+c}\right)^2$$
$$(a+c)^2 c^2 - a^2 c^2 = (a+c)^2 b^2 - a^4$$
$$a^4 - a^2 c^2 = (b^2 - c^2)(a+c)^2$$
$$a^2(a-c) = (b^2 - c^2)(a+c).$$

Therefore the three lines are concurrent if and only if the given equation holds, as desired.

Alternatively, applying the law of cosines gives
$$\frac{BD}{DC} = \frac{c \cos B}{b \cos C} = \frac{c}{b} \cdot \frac{a^2 + c^2 - b^2}{2ac} \cdot \frac{2ab}{a^2 + b^2 - c^2} = \frac{a^2 + c^2 - b^2}{a^2 + b^2 - c^2}.$$
Again, this equals $\frac{c}{a}$ exactly when the given equation holds.

Problem 4 A 100 by 100 square floor is to be tiled. The only available tiles are rectangular 1 by 3 tiles, fitting exactly over three squares of the floor.

(a) If a 2 by 2 square is removed from the center of the floor, prove that the remaining part of the floor can be tiled with available tiles.

(b) If, instead, a 2 by 2 square is removed from the corner, prove that the remaining part of the floor cannot be tiled with the available tiles.

Solution. Choose a coordinate system so that the corners of the square floor lie along the lattice points $\{(x,y) \mid 0 \leq x, y \leq 100, x, y \in \mathbb{Z}\}$.

Denote the rectangular region $\{(x,y) \mid a \leq x \leq b, c \leq y \leq d\}$ by $[(a,c) - (b,d)]$.

(a) It is evident that any rectangle with at least one dimension divisible by 3 can be tiled. First tile the four rectangles

$$[(0,0) - (48, 52)], \quad [(0, 52) - (52, 100)],$$
$$[(52, 48) - (100, 100)], \quad \text{and} \quad [(48, 0) - (100, 52)].$$

The only part of the board left untiled is $[(48, 48) - (52, 52)]$. Also recall that the central region $[(49, 49) - (51, 51)]$ has been removed. It is obvious that the remaining portion can be tiled.

(b) Assume without loss of generality that $[(0,0) - (2,2)]$ is the 2 by 2 square which is removed. Label each remaining square $[(x, y) - (x+1, y+1)]$ with the number $L(x, y) \in \{0, 1, 2\}$ such that $L(x, y) \equiv x+y \pmod{3}$. There are 3333 squares labelled 0, 3331 squares labelled 1, and 3332 squares labelled 2. However, each 1 by 3 tile covers an equal number of squares of each label. Therefore, the floor cannot be tiled.

Problem 5 Define a sequence u_n, $n = 0, 1, 2, \ldots$ as follows: $u_0 = 0$, $u_1 = 1$, and for each $n \geq 1$, u_{n+1} is the smallest positive integer such that $u_{n+1} > u_n$ and $\{u_0, u_1, \ldots, u_{n+1}\}$ contains no three elements which are in arithmetic progression. Find u_{100}.

Solution. Take any nonnegative integer n (e.g., 100) and express it in base 2 (e.g., $100 = 1100100_2$). Now interpret that sequence of 1's and 0's as an integer in base 3 (e.g., $1100100_3 = 981$). Call that integer t_n (e.g., $t_{100} = 981$).

We now prove that $t_n = u_n$ by strong induction on n. It is obvious that $t_0 = u_0$ and that $t_1 = u_1$. Now assume that $t_k = u_k$ for all $k < n$. We shall show that $t_n = u_n$.

First we show that $u_n \leq t_n$ by proving that, in the sequence $t_0, t_1, t_2, \ldots, t_n$, no three numbers form an arithmetic progression. Pick any three numbers $0 \leq \alpha < \beta < \gamma \leq n$, and consider t_α, t_β, and t_γ in base 3. t_α and t_γ contain no 2's, so in the addition $t_\alpha + t_\gamma$ no carrying can occur. Also note that t_α and t_γ differ in at least one digit, and in that digit the number $t_\alpha + t_\gamma$ must contain a 1. On the other hand, $2t_\beta$ consists of only 2's and 0's, so $t_\alpha + t_\gamma \neq 2t_\beta$ for any choice of α, β, γ. Therefore, among $t_0, t_1, t_2, \ldots, t_n$, no three numbers are in arithmetic progression. Hence $u_n \leq t_n$.

Next we show that $u_n \geq t_n$ by showing that for all $k \in \{t_{n-1} + 1, t_{n-1} + 2, \ldots, t_n - 1\}$, there exist numbers a and b such that $t_a + k = 2t_b$.

First note that k must contain a 2 in its base 3 representation, because the t_i are the only nonnegative integers consisting of only 1's and 0's in base 3. Therefore, we can find two numbers a and b with $0 \leq t_a < t_b < k$ such that:

- whenever k has a 0 or a 1 in its base 3 representation, t_a and t_b each also have the same digit in the corresponding positions in their base 3 representations;
- whenever k has a 2 in its base 3 representation, t_a has a 0 in the corresponding position in its base 3 representation, but t_b has a 1 in the corresponding position in its base 3 representation.

The t_a and t_b, thus constructed, satisfy $t_a < t_b < k$ while $t_a + k = 2t_b$, so t_a, t_b, k form an arithmetic progression. Thus, $u_n \geq t_n$. Putting this result together with the previous result, we have forced $u_n = t_n$. Hence $u_{100} = t_{100} = 981$.

Problem 6 Solve the system of equations

$$y^2 - (x+8)(x^2+2) = 0$$
$$y^2 - (8+4x)y + (16+16x-5x^2) = 0.$$

Solution. We first note that the solutions $(x,y) = (-2,-6)$ and $(-2,6)$ both work and are the only solutions with $x = -2$.

Assume that $x \neq -2$. We substitute $y^2 = (x+8)(x^2+2)$ into $y^2 + 16 + 16x - 5x^2 = 4(x+2)y$ to get

$$4(x+2)y = x^3 + 3x^2 + 18x + 32 = (x+2)(x^2+x+16).$$

Dividing by $x+2$, we find that $4y = x^2 + x + 16$. Solving for y and substituting the resulting expression into the first given equation, we obtain

$$x^4 - 14x^3 - 95x^2 = 0$$

or $x^2(x+5)(x-19) = 0$. Thus, $x \in \{0, -5, 19\}$. Using the equation $4y = x^2 + x + 16$, we find that $(x,y) = (0,4), (-5,9)$, or $(19,99)$. It can be easily checked that each of these pairs works.

Problem 7 A function $f : \mathbb{N} \to \mathbb{N}$ satisfies

(i) $f(ab) = f(a)f(b)$ whenever the greatest common divisor of a and b is 1;

(ii) $f(p+q) = f(p) + f(q)$ for all prime numbers p and q.

Prove that $f(2) = 2$, $f(3) = 3$, and $f(1999) = 1999$.

Solution. We shall write (i)$_{a,b}$ when we plug (a,b) (where a and b are relatively prime) into (i), and (ii)$_{p,q}$ when we plug (p,q) (where p and q are primes) into (ii).

First we find $f(1)$, $f(2)$, and $f(4)$. By (i)$_{1,b}$ we find $f(1) = 1$. By (i)$_{2,3}$ we find $f(6) = f(2)f(3)$, and by (ii)$_{3,3}$ we get $f(6) = 2f(3)$. Thus, $f(2) = 2$. By (ii)$_{2,2}$, we then also find that $f(4) = 4$.

From (ii)$_{3,2}$ and (ii)$_{5,2}$, respectively, we obtain

$$f(5) = f(3) + 2, \qquad f(7) = f(5) + 2 = f(3) + 4.$$

By (ii)$_{5,7}$ we have $f(12) = f(5) + f(7) = 2f(3) + 6$, and from (i)$_{4,3}$ we have $f(12) = 4f(3)$. Solving for $f(3)$ yields $f(3) = 3$. Then, using the above equations, we find that $f(5) = 5$ and $f(7) = 7$.

We proceed to find $f(13)$ and $f(11)$. By (i)$_{3,5}$, we have $f(15) = 15$. From (ii)$_{13,2}$ and (ii)$_{11,2}$, respectively, we find

$$f(13) = f(15) - f(2) = 13, \qquad f(11) = f(13) - f(2) = 11.$$

Finally, we can calculate $f(1999)$. By applying (i) repeatedly, we find $f(2002) = f(2 \cdot 7 \cdot 11 \cdot 13) = f(2)f(7)f(11)f(13) = 2002$. Noting that 1999 is a prime number, from (ii)$_{1999,3}$ we obtain

$$f(1999) = f(2002) - f(3) = 1999,$$

as desired.

Problem 8 Let a, b, c, d be positive real numbers whose sum is 1. Prove that
$$\frac{a^2}{a+b} + \frac{b^2}{b+c} + \frac{c^2}{c+d} + \frac{d^2}{d+a} \geq \frac{1}{2}$$
with equality if and only if $a = b = c = d = 1/4$.

Solution. Applying the Cauchy-Schwarz inequality, we find

$$[(a+b) + (b+c) + (c+d) + (d+a)]$$
$$\times \left(\frac{a^2}{a+b} + \frac{b^2}{b+c} + \frac{c^2}{c+d} + \frac{d^2}{d+a} \right) \geq (a+b+c+d)^2.$$

Hence
$$\frac{a^2}{a+b} + \frac{b^2}{b+c} + \frac{c^2}{c+d} + \frac{d^2}{d+a} \geq \frac{1}{2}(a+b+c+d) = \frac{1}{2},$$
with equality if and only if
$$\frac{a+b}{a} = \frac{b+c}{b} = \frac{c+d}{d} = \frac{d+a}{a},$$
i.e., if and only if $a = b = c = d = \frac{1}{4}$.

Problem 9 Find all positive integers m such that the fourth power of the number of positive divisors of m equals m.

Solution. If the given condition holds for some integer m, then m must be a perfect fourth power and we may write its prime factorization as $m = 2^{4a_2} 3^{4a_3} 5^{4a_5} 7^{4a_7} \cdots$ for nonnegative integers $a_2, a_3, a_5, a_7, \ldots$. The number of positive divisors of m equals

$$(4a_2 + 1)(4a_3 + 1)(4a_5 + 1)(4a_7 + 1) \cdots.$$

This number is odd, so m is odd and $a_2 = 0$. Thus,

$$1 = \frac{4a_3 + 1}{3^{a_3}} \cdot \frac{4a_5 + 1}{5^{a_5}} \cdot \frac{4a_7 + 1}{7^{a_7}} \cdots = x_3 x_5 x_7 \cdots,$$

where we write $x_p = \frac{4a_p+1}{p^{a_p}}$ for each p. We proceed to examine x_p through three cases: $p = 3$, $p = 5$, and $p > 5$.

When $a_3 = 1$, $x_3 = \frac{5}{3}$; when $a_3 = 0$ or 2, $x_3 = 1$. When $a_3 > 2$, by Bernoulli's inequality we have

$$3^{a_3} = (8+1)^{a_3/2} > 8(a_3/2) + 1 = 4a_3 + 1$$

so that $x_3 < 1$.

When $a_5 = 0$ or 1, $x_5 = 1$; when $a_5 \geq 2$, by Bernoulli's inequality we have

$$5^{a_5} = (24+1)^{a_5/2} \geq 24 a_5/2 + 1 = 12 a_5 + 1$$

so that $x_5 \leq \frac{4a_5+1}{12a_5+1} \leq \frac{9}{25}$.

Finally, for any $p > 5$ when $a_p = 0$ we have $x_p = 1$; when $a_p = 1$ we have $p^{a_p} = p > 5 = 4a_p + 1$ so that $x_p < 1$; and when $a_p > 1$ then again by Bernoulli's inequality we have

$$p^{a_p} > 5^{a_p} > 12 a_p + 1$$

so that as above $x_p < \frac{9}{25}$.

Now if $a_3 \neq 1$ then we have $x_p \leq 1$ for all p. Because $1 = x_2 x_3 x_5 \cdots$ we must actually have $x_p = 1$ for all p. This means that $a_3 \in \{0, 2\}$, $a_5 \in \{0, 1\}$, and $a_7 = a_{11} = \cdots = 0$. Hence $m = 1^4, (3^2)^4, 5^4$, or $(3^2 \cdot 5)^4$.

Otherwise, if $a_3 = 1$ then $3 \mid m = 5^4(4a_5 + 1)^4(4a_7 + 1)^4 \cdots$. Then for some prime $p' \geq 5$, $3 \mid 4a_{p'} + 1$ so that $a_{p'} \geq 2$; from above, we have $x_{p'} \leq \frac{9}{25}$. Then $x_3 x_5 x_7 \cdots \leq \frac{5}{3} \frac{9}{25} < 1$, a contradiction.

Thus, the only such integers m are 1, 5^4, 3^8, and $3^8 \cdot 5^4$, and it is easily verified that these integers work.

Problem 10 Let $ABCDEF$ be a convex hexagon such that $AB = BC$, $CD = DE$, $EF = FA$, and

$$\angle ABC + \angle CDE + \angle EFA = 360°.$$

Prove that the respective perpendiculars from A, C, E to FB, BD, DF are concurrent.

First Solution. The result actually holds even without the given angle condition. Let $\mathcal{C}_1$ be the circle with center B and radius $AB = BC$, $\mathcal{C}_2$ the circle with center D and radius $CD = DE$, and $\mathcal{C}_3$ the circle with center F and radius $EF = FA$. The line through A and perpendicular to line FB is the radical axis of circles $\mathcal{C}_3$ and $\mathcal{C}_1$, the line through C and perpendicular to line BD is the radical axis of circles $\mathcal{C}_1$ and $\mathcal{C}_2$, and the line through E and perpendicular to line DF is the radical axis of circles $\mathcal{C}_2$ and $\mathcal{C}_3$. The result follows because these three radical axes meet at the radical center of the three circles.

Second Solution. We first establish a lemma.

Lemma. *Given points $W \neq X$ and $Y \neq Z$, lines WX and YZ are perpendicular if and only if*

$$YW^2 - WZ^2 = XY^2 - XZ^2. \tag{1}$$

Proof: Introduce Cartesian coordinates such that $W = (0,0)$, $X = (1,0)$, $Y = (x_1, y_1)$, and $Z = (x_2, y_2)$. Then (1) becomes

$$x_1^2 + y_1^2 - x_2^2 - y_2^2 = (x_1 - 1)^2 + y_1^2 - (x_2 - 1)^2 - y_2^2,$$

which upon cancellation yields $x_1 = x_2$. This is true if and only if line YZ is perpendicular to the x-axis WX. ∎

If P is the intersection of the perpendiculars from A and C to lines FB and BD, respectively, then the lemma implies that

$$PF^2 - PB^2 = AF^2 - AB^2 \quad \text{and} \quad PB^2 - PD^2 = CB^2 - CD^2.$$

From the given isosceles triangles, we have $EF = FA$, $AB = BC$, and $CD = DE$. Adding the two equations then gives

$$PF^2 - PD^2 = EF^2 - ED^2.$$

Hence, line PE is also perpendicular to line DF, which completes the proof.

12 Italy

Problem 1 Given a rectangular sheet with sides a and b, with $a > b$, fold it along a diagonal. Determine the area of the triangle that passes over the edge of the paper.

Solution. Let $ABCD$ be a rectangle with $AD = a$ and $AB = b$. Let D' be the reflection of D across line AC, and let $E = \overline{AD'} \cap \overline{BC}$. We wish to find $[CD'E]$. Because $AB = CD'$, $\angle ABE = \angle CD'E = 90°$, and $\angle BEA = \angle D'EC$, triangles ABE and $CD'E$ are congruent. Thus $AE = EC$ and $CE^2 = AE^2 = AB^2 + BE^2 = b^2 + (a - CE)^2$. Hence $CE = \frac{a^2+b^2}{2a}$. It follows that

$$[CD'E] = [ACD'] - [ACE] = \frac{ab}{2} - \frac{b}{2} \cdot CE = \frac{b(a^2 - b^2)}{4a}.$$

Problem 2 A positive integer is said to be *balanced* if the number of its decimal digits equals the number of its distinct prime factors. For instance, 15 is balanced, while 49 is not. Prove that there are only finitely many balanced numbers.

Solution. For $n > 15$, consider the product of the first n primes. The first sixteen primes have product

$$(2 \cdot 53)(3 \cdot 47)(5 \cdot 43)(7 \cdot 41) \cdot 11 \cdot 13 \cdot 17 \cdot 19 \cdot 23 \cdot 29 \cdot 31 \cdot 37 > 100^4 \cdot 10^8 = 10^{16},$$

while the other $n - 16$ primes are each at least 10. Thus, the product of the first n primes is greater than 10^n.

If x has n digits and is balanced, then it is greater or equal to the product of the first n primes. If $n \geq 16$, then from the previous paragraph x would be greater than 10^n and would have at least $n + 1$ digits, a contradiction. Thus x can have at most 15 digits, implying that the number of balanced numbers is finite.

Problem 3 Let $\omega, \omega_1, \omega_2$ be three circles with radii r, r_1, r_2, respectively, with $0 < r_1 < r_2 < r$. The circles ω_1 and ω_2 are internally tangent to ω at two distinct points A and B and meet in two distinct points. Prove that $\overline{AB}$ contains an intersection point of ω_1 and ω_2 if and only if $r_1 + r_2 = r$.

Solution. Let O be the center of ω, and note that the centers C, D of ω_1, ω_2 lie on $\overline{OA}$ and $\overline{OB}$, respectively. Let E be a point on $\overline{AB}$ such that $CE \parallel OB$. Then $\triangle ACE \sim \triangle AOB$. Hence $AC = CE$ and E is on ω_1. We need to prove that $r = r_1 + r_2$ if and only if E is on ω_2.

Note that the following conditions are equivalent: $r = r_1+r_2$; $r-r_2 = r_1$; $OB-BD = AC$; $OD = CE$; $CEDO$ is a parallelogram; $DE \parallel AO$; $\triangle BDE \sim \triangle BOA$; $BD = DE$; E is on ω_2. This completes the proof.

Problem 4 Albert and Barbara play the following game. On a table there are 1999 sticks: each player in turn must remove from the table some sticks, provided that the player removes at least one stick and at most half of the sticks remaining on the table. The player who leaves just one stick on the table loses the game. Barbara moves first. Determine for which of the players there exists a winning strategy.

Solution. Call a number k *hopeless* if a player faced with k sticks has no winning strategy. If k is hopeless, then so is $2k+1$: a player faced with $2k+1$ sticks can only leave a pile of $k+1, k+2, \ldots$, or $2k$ sticks, from which the other player can leave k sticks. Because 2 is hopeless, so are $5, 11, \ldots, 3 \cdot 2^n - 1$ for all $n \geq 0$. Conversely, if $3 \cdot 2^n - 1 < k < 3 \cdot 2^{n+1} - 1$, then given k sticks a player can leave $3 \cdot 2^n - 1$ sticks and force a win. Because 1999 is not of the form $3 \cdot 2^n - 1$, it is not hopeless and hence Barbara has a winning strategy.

Problem 5 On a lake there is a village of pile-built dwellings, set on the nodes of an $m \times n$ rectangular array. Each dwelling is an endpoint of exactly p bridges which connect the dwelling with one or more of the adjacent dwellings (here adjacent means with respect to the array, hence diagonal connection is not allowed). Determine for which values of m, n, p it is possible to place the bridges so that from any dwelling one can reach any other dwelling. (Given two adjacent dwellings A and B, there may be several bridges, each of which connects A to B.)

Solution. Suppose it is possible to place the bridges in this manner, and set the villages along the lattice points $\{(a,b) \mid 1 \leq a \leq m, 1 \leq b \leq n\}$. Color the dwellings cyan and magenta in a checkerboard fashion, so that every bridge connects a cyan dwelling with a magenta dwelling. Because each dwelling is at the end of the same number of bridges (exactly p of them), the number of cyan dwellings must equal the number of magenta dwellings. Thus, $2 \mid mn$.

We first examine the case when at least one of m, n, and p equals 1. If $(m,n) = (1,2)$ or $(2,1)$, all values of p work. If $m=1$ and $n > 2$ (or $n=1$ and $m > 2$) then p bridges must connect $(1,1)$ and $(1,2)$ (or $(1,1)$ and $(2,1)$). Then neither of these dwellings is connected to any other dwellings, a contradiction. Finally, we cannot have $p=1$ and $mn > 2$, because otherwise if any two dwellings A and B are connected, then they cannot be connected to any other dwellings.

Now assume that $2 \mid mn$ with m, n, and p greater than 1. Assume without loss of generality that $2 \mid m$. Build a sequence of bridges starting at $(1,1)$, going up to $(1,n)$, right to (m,n), down to $(m,1)$, and left to $(m-1,1)$; and then weaving back to $(1,1)$ by repeatedly going from $(k,1)$ up to $(k,n-1)$ left to $(k-1,n-1)$ down to $(k-1,1)$ and left to $(k-2,1)$ for $k = m-1, m-3, \ldots, 3$. (The sideways E below shows this construction for $m = 6$, $n = 4$.)

So far we have built two bridges leading out of every dwelling, and any dwelling can be reached from any other dwelling. For the remaining $p-2$ bridges needed for each dwelling, note that our sequence contains exactly mn bridges, an even number. Thus if we build *every other* bridge in our sequence, and do this $p-2$ times, then exactly p bridges come out of every dwelling.

Thus either $mn = 2$ while p equals any positive integer, or else $2 \mid mn$ with $m, n, p > 1$.

Problem 6 Determine all triples (x, k, n) of positive integers such that
$$3^k - 1 = x^n.$$

Solution. All triples of the form $(3^k - 1, k, 1)$ for positive integers k, and $(2, 2, 3)$.

The solutions when $n = 1$ are obvious. Now, n cannot be even because then 3 could not divide $3^k = \left(x^{\frac{n}{2}}\right)^2 + 1$ (because no square is congruent to 2 modulo 3). Also, we must have $x \neq 1$.

Assume that $n > 1$ is odd and $x \geq 2$. Then $3^k = (x+1)\sum_{i=0}^{n-1}(-x)^i$, implying that both $x+1$ and $\sum_{i=0}^{n-1}(-x)^i$ are powers of 3. Because $x + 1 \leq x^2 - x + 1 \leq \sum_{i=0}^{n-1}(-x)^i$, we must have $0 \equiv \sum_{i=0}^{n-1}(-x)^i \equiv n \pmod{x+1}$, so that $x + 1 \mid n$. Specifically, this means that $3 \mid n$.

Writing $x' = x^{\frac{n}{3}}$, we have $3^k = x'^3 + 1 = (x'+1)(x'^2 - x' + 1)$. As before $x'+1$ must equal some power of 3, say 3^t. Then $3^k = (3^t-1)^3 + 1 = 3^{3t} - 3^{2t+1} + 3^{t+1}$, which is strictly between 3^{3t-1} and 3^{3t} for $t > 1$. Therefore we must have $t = 1$, $x' = 2$, and $k = 2$, giving the solution $(x, k, n) = (2, 2, 3)$.

Problem 7 Prove that for each prime p the equation
$$2^p + 3^p = a^n$$
has no integer solutions (a, n) with $a, n > 1$.

Solution. When $p = 2$ we have $a^n = 13$, which is impossible. Otherwise, p is odd and $5 \mid 2^p + 3^p$. Because $n > 1$, we must have $25 \mid 2^p + 3^p$. Hence
$$2^p + (5-2)^p \equiv 2^p + \left(\binom{p}{1} 5 \cdot (-2)^{p-1} + (-2)^p\right) \equiv 5p \cdot 2^{p-1} \pmod{25},$$
so $5 \mid p$. Thus $p = 5$, but the equation $a^n = 2^5 + 3^5 = 5^2 \cdot 11$ has no solutions.

Problem 8 Points D and E are given on the sides AB and AC of triangle ABC such that $DE \parallel BC$ and $\overline{DE}$ is tangent to the incircle of ABC. Prove that
$$DE \leq \frac{AB + BC + CA}{8}.$$

Solution. Let $BC = a$, $CA = b$, $AB = c$. Also let $h = \frac{2[ABC]}{a}$ be the distance from A to line BC, and let $r = \frac{2[ABC]}{a+b+c}$ the inradius of triangle ABC. Note that
$$\frac{h - 2r}{h} = \frac{b + c - a}{a + b + c}.$$
Let $x = b + c - a$, $y = c + a - b$, $z = a + b - c$. Then
$$(x + y + z)^2 \geq \left(2\sqrt{x(y+z)}\right)^2 = 4x(y+z)$$
by the AM-GM inequality. Therefore, $(a + b + c)^2 \geq 8(b + c - a)a$, or
$$\frac{b+c-a}{a+b+c} \cdot a \leq \frac{a+b+c}{8} \implies \frac{h-2r}{h} \cdot BC \leq \frac{AB + BC + CA}{8}.$$
Because $DE \parallel BC$, we have $\frac{DE}{BC} = \frac{h-2r}{h}$. Substituting this into the above inequality gives the desired result.

Problem 9

(a) Find all the strictly monotonic functions $f : \mathbb{R} \to \mathbb{R}$ such that
$$f(x + f(y)) = f(x) + y, \qquad \text{for all } x, y \in \mathbb{R}.$$

(b) Prove that for every integer $n > 1$ there do not exist strictly monotonic functions $f : \mathbb{R} \to \mathbb{R}$ such that
$$f(x + f(y)) = f(x) + y^n, \qquad \text{for all } x, y \in \mathbb{R}.$$

Solution. (a) The only such functions are $f(x) = x$ and $f(x) = -x$. Setting $x = y = 0$ gives $f(f(0)) = f(0)$, while setting $x = -f(0), y = 0$ gives $f(-f(0)) = f(0)$. Because f is strictly monotonic it is injective, so $f(0) = -f(0)$ and thus $f(0) = 0$. Next, setting $x = 0$ gives $f(f(y)) = y$ for all y.

Suppose f is increasing. If $f(x) > x$ then $x = f(f(x)) > f(x)$, a contradiction; if $f(x) < x$ then $x = f(f(x)) < f(x)$, a contradiction. Thus $f(x) = x$ for all x.

Next suppose that f is decreasing. Plugging in $x = -f(t), y = t$, and then $x = 0, y = -t$ shows that $f(-f(t)) = f(f(-t)) = -t$, so $f(t) = -f(-t)$ for all t. Now given x, if $f(x) < -x$, then

$$x = f(f(x)) > f(-x) = -f(x),$$

a contradiction. If instead $f(x) > -x$, then

$$x = f(f(x)) < f(-x) = -f(x),$$

a contradiction. Hence we must have $f(x) = -x$ for all x.

Therefore either $f(x) = x$ for all x or $f(x) = -x$ for all x, and it is easy to check that these two functions work.

(b) Because f is strictly monotonic, it is injective. Then for $y \neq 0$ we have $f(y) \neq f(-y)$ so that $f(x + f(y)) \neq f(x + f(-y))$ and hence $f(x) + y^n \neq f(x) + (-y)^n$. Thus, n cannot be even.

Now suppose there is such an f for odd n. By arguments similar to those in part (a), we find that $f(0) = 0$ and $f(f(y)) = y^n$. Specifically, $f(f(1)) = 1$. If f is increasing then as in part (a) we have $f(1) = 1$; then $f(2) = f(1 + f(1)) = f(1) + 1^n = 2$ and $2^n = f(f(2)) = f(2) = 2$, a contradiction. If f is decreasing, then as in part (a) we have $f(1) = -1$; then

$$f(2) = f(1 + f(-1)) = f(1) + (-1)^n = -2$$

and

$$2^n = f(f(2)) = f(-2) = -f(2) = 2,$$

a contradiction.

Problem 10 Let X be a set with $|X| = n$, and let $A_1, A_2, \ldots, A_m$ be subsets of X such that

(a) $|A_i| = 3$ for $i = 1, 2, \ldots, m$.

(b) $|A_i \cap A_j| \leq 1$ for all $i \neq j$.

Prove that there exists a subset of X with at least $\lfloor\sqrt{2n}\rfloor$ elements, which does not contain A_i for $i = 1, 2, \ldots, m$.

Solution. Let A be a subset of X containing no A_i, and having the maximum number of elements subject to this condition. Let k be the size of A. By assumption, for each $x \in X - A$, there exists $i(x) \in \{1, \ldots, m\}$ such that $A_{i(x)} \subseteq A \cup \{x\}$. Let $L_x = A \cap A_{i(x)}$, which by the previous observation must have 2 elements. Because $|A_i \cap A_j| \leq 1$ for $i \neq j$, the L_x must all be distinct. Now there are $\binom{k}{2}$ 2-element subsets of A, so there can be at most $\binom{k}{2}$ sets L_x. Thus $n - k \leq \binom{k}{2}$ or $k^2 + k \geq 2n$. It follows that
$$k \geq \frac{1}{2}(-1 + \sqrt{1+8n}) > \sqrt{2n} - 1,$$
that is, $k \geq \lfloor\sqrt{2n}\rfloor$.

13 Japan

Problem 1 Each square of a 1999×1999 grid contains one or zero stones. Find the minimum number of stones needed such that, when an arbitrary blank square is selected, the total number of stones in the corresponding row and column is at least 1999.

Solution. Place stones in a checkerboard pattern on the grid, so that stones are placed on the four corner squares. This placement satisfies the condition and contains $1000 \times 1000 + 999 \times 999 = 1998001$ stones. We now prove this number is minimal.

Suppose the condition is satisfied. Assume that the minimum number of stones in any row or column is k. Without loss of generality, assume that some column contains k stones. For each of the k stones in that column, the row containing that stone must contain at least k stones by our minimal choice of k. For the $1999 - k$ blank squares in our column, to satisfy the given condition there must be at least $1999 - k$ stones in the row containing that square. Thus total number of stones is at least

$$k^2 + (1999-k)^2 = 2\left(k - \frac{1999}{2}\right)^2 + \frac{1999^2}{2} \geq \frac{1999^2}{2} = 1998000.5,$$

and it follows that there indeed must be at least 1998001 stones.

Problem 2 Let $f(x) = x^3 + 17$. Prove that for each natural number n, $n \geq 2$, there is a natural number x for which $f(x)$ is divisible by 3^n but not by 3^{n+1}.

Solution. We prove the result by induction on n. If $n = 2$, then $x = 1$ suffices. Now suppose that the claim is true for $n \geq 2$ — that is, there is a natural number y such that $y^3 + 17$ is divisible by 3^n but not 3^{n+1}. We prove that the claim is true for $n + 1$.

Suppose we have integers a, m such that a is not divisible by 3 and $m \geq 2$. Then $a^2 \equiv 1 \pmod{3}$ and thus $3^m a^2 \equiv 3^m \pmod{3^{m+1}}$. Also, because $m \geq 2$ we have $3m - 3 \geq 2m - 1 \geq m + 1$. Hence

$$(a + 3^{m-1})^3 \equiv a^3 + 3^m a^2 + 3^{2m-1} a + 3^{3m-3} \equiv a^3 + 3^m \pmod{3^{m+1}}.$$

Because $y^3 + 17$ is divisible by 3^n, it is congruent to either 0, 3^n, or $2 \cdot 3^n$ modulo 3^{n+1}. Because 3 does not divide 17, 3 cannot divide y either. Hence applying our result from the previous paragraph twice — once with $(a, m) = (y, n)$ and once with $(a, m) = (y + 3^{n-1}, n)$ — we find that 3^{n+1} must divide either $(y + 3^{n-1})^3 + 17$ or $(y + 2 \cdot 3^{n-1})^3 + 17$.

Hence there exists a natural number x' not divisible by 3 such that $3^{n+1} \mid x'^3 + 17$. If 3^{n+2} does not divide $x'^3 + 17$, we are done. Otherwise, we claim the number $x = x' + 3^n$ suffices. Because $x = x' + 3^{n-1} + 3^{n-1} + 3^{n-1}$, the result from two paragraphs ago tells us that $x^3 \equiv x'^3 + 3^n + 3^n + 3^n \equiv x'^3 \pmod{3^{n+1}}$. Thus $3^{n+1} \mid x^3 + 17$ as well. On the other hand, because $x = x' + 3^n$, we have $x^3 \equiv x'^3 + 3^{n+1} \not\equiv x'^3 \pmod{3^{n+2}}$. It follows that 3^{n+2} does *not* divide $x^3 + 17$, as desired. This completes the inductive step.

Problem 3 From a set of $2n+1$ weights (where n is a natural number), if any one weight is excluded, then the remaining $2n$ weights can be divided into two sets of n weights that balance each other. Prove that all the weights are equal.

Solution. Label the weights $a_1, a_2, \ldots, a_{2n+1}$. Then for each j, $1 \leq j \leq 2n$, we have

$$c_1^{(j)} a_1 + c_2^{(j)} a_2 + \cdots + c_{2n}^{(j)} a_{2n} = a_{2n+1}$$

where $c_j^{(j)} = 0$, n of the other $c_i^{(j)}$ equal 1, and the remaining $c_i^{(j)}$ equal -1.

Thus we have $2n$ equations in the variables $a_1, a_2, \ldots, a_{2n}$. Clearly $(a_1, a_2, \ldots, a_{2n}) = (a_{2n+1}, a_{2n+1}, \ldots, a_{2n+1})$ is a solution to this system of equations. By Cramer's Rule, this solution is unique if and only if the determinant of the matrix

$$M = \begin{bmatrix} c_1^{(1)} & c_2^{(1)} & \cdots & c_{2n}^{(1)} \\ c_1^{(2)} & c_2^{(2)} & \cdots & c_{2n}^{(2)} \\ \vdots & \vdots & \ddots & \vdots \\ c_1^{(2n)} & c_2^{(2n)} & \cdots & c_{2n}^{(2n)} \end{bmatrix}$$

is nonzero. The diagonal entries of M are all even, and its off-diagonal entries are all odd. Hence if we multiply M by itself, the resulting matrix has odd diagonal entries and even off-diagonal entries. This implies that $(\det M)^2 = \det(M^2)$ is odd, so $\det M$ must indeed be nonzero.

Problem 4 Prove that

$$f(x) = (x^2 + 1^2)(x^2 + 2^2)(x^2 + 3^2) \cdots (x^2 + n^2) + 1$$

cannot be expressed as a product of two integral-coefficient polynomials with degree greater than 0.

Solution. The claim is obvious when $n = 1$. Now assume $n \geq 2$ and suppose, by way of contradiction, that $f(x)$ *could* be expressed as such a product $g(x)h(x)$ with

$$g(x) = a_0 + a_1 x + \cdots + a_\ell x^\ell,$$
$$h(x) = b_0 + b_1 x + \cdots + b_{\ell'} x^{\ell'},$$

where $\ell, \ell' > 0$ and the coefficients a_i and b_i are integers.

For $m = \pm 1, \pm 2, \ldots, \pm n$, because $(mi)^2 + m^2 = 0$ we have $1 = f(mi) = g(mi)h(mi)$. Because g and h have integer coefficients, $g(mi)$ equals either 1, -1, i, or $-i$. Moreover, because the imaginary part of

$$g(mi) = (a_0 - a_2 m^2 + a_4 m^4 - \cdots) + m(a_1 - a_3 m^2 + a_5 m^4 - \cdots)i$$

is a multiple of m, $g(mi)$ must equal ± 1 for $m \neq \pm 1$. Going further, because $1 = g(mi)h(mi)$ we have $g(mi) = h(mi) = \pm 1$ for $m \neq \pm 1$.

Then by the factor theorem,

$$g(x) - h(x) = (x^2 + 2^2)(x^2 + 3^2) \cdots (x^2 + n^2) k(x)$$

for some integer-coefficient polynomial $k(x)$ with degree at most 1. Because $(g(i), h(i))$ equals $(1, 1)$, $(-1, -1)$, $(i, -i)$, or $(-i, i)$, we have

$$2 \geq |g(i) - h(i)| = (-1 + 2^2)(-1 + 3^2) \cdots (-1 + n^2) |k(i)|,$$

and hence we must have $k(i) = 0$. Because $k(x)$ has degree at most 1, this implies that $k(x) = 0$ for *all* x and that $g(x) = h(x)$ for all x. Then $a_0^2 = g(0)h(0) = f(0) = (1^2)(2^2) \cdots (n^2) + 1$, which is impossible.

Problem 5 For a convex hexagon $ABCDEF$ whose side lengths are all 1, let M and m be the maximum and minimum values of the three diagonals AD, BE, and CF. Find all possible values of m and M.

Solution. We claim that the possible values are $\sqrt{3} \leq M \leq 3$ and $1 \leq m \leq 2$.

First we show all such values are attainable. Continuously transform $ABCDEF$ from an equilateral triangle ACE of side length 2, into a regular hexagon of side length 1, and finally into a segment of length 3 (for instance, by enlarging the diagonal AD of the regular hexagon while bringing B, C, E, F closer to line AD). Then M continuously varies from $\sqrt{3}$ to 2 to 3. Similarly, by continuously transforming $ABCDEF$ from a 1×2 rectangle into a regular hexagon, we can make m vary continuously from 1 to 2.

Now we prove no other values are attainable. First, we have $AD \leq AB + BC + CD = 3$, and similarly $BE, CF \leq 3$ so that $M \leq 3$.

Next suppose, by way of contradiction, that $m < 1$ and assume without loss of generality that $AD < 1$. Because $AD < AB = BC = CD = 1$,

$$\angle DCA < \angle DAC, \quad \angle ABD < \angle ADB,$$
$$\angle CBD = \angle CDB, \quad \angle BCA = \angle BAC.$$

Therefore,

$$\angle CDA + \angle BAD = \angle CDB + \angle BDA + \angle BAC + \angle CAD$$
$$> \angle CBD + \angle DBA + \angle BCA + \angle ACD$$
$$= \angle CBA + \angle BCD.$$

Consequently $\angle CDA + \angle BAD > 180°$ and likewise $\angle EDA + \angle FAD > 180°$. Then

$$\angle CDE + \angle BAF = \angle CDA + \angle EDA + \angle BAD + \angle FAD > 360°,$$

which is impossible because $ABCDEF$ is convex. Hence $m \geq 1$.

Next we demonstrate that $M \geq \sqrt{3}$ and $m \leq 2$. Because the sum of the six interior angles in $ABCDEF$ is $720°$, some pair of adjacent angles has sum greater than or equal to $240°$ and some pair has sum less than or equal to $240°$. Thus it suffices to prove that $CF \geq \sqrt{3}$ when $\angle A + \angle B \geq 240°$, and that $CF \leq 2$ when $\angle A + \angle B \leq 240°$.

By the law of cosines,

$$CF^2 = BC^2 + BF^2 - 2BC \cdot BF \cos \angle FBC.$$

Thus if we fix A, B, F and decrease $\angle ABC$, we decrease $\angle FBC$ and CF. Similarly, by fixing A, B, C and decreasing $\angle BAF$, we decrease CF. Therefore, it suffices to prove that $\sqrt{3} \geq CF$ when $\angle A + \angle B = 240°$. Likewise, it suffices to prove that $CF \leq 2$ when $\angle A + \angle B = 240°$.

Now suppose that $\angle A + \angle B$ *does* equal $240°$. Let lines AF and BC intersect at P, and set $x = PA$ and $y = PB$. Because $\angle A + \angle B = 240°$, $\angle P = 60°$. Then applying the law of cosines to triangles PAB and PCF yields

$$1 = AB^2 = x^2 + y^2 - xy$$

and

$$CF^2 = (x+1)^2 + (y+1)^2 - (x+1)(y+1) = 2 + x + y.$$

Therefore, we need only find the possible values of $x+y$ given that $x^2+y^2-xy=1$ and $x,y \geq 0$. These conditions imply that
$$(x+y)^2 + 3(x-y)^2 = 4, \quad x+y \geq 0,$$
and $|x-y| \leq x+y$. Hence
$$1 = \frac{1}{4}(x+y)^2 + \frac{3}{4}(x-y)^2$$
$$\leq (x+y)^2 \leq (x+y)^2 + 3(x-y)^2 = 4,$$
so $1 \leq x+y \leq 2$ and $\sqrt{3} \leq CF \leq 2$. This completes the proof.

14 Korea

Problem 1 Let R and r be the circumradius and inradius, respectively, of triangle ABC, and let R' and r' be the circumradius and inradius, respectively, of triangle $A'B'C'$. Prove that if $\angle C = \angle C'$ and $Rr' = R'r$, then the triangles are similar.

Solution. Let ω be the circumcircle of triangle ABC. By scaling, rotating, translating, and reflecting, we may assume that $A = A'$, $B = B'$, $R = R'$, $r = r'$ and that C, C' lie on the same arc $\widehat{AB}$ of ω. If the triangles were similar before these transformations, they still remain similar, so it suffices to prove they are now congruent.

Because
$$r = \frac{1}{2}(AC + BC - AB)\cot(\angle C)$$
and
$$r' = \frac{1}{2}(A'C' + B'C' - A'B')\cot(\angle C')$$
$$= \frac{1}{2}(A'C' + B'C' - AB)\cot(\angle C),$$
we must have $AC + BC = A'C' + B'C'$ and hence $AB + BC + CA = A'B' + B'C' + C'A'$. Then the area of triangle ABC is $\frac{1}{2}r(AB + BC + CA)$, which thus equals the area of triangle $A'B'C'$, $\frac{1}{2}r'(A'B' + B'C' + C'A')$. Because these triangles share the same base $\overline{AB}$, we know that the altitudes from C and C' to $\overline{AB}$ are equal. This implies that $\triangle ABC$ is congruent to either $\triangle A'B'C'$ or $\triangle B'A'C'$, as desired.

Problem 2 Suppose $f : \mathbb{Q} \to \mathbb{R}$ is a function satisfying
$$|f(m+n) - f(m)| \leq \frac{n}{m}$$
for all positive rational numbers m and n. Show that for all positive integers k,
$$\sum_{i=1}^{k} |f(2^k) - f(2^i)| \leq \frac{k(k-1)}{2}.$$

Solution. It follows from the condition $|f(m+n) - f(m)| \leq \frac{n}{m}$ that
$$|f(2^{i+1}) - f(2^i)| \leq \frac{2^{i+1} - 2^i}{2^i} = 1.$$

Therefore, for $k > i$,

$$|f(2^k) - f(2^i)| \leq \sum_{j=i}^{k-1} |f(2^{j+1}) - f(2^j)| \leq k - i.$$

From the above inequality, we obtain

$$\sum_{i=1}^{k} |f(2^k) - f(2^i)| = \sum_{i=1}^{k-1} |f(2^k) - f(2^i)| \leq \sum_{i=1}^{k-1} (k-i) = \frac{k(k-1)}{2}.$$

This completes the proof.

Problem 3 Find all positive integers n such that $2^n - 1$ is a multiple of 3 and $\frac{2^n-1}{3}$ is a divisor of $4m^2 + 1$ for some integer m.

Solution. The answer is all $n = 2^k$ where $k = 1, 2, \ldots$.

First observe that $2 \equiv -1 \pmod{3}$. Hence $3 \mid 2^n - 1$ if and only if n is even.

Suppose, by way of contradiction, that $\ell \geq 3$ is a positive odd divisor of n. Then $2^\ell - 1$ is not divisible by 3 but it *is* a divisor of $2^n - 1$, so it is a divisor of $4m^2 + 1$ as well. On the other hand, $2^\ell - 1$ has a prime divisor p of the form $4r + 3$. Then $(2m)^2 \equiv -1 \pmod{4r+3}$, but a standard number theory result states that a square cannot be congruent to -1 modulo a prime of the form $4r + 3$.

Therefore, n is indeed of the form 2^k for $k \geq 1$. For such n, we have

$$\frac{2^n - 1}{3} = (2^{2^1} + 1)(2^{2^2} + 1)(2^{2^3} + 1) \cdots (2^{2^{k-1}} + 1).$$

The factors on the right side are all relatively prime to 2 because they are all odd. They are also Fermat numbers, and another result from number theory states that they are relatively prime. (Suppose that some prime p divided both $2^{2^a} + 1$ and $2^{2^b} + 1$ for $a < b$. Then $2^{2^a} \equiv 2^{2^b} \equiv -1 \pmod{p}$. Hence $-1 \equiv 2^{2^b} = \left(2^{2^a}\right)^{2^{b-a}} \equiv \left((-1)^2\right)^{2^{b-a-1}} \equiv 1 \pmod{p}$, implying that $p = 2$. This is impossible.) Therefore by the Chinese Remainder Theorem, there is a positive integer c simultaneously satisfying

$$c \equiv 2^{2^{i-1}} \pmod{2^{2^i} + 1} \quad \text{for all } i = 1, 2, \ldots, k-1$$

and $c \equiv 0 \pmod{2}$. Putting $c = 2m$, $4m^2 + 1$ is a multiple of $\frac{2^n-1}{3}$, as desired.

Problem 4 Suppose that for any real x with $|x| \neq 1$, a function $f(x)$ satisfies
$$f\left(\frac{x-3}{x+1}\right) + f\left(\frac{3+x}{1-x}\right) = x.$$
Find all possible $f(x)$.

Solution. Set $t = \frac{x-3}{x+1}$ so that $x = \frac{3+t}{1-t}$. Then the given equation can be rewritten as
$$f(t) + f\left(\frac{t-3}{t+1}\right) = \frac{3+t}{1-t}.$$
Similarly, set $t = \frac{3+x}{1-x}$ so that $x = \frac{t-3}{t+1}$ and $\frac{x-3}{x+1} = \frac{3+t}{1-t}$. Again we can rewrite the given equation, this time as
$$f\left(\frac{3+t}{1-t}\right) + f(t) = \frac{t-3}{t+1}.$$
Adding these two equations we have
$$\frac{8t}{1-t^2} = 2f(t) + f\left(\frac{t-3}{t+1}\right) + f\left(\frac{3+t}{1-t}\right) = 2f(t) + t,$$
so that
$$f(t) = \frac{4t}{1-t^2} - \frac{t}{2},$$
and some algebra verifies that this solution works.

Problem 5 Consider a permutation $(a_1, a_2, \ldots, a_6)$ of $1, 2, \ldots, 6$ such that the minimum number of transpositions needed to transform $(a_1, a_2, a_3, a_4, a_5, a_6)$ to $(1, 2, 3, 4, 5, 6)$ is four. Find the number of such permutations.

Solution. Suppose we are given distinct numbers $b_1, b_2, \ldots, b_k$ between 1 and n. In a k-cycle involving these numbers, b_1 is mapped to one of the other $k - 1$ numbers; its image is mapped to one of the $k - 2$ remaining numbers; and so on until the remaining number is mapped to b_1. Hence there are $(k - 1)(k - 2) \cdots (1) = (k - 1)!$ cycles of length k involving these numbers.

Any permutation which can be achieved with four transpositions is even, so a permutation satisfying the given conditions must be either (i) the identity permutation, (ii) a composition of two transpositions, (iii) a 3-cycle, (iv) a composition of a 2-cycle and a 4-cycle, (v) a composition of two 3-cycles, or (vi) a 5-cycle. Permutations of type (i), (ii), and (iii) can be attained with fewer transpositions from our observations above.

Conversely, any even permutation that can be achieved with zero or two transpositions is of these three types. Hence the permutations described in the problem statement are precisely those of types (iv), (v), and (vi).

For type-(iv) permutations, there are $\binom{6}{2} = 15$ ways to assign which cycle each of $1, 2, \ldots, 6$ belongs. There are also $(2-1)!(4-1)! = 6$ ways to rearrange the numbers within the cycles, for a total of $15 \cdot 6 = 90$ permutations.

For type-(v) permutations, there are $\frac{1}{2}\binom{6}{3} = 10$ ways to assign which cycle each number belongs to (because $\binom{6}{3}$ counts each such permutation twice, once in the form $(a\,b\,c)(d\,e\,f)$ and again in the form $(d\,e\,f)(a\,b\,c)$). There are $(3-1)!(3-1)! = 4$ ways to rearrange the numbers within these two cycles for a total of $10 \cdot 4 = 40$ type-(v) permutations.

Finally, for type-(v) permutations there are $\binom{6}{5} = 6$ ways to choose which five numbers are cycled, and $(5-1)! = 24$ different cycles among any five numbers. This gives a total of $6 \cdot 24 = 144$ type-(v) permutations, and altogether

$$90 + 40 + 144 = 274$$

permutations which can be attained with four permutations, but no fewer.

Problem 6 Let $a_1, a_2, \cdots, a_{1999}$ be nonnegative real numbers satisfying the following two conditions:

(a) $a_1 + a_2 + \cdots + a_{1999} = 2$;

(b) $a_1 a_2 + a_2 a_3 + \cdots + a_{1998} a_{1999} + a_{1999} a_1 = 1$.

Let $S = a_1^2 + a_2^2 + \cdots + a_{1999}^2$. Find the maximum and minimum possible values of S.

Solution. Without loss of generality assume that a_{1999} is the minimum a_i. We may also assume that $a_1 > 0$. From the given equations we have

$$4 = (a_1 + a_2 + \cdots + a_{1999})^2$$
$$\geq (a_1 + a_2 + \cdots + a_{1999})^2 - (a_1 - a_2 + a_3 - \cdots - a_{1998} + a_{1999})^2$$
$$= 4(a_1 + a_3 + \cdots + a_{1999})(a_2 + a_4 + \cdots + a_{1998})$$
$$\geq 4(a_1 a_2 + a_2 a_3 + \cdots + a_{1998} a_{1999})$$
$$\quad + 4(a_1 a_4 + a_2 a_5 + \cdots + a_{1996} a_{1999}) + 4a_1(a_6 + a_8 + \cdots + a_{1998})$$
$$= 4(1 - a_{1999} a_1) + 4(a_1 a_4 + a_2 a_5 \cdots + a_{1996} a_{1999})$$
$$\quad + 4a_1(a_6 + a_8 + \cdots + a_{1998})$$

$$= 4 + 4(a_1 a_4 + a_2 a_5 + \cdots + a_{1996} a_{1999})$$
$$+ 4a_1(a_6 + a_8 + \cdots + a_{1998} - a_{1999})$$
$$\geq 4.$$

Hence equality must hold in the first and third inequality. Thus we must have

(i) $a_1 + a_3 + \cdots + a_{1999} = a_2 + a_4 + \cdots + a_{1998} = 1$
(ii) $a_1 a_4 = a_2 a_5 = \cdots = a_{1996} a_{1999} = 0$
(iii) $a_6 + a_8 + \cdots + a_{1998} = a_{1999}$.

Condition (ii) implies $a_4 = 0$. From (iii) we get $a_6 = a_8 = \cdots = a_{1998} = 0$. Thus from (i), we have $a_2 = 1$, and from (b), we have $a_1 + a_3 = 1$. Applying these to the first given condition (a), we have

$$a_4 + a_5 + \cdots + a_{1999} = 0,$$

so that $a_4 = a_5 = \cdots = a_{1999} = 0$. Therefore

$$S = a_1^2 + a_2^2 + a_3^2$$
$$= a_1^2 + 1 + (1 - a_1)^2 \quad \text{because } a_2 = a_1 + a_3 = 1$$
$$= 2(a_1^2 - a_1 + 1)$$
$$= 2\left(a_1 - \frac{1}{2}\right)^2 + \frac{3}{2}.$$

Thus S has maximum value 2 attained when $a_1 = 1$, and minimum value $\frac{3}{2}$ when $a_1 = \frac{1}{2}$.

15 Poland

Problem 1 Let D be a point on side BC of triangle ABC such that $AD > BC$. Point E on side AC is defined by the equation
$$\frac{AE}{EC} = \frac{BD}{AD - BC}.$$
Show that $AD > BE$.

First Solution. Fix the points B, C, D and the distance AD, and let A vary. The locus of A is a circle with center D. From the equation, the ratio $\frac{AE}{EC}$ is fixed. Therefore, $\lambda = \frac{EC}{AC}$ is also fixed. Because E is the image of A under a homothety about C with ratio λ, the locus of all points E is the image of the locus of A under this homothety — a circle centered on $\overline{BC}$. Then BE has its unique maximum when E is the intersection of the circle with line BC farther from B. If we show that $AD = BE$ in this case then we are done (the original inequality $AD > BE$ will be strict because equality can only hold in this degenerate case). Indeed, in this case the points B, D, C, E, A are collinear in that order, and our equation gives
$$AE \cdot (AC - BD) = AE \cdot (AD - BC) = EC \cdot BD \Rightarrow AE \cdot AC$$
$$= (AE + EC) \cdot BD = AC \cdot BD \Rightarrow AE$$
$$= BD \Rightarrow AD = BE.$$

Second Solution. Let F be the point on $\overline{AD}$ such that $FA = BC$, and let line BF hit side AC at E'. By the law of sines we have
$$AE' = FA \cdot \frac{\sin \angle AFE'}{\sin \angle FE'A} = CB \cdot \frac{\sin \angle DFB}{\sin \angle CE'F}$$
and
$$E'C = CB \cdot \frac{\sin \angle E'BC}{\sin \angle CE'B} = CB \cdot \frac{\sin \angle FBD}{\sin \angle CE'F}.$$
Hence
$$\frac{AE'}{E'C} = \frac{\sin \angle DFB}{\sin \angle FBD} = \frac{DB}{FD} = \frac{BD}{AD - BC} = \frac{AE}{EC},$$
and $E' = E$.

Let ℓ be the line passing through A parallel to side BC. Draw G on ray BC such that $BG = AD$ and $CG = FD$, and let lines GE and ℓ intersect at H. Triangles ECG and EAH are similar, so $AH = CG \cdot \frac{AE}{EC} = FD \cdot \frac{AE}{EC}$.

By Menelaus' Theorem applied to triangle CAD and line EFB, we have
$$\frac{CE \cdot AF \cdot DB}{EA \cdot FD \cdot BC} = 1.$$
Thus
$$AH = FD \cdot \frac{AE}{EC} = FD \cdot \frac{AF \cdot DB}{FD \cdot BC} = DB \cdot \frac{AF}{BC} = DB,$$
implying that quadrilateral $BDAH$ is a parallelogram and that $BH = AD$. It follows that triangle BHG is isosceles with $BH = BG = AD$. Because $\overline{BE}$ in a cevian in this triangle, we must have $BE < AD$, as desired.

Problem 2 Given are nonnegative integers $a_1 < a_2 < \cdots < a_{101}$ smaller than 5050. Show that one can choose four distinct integers a_k, a_l, a_m, a_n such that
$$5050 \mid (a_k + a_l - a_m - a_n).$$

Solution. First observe that the a_i are all distinct modulo 5050 because they are all between 0 and 5050. Now consider all sums $a_i + a_j, i < j$; there are $\binom{101}{2} = 5050$ such sums. If any two such sums, $a_k + a_l$ and $a_m + a_n$, are congruent mod 5050, we are done. (In this case, all four indices would indeed be distinct: if, for example, $k = m$, then we would also have $l = n$ because all a_i are different mod 5050, but we chose the pairs $\{k,l\}$ and $\{m,n\}$ to be distinct.) The only other possibility is that these sums occupy every possible congruence class mod 5050. Then, adding all such sums gives
$$100(a_1 + a_2 + \cdots + a_{101}) \equiv 0 + 1 + \cdots + 5049 = 2525 \cdot 5049 \pmod{5050}.$$
Because the number on the left side is even but $2525 \cdot 5049$ is odd, we get a contradiction.

Problem 3 For a positive integer n, let $S(n)$ denote the sum of its digits. Prove that there exist distinct positive integers $\{n_i\}_{1 \leq i \leq 50}$ such that
$$n_1 + S(n_1) = n_2 + S(n_2) = \cdots = n_{50} + S(n_{50}).$$

Solution. We show by induction on k that there exist positive integers $n_1, \ldots, n_k$ with the desired property. For $k = 1$ the statement is obvious. For $k > 1$, let $m_1 < \cdots < m_{k-1}$ satisfy the induction hypothesis for $k-1$. Note that we can make all the m_i arbitrarily large by adding some large power of 10 to all of them, which preserves the described property. Then, choose m with $1 \leq m \leq 9$ and $m \equiv m_1 + 1 \pmod{9}$. Observing that

$S(x) \equiv x \pmod 9$, we have $m_1 - m + S(m_1) - S(m) + 11 = 9\ell$ for some integer ℓ. By choosing the m_i large enough we can ensure $10^\ell > m_{k-1}$. Now let $n_i = 10^{\ell+1} + m_i$ for $i < k$ and $n_k = m + 10^{\ell+1} - 10$. It is obvious that $n_i + S(n_i) = n_j + S(n_j)$ for $i, j < k$, and

$$\begin{aligned} n_1 + S(n_1) &= (10^{l+1} + m_1) + (1 + S(m_1)) \\ &= (m_1 + S(m_1) + 1) + 10^{l+1} \\ &= (9\ell + S(m) + m - 10) + 10^{\ell+1} \\ &= (m + 10^{l+1} - 10) + (9\ell + S(m)) \\ &= n_k + S(n_k), \end{aligned}$$

as needed.

Problem 4 Find all integers $n \geq 2$ for which the system of equations

$$x_1^2 + x_2^2 + 50 = 16x_1 + 12x_2$$
$$x_2^2 + x_3^2 + 50 = 16x_2 + 12x_3$$
$$\dots\dots\dots$$
$$x_{n-1}^2 + x_n^2 + 50 = 16x_{n-1} + 12x_n$$
$$x_n^2 + x_1^2 + 50 = 16x_n + 12x_1$$

has a solution in integers $(x_1, x_2, \ldots, x_n)$.

Solution. Answer: $n \equiv 0 \pmod 3$.

We rewrite the equation $x^2 + y^2 + 50 = 16x + 12y$ as $(x-8)^2 + (y-6)^2 = 50$, whose integer solutions are

$(7, -1), (7, 13), (9, -1), (9, 13), (3, 1), (3, 11)$

$(13, 1), (13, 11), (1, 5), (1, 7), (15, 5), (15, 7).$

Thus every pair (x_i, x_{i+1}) (where $x_{n+1} = x_1$) must be one of these. If $3 \mid n$ then let $x_{3i} = 1$, $x_{3i+1} = 7$, $x_{3i+2} = 13$ for each i. Conversely, if a solution exists, consider the pairs (x_i, x_{i+1}) which occur. Every pair's first coordinate is the second coordinate of another pair, and vice versa, which reduces the above possibilities to $(1, 7), (7, 13), (13, 1)$. It follows that the x_i must form a repeating sequence $1, 7, 13, 1, 7, 13, \ldots$, which is only possible when $3 \mid n$.

Problem 5 Let $a_1, a_2, \ldots, a_n, b_1, b_2, \ldots, b_n$ be integers. Prove that

$$\sum_{1 \leq i < j \leq n} (|a_i - a_j| + |b_i - b_j|) \leq \sum_{1 \leq i,j \leq n} |a_i - b_j|.$$

Solution. Define $f_{\{a,b\}}(x) = 1$ if either $a \leq x < b$ or $b \leq x < a$, and 0 otherwise. Observe that when a, b are integers, $|a - b| = \sum_x f_{\{a,b\}}(x)$ where the sum is over all integers (the sum is well-defined as only finitely many terms are nonzero). Now fix an arbitrary integer x and suppose $a_\leq$ is the number of values of i for which $a_i \leq x$. Define $a_>, b_\leq, b_>$ analogously. We have

$$(a_\leq - b_\leq) + (a_> - b_>) = (a_\leq + a_>) - (b_\leq + b_>)$$
$$= n - n = 0 \Rightarrow (a_\leq - b_\leq)(a_> - b_>) \leq 0.$$

Thus $a_\leq a_> + b_\leq b_> \leq a_\leq b_> + a_> b_\leq$. Now $a_\leq a_> = \sum_{i<j} f_{\{a_i,a_j\}}(x)$ because both sides count the same set of pairs, and the other terms reduce similarly, giving

$$\sum_{1 \leq i < j \leq n} f_{\{a_i,a_j\}}(x) + f_{\{b_i,b_j\}}(x) \leq \sum_{1 \leq i,j \leq n} f_{\{a_i,b_j\}}(x).$$

Because x was an arbitrary integer, this last inequality holds for all integers x. Summing over all integers x and using our first observation, we get the desired inequality. Equality holds if and only if the above inequality is an equality for all x, which is true precisely when the a_i equal the b_i in some order.

Problem 6 In a convex hexagon $ABCDEF$, $\angle A + \angle C + \angle E = 360°$ and

$$AB \cdot CD \cdot EF = BC \cdot DE \cdot FA.$$

Prove that $AB \cdot FD \cdot EC = BF \cdot DE \cdot CA$.

First Solution. Construct point G so that triangle GBC is similar to triangle FBA (and with the same orientation). Then $\angle DCG = 360° - (\angle GCB + \angle BCD) = \angle DEF$ and

$$\frac{GC}{CD} = \frac{FA \cdot \frac{BC}{AB}}{CD} = \frac{FE}{ED},$$

so $\triangle DCG \sim \triangle DEF$.

Now $\frac{AB}{BF} = \frac{CB}{BG}$ by similar triangles, and $\angle ABC = \angle ABF + \angle FBC = \angle CBG + \angle FBC = \angle FBG$. Thus $\triangle ABC \sim \triangle FBG$, and likewise

$\triangle EDC \sim \triangle FDG$. Then

$$\frac{AB}{CA} \cdot \frac{EC}{DE} \cdot \frac{FD}{BF} = \frac{FB}{GF} \cdot \frac{FG}{DF} \cdot \frac{FD}{BF} = 1$$

as needed.

Second Solution. Invert about F with some radius r. The original equality becomes

$$\frac{A'B' \cdot r^2}{A'F \cdot B'F} \cdot \frac{C'D' \cdot r^2}{C'F \cdot D'F} \cdot \frac{r^2}{E'F} = \frac{B'C' \cdot r^2}{B'F \cdot C'F} \cdot \frac{D'E' \cdot r^2}{D'F \cdot E'F} \cdot \frac{r^2}{A'F}$$

or

$$\frac{A'B'}{B'C'} = \frac{E'D'}{D'C'}.$$

The original angle condition is

$$\angle FAB + \angle BCF + \angle FCD + \angle DEF = 360°.$$

Using directed angles, this condition turns into $\angle A'B'F + \angle FB'C' + \angle C'D'F + \angle FD'E' = 360°$, or $\angle A'B'C' = \angle E'D'C'$. Thus triangles $A'B'C'$, $E'D'C'$ are similar, giving

$$\frac{A'B'}{A'C'} = \frac{E'D'}{E'C'}$$

or, equivalently,

$$\frac{A'B' \cdot r^2}{A'F \cdot B'F} \cdot \frac{r^2}{D'F} \cdot \frac{E'C' \cdot r^2}{C'F \cdot E'F} = \frac{r^2}{B'F} \cdot \frac{D'E' \cdot r^2}{D'F \cdot E'F} \cdot \frac{C'A' \cdot r^2}{A'F \cdot C'F}.$$

Inverting back yields the desired result.

Third Solution. Position the hexagon in the complex plane and let $a = B - A, b = C - B, \ldots, f = A - F$. The product identity implies that $|ace| = |bdf|$, and the angle equality implies $\frac{-b}{a} \cdot \frac{-d}{c} \cdot \frac{-f}{e}$ is real and positive. Hence, $ace = -bdf$. Also, $a+b+c+d+e+f = 0$. Multiplying this by ad and adding $ace + bdf = 0$ gives

$$a^2d + abd + acd + ad^2 + ade + adf + ace + bdf = 0$$

which factors to $a(d + e)(c + d) + d(a + b)(f + a) = 0$. Thus

$$|a(d+e)(c+d)| = |d(a+b)(f+a)|,$$

which is what we wanted.

16 Romania

National Olympiad

Problem 7.1 Determine the side lengths of a right triangle if they are integers and the product of the leg lengths is equal to three times the perimeter.

Solution. One of the leg lengths must be divisible by 3. Let the legs have lengths $3a$ and b and let the hypotenuse have length c, where a, b, and c are positive integers. From the given condition we have $3ab = 3(3a + b + c)$, or $c = ab - 3a - b$. By the Pythagorean theorem, we have $(3a)^2 + b^2 = c^2 = (ab - 3a - b)^2$, which simplifies to

$$ab[(a-2)(b-6) - 6] = 0.$$

Because $a, b > 0$, we have $(a, b) \in \{(3, 12), (4, 9), (5, 8), (8, 7)\}$, and therefore the side lengths of the triangle are either $(9, 12, 15)$, $(8, 15, 17)$, or $(7, 24, 25)$.

Problem 7.2 Let a, b, c be nonzero integers, $a \neq c$, such that

$$\frac{a}{c} = \frac{a^2 + b^2}{c^2 + b^2}.$$

Prove that $a^2 + b^2 + c^2$ cannot be a prime number.

Solution. Cross-multiplying and factoring, we have $(a-c)(b^2 - ac) = 0$. Because $a \neq c$, we have $ac = b^2$. Now,

$$a^2 + b^2 + c^2 = a^2 + (2ac - b^2) + c^2$$
$$= (a+c)^2 - b^2 = (a+b+c)(a-b+c).$$

Also, $|a|$, $|c|$ cannot both be 1. Then $a^2 + b^2 + c^2 > |a| + |b| + |c| \geq |a+b+c|, |a-b+c|$, whence $a^2 + b^2 + c^2$ cannot be a prime number.

Problem 7.3 Let $ABCD$ be a convex quadrilateral with $\angle BAC = \angle CAD$ and $\angle ABC = \angle ACD$. Rays AD and BC meet at E and rays AB and DC meet at F. Prove that

(a) $AB \cdot DE = BC \cdot CE$;

(b) $AC^2 < \frac{1}{2}(AD \cdot AF + AB \cdot AE)$.

Solution.

(a) Because $\angle BAC + \angle CBA = \angle ECA$, we have $\angle ECD = \angle BAC$. Then $\triangle CDE \sim \triangle ACE$, and $\frac{CE}{DE} = \frac{AE}{CE}$. Because $\overline{AC}$ is the angle bisector of $\angle A$ in triangle ABE, we also have $\frac{AE}{CE} = \frac{AB}{BC}$. Thus $\frac{CE}{DE} = \frac{AB}{BC}$, whence $AB \cdot DE = BC \cdot CE$.

(b) Note that $\overline{AC}$ is an angle bisector of both triangle ADF and triangle AEB. Thus it is enough to prove that if $\overline{XL}$ is an angle bisector in an arbitrary triangle XYZ, then $XL^2 < XY \cdot XZ$. Let M be the intersection of $\overrightarrow{XL}$ and the circumcircle of triangle XYZ. Because $\triangle XYL \sim \triangle XMZ$, we have $XL^2 < XL \cdot XM = XY \cdot XZ$, as desired.

Problem 7.4 In triangle ABC, D and E lie on sides BC and AB, respectively, F lies on side AC such that $EF \parallel BC$, G lies on side BC such that $EG \parallel AD$. Let M and N be the midpoints of $\overline{AD}$ and $\overline{BC}$, respectively. Prove that

(a) $\dfrac{EF}{BC} + \dfrac{EG}{AD} = 1;$

(b) the midpoint of $\overline{FG}$ lies on line MN.

Solution.

(a) Because $EF \parallel BC$, $\triangle AEF \sim \triangle ABC$ and $\frac{EF}{BC} = \frac{AE}{AB}$. Similarly, because $EG \parallel AD$, $\triangle BEG \sim \triangle BAD$ and $\frac{EG}{AD} = \frac{EB}{AB}$. Hence $\frac{EF}{BC} + \frac{EG}{AD} = 1$.

(b) Let lines AN, EF intersect at point P, and let Q be the point on line BC such that $PQ \parallel AD$. Because $BC \parallel EF$ and N is the midpoint of $\overline{BC}$, P is the midpoint of $\overline{EF}$. Then vector EP equals both vectors PF and GQ, and $PFQG$ is a parallelogram. Thus the midpoint X of $\overline{FG}$ must also be the midpoint of $\overline{PQ}$. Because M is the midpoint of $\overline{AD}$ and $AD \parallel PQ$, it follows that points M, X, N must be collinear.

Problem 8.1 Let $p(x) = 2x^3 - 3x^2 + 2$, and let

$$S = \{p(n) \mid n \in \mathbb{N}, n \leq 1999\},$$
$$T = \{n^2 + 1 \mid n \in \mathbb{N}\},$$
$$U = \{n^2 + 2 \mid n \in \mathbb{N}\}.$$

Prove that $S \cap T$ and $S \cap U$ have the same number of elements.

Solution. Note that $|S \cap T|$ is the number of squares of the form $2n^3 - 3n^2 + 1 = (n-1)^2(2n+1)$ where $n \in \mathbb{N}, n \leq 1999$. For $n \leq 1999$, $(n-1)^2(2n+1)$ is a square precisely when either $n = 1$ or when $n \in \{\frac{1}{2}(k^2 - 1) \mid k = 1, 3, 5, \ldots, 63\}$. Thus, $|S \cap T| = 33$.

Next, $|S \cap U|$ is the number of squares of the form $2n^3 - 3n^2 = n^2(2n-3)$ where $n \in \mathbb{N}, n \leq 1999$. For $n \leq 1999$, $n^2(2n-3)$ is a square precisely when either $n = 0$ or when $n \in \{\frac{1}{2}(k^2+3) \mid k = 1, 3, 5, \ldots, 63\}$. Thus $|S \cap U| = 33$ as well, and we are done.

Problem 8.2

(a) Let $n \geq 2$ be a positive integer and

$$x_1, y_1, x_2, y_2, \ldots, x_n, y_n$$

be positive real numbers such that

$$x_1 + x_2 + \cdots + x_n \geq x_1 y_1 + x_2 y_2 + \cdots + x_n y_n.$$

Prove that

$$x_1 + x_2 + \cdots + x_n \leq \frac{x_1}{y_1} + \frac{x_2}{y_2} + \cdots + \frac{x_n}{y_n}.$$

(b) Let a, b, c be positive real numbers such that

$$ab + bc + ca \leq 3abc.$$

Prove that

$$a^3 + b^3 + c^3 \geq a + b + c.$$

Solution.

(a) Applying the Cauchy-Schwarz inequality and then the given inequality, we have

$$\left(\sum_{i=1}^n x_i\right)^2 \leq \sum_{i=1}^n x_i y_i \cdot \sum_{i=1}^n \frac{x_i}{y_i} \leq \sum_{i=1}^n x_i \cdot \sum_{i=1}^n \frac{x_i}{y_i}.$$

Dividing both sides by $\sum_{i=1}^n x_i$ yields the desired inequality.

(b) By the AM-HM inequality on a, b, c we have

$$a + b + c \geq \frac{9abc}{ab + bc + ca} \geq \frac{9abc}{3abc} = 3.$$

The given condition is equivalent to $\frac{1}{a} + \frac{1}{b} + \frac{1}{c} \leq 3$, implying that $a + b + c \geq \frac{1}{a} + \frac{1}{b} + \frac{1}{c}$. Hence setting $x_1 = a, x_2 = b, x_3 = c$

and $y_1 = \frac{1}{a^2}, y_2 = \frac{1}{b^2}, y_3 = \frac{1}{c^2}$ in the result from part (a) gives $a + b + c \leq a^3 + b^3 + c^3$, as desired.

Problem 8.3 Let $ABCDA'B'C'D'$ be a rectangular box, let E and F be the feet of perpendiculars from A to lines $A'D$ and $A'C$ respectively, and let P and Q be the feet of perpendiculars from B' to lines $A'C'$ and $A'C$ respectively. Prove that

(a) planes AEF and $B'PQ$ are parallel;

(b) triangles AEF and $B'PQ$ are similar.

Solution.

(a) Let $(P_1 P_2 \ldots P_k)$ denote the plane containing points $P_1, P_2, \ldots, P_k$. First observe that quadrilateral $A'B'CD$ is a parallelogram and thus lies in a single plane.

We are given that $AE \perp A'D$. Also, line AE is contained in plane $(ADD'A)$, which is perpendicular to line CD. Hence $AE \perp CD$ as well, and therefore $AE \perp (A'B'CD)$ and $AE \perp A'C$. Because we know that $A'C \perp AF$, we have $A'C \perp (AEF)$ and $A'C \perp EF$.

Likewise, $B'Q \perp A'C$. Because lines $EF, B'Q$, and $A'C$ all lie in plane $(A'B'CD)$, it follows that $EF \parallel B'Q$. In a similar way we deduce that $AF \parallel PQ$. Hence the planes (AEF) and $(B'PQ)$ are parallel, as desired.

(b) Because $EF \parallel B'Q$ and $FA \parallel QP$, we have $\angle EFA = \angle PQB'$. Furthermore, from above $AE \perp EF$ and likewise $B'P \perp PQ$, implying that $\angle AEF = \angle B'PQ = 90°$ as well. Therefore $\triangle AEF \sim \triangle B'PQ$, as desired.

Problem 8.4 Let $SABC$ be a right pyramid with equilateral base ABC, let O be the center of ABC, and let M be the midpoint of $\overline{BC}$. If $AM = 2SO$ and N is a point on edge SA such that $SA = 25SN$, prove that planes ABP and SBC are perpendicular, where P is the intersection of lines SO and MN.

Solution. Let $AB = BC = CA = s$. Then some quick calculations show that $AO = \frac{\sqrt{3}}{3}s$, $AM = \frac{\sqrt{3}}{2}s$, $AS = \frac{5}{\sqrt{48}}s$, and $AN = \frac{24}{5\sqrt{48}}s$. Then $AO \cdot AM = AN \cdot AS = \frac{1}{2}s^2$, whence $MONS$ is a cyclic quadrilateral. Thus, $\angle MNS = 90°$, and P is the orthocenter of triangle AMS. Let Q be the intersection of lines AP and MS. Note that $\angle AMB = \angle AQM = \angle QMB = 90°$. From repeated applications of the Pythagorean theorem, we have $AB^2 = AM^2 + MB^2 = AQ^2 + QM^2 + MB^2 = AQ^2 + QB^2$, whence $\angle AQB = 90°$. Now $AQ \perp QB$ and $AQ \perp QM$, so line AQ must be perpendicular to plane (SBC). Because plane (ABP) contains line AQ, planes (ABP) and (SBC) must be perpendicular.

Problem 9.1 Let ABC be a triangle with angle bisector $\overline{AD}$. One considers the points M, N on rays AB and AC respectively, such that $\angle MDA = \angle ABC$ and $\angle NDA = \angle BCA$. Lines AD and MN meet at P. Prove that
$$AD^3 = AB \cdot AC \cdot AP.$$

Solution. Because $\triangle ADB \sim \triangle AMD$, $\frac{AD}{AB} = \frac{AM}{AD}$. Also, $\angle MAN + \angle NDM = 180°$, whence $AMDN$ is cyclic. Because $\angle DCA = \angle ADN = \angle AMN$, $\triangle ADC \sim \triangle APM$, and $\frac{AD}{AP} = \frac{AC}{AM}$. Therefore,
$$\frac{AD}{AB}\frac{AD}{AC}\frac{AD}{AP} = \frac{AM}{AD}\frac{AD}{AC}\frac{AC}{AM} = 1,$$
as desired.

Problem 9.2 For $a, b > 0$, denote by $t(a, b)$ the positive root of the equation
$$(a+b)x^2 - 2(ab-1)x - (a+b) = 0.$$
Let $M = \{(a, b) \mid a \neq b,\ t(a, b) \leq \sqrt{ab}\}$. Determine, for $(a, b) \in M$, the minimum value of $t(a, b)$.

Solution. Consider the polynomial
$$P(x) = (a+b)x^2 - 2(ab-1)x - (a+b) = 0.$$
Because $a+b \neq 0$, the product of its roots is $-\frac{a+b}{a+b} = -1$. Hence P must have a unique positive root $t(a, b)$ and a unique negative root. Because the leading coefficient of $P(x)$ is positive, the graph of $P(x)$ is positive for $x > t(a, b)$ and negative for $0 \leq x < t(a, b)$ (because in the latter case, x is between the two roots). Thus, the condition $t(a, b) \leq \sqrt{ab}$ is equivalent to $P(\sqrt{ab}) \geq 0$, or
$$(ab-1)(a+b-2\sqrt{ab}) \geq 0.$$
$a + b > 2\sqrt{ab}$ by the AM-GM inequality, where the inequality is strict because $a \neq b$. Thus $t(a, b) \leq \sqrt{ab}$ exactly when $ab \geq 1$.

Now using the quadratic formula, we find that

$$t(a,b) = \frac{ab-1}{a+b} + \sqrt{\left(\frac{ab-1}{a+b}\right)^2 + 1}.$$

Thus given $ab \geq 1$, we have $t(a,b) \geq 1$ with equality when $ab = 1$.

Problem 9.3 In the convex quadrilateral $ABCD$ the bisectors of angles A and C meet at I. Prove that there exists a circle inscribed in $ABCD$ if and only if

$$[AIB] + [CID] = [AID] + [BIC].$$

Solution. It is well known that a circle can be inscribed in a convex quadrilateral $ABCD$ if and only if $AB + CD = AD + BC$. The bisector of angle A consists of those points lying inside $\angle BAD$ equidistant from lines AB and AD. Similarly, the bisector of angle C consists of those points lying inside $\angle BCD$ equidistant from lines BC and BD.

Suppose $ABCD$ has an incircle. Then its center is equidistant from all four sides of the quadrilateral, so it lies on both bisectors and hence equals I. If we let r denote the radius of the incircle, then we have

$$[AIB] + [CID] = r(AB + CD) = r(AD + BC) = [AID] + [BIC].$$

Conversely, suppose that $[AIB] + [CID] = [AID] + [BIC]$. Let $d(I, \ell)$ denote the distance from I to any line ℓ, and write $x = d(I, AB) = d(I, AD)$ and $y = d(I, BC) = d(I, CD)$. Then

$$[AIB] + [CID] = [AID] + [BIC]$$
$$AB \cdot x + CD \cdot y = AD \cdot x + BC \cdot y$$
$$x(AB - AD) = y(BC - CD).$$

If $AB = AD$, then $BC = CD$ and it follows that $AB + CD = AD + BC$. Otherwise, suppose that $AB > AD$; then $BC > CD$ as well. Consider the points $A' \in \overline{AB}$ and $C' \in \overline{BC}$ such that $AD = AA'$ and $CD = CC'$. By SAS, we have $\triangle AIA' \cong \triangle AID$ and $\triangle DCI \cong \triangle C'IC$. Hence $IA' = ID = IC'$. Furthermore, subtracting $[AIA'] + [DCI] = [AID] + [C'IC]$ from both sides of our given condition, we have $[A'IB] = [C'IB]$ or $IA' \cdot IB \cdot \sin \angle A'IB = IC' \cdot IB \cdot \sin \angle CIB$. Thus $\angle A'IB = \angle C'IB$, and hence $\triangle A'IB \cong C'IB$ by SAS.

Thus $\angle IBA' = \angle IBC'$, implying that I lies on the angle bisector of $\angle ABC$. Therefore $x = d(I, AB) = d(I, BC) = y$, and the circle centered at I with radius $x = y$ is tangent to all four sides of the quadrilateral.

Problem 9.4

(a) Let $a, b \in \mathbb{R}$, $a < b$. Prove that $a < x < b$ if and only if there exists $0 < \lambda < 1$ such that $x = \lambda a + (1 - \lambda)b$.

(b) The function $f : \mathbb{R} \to \mathbb{R}$ has the property:
$$f(\lambda x + (1 - \lambda)y) < \lambda f(x) + (1 - \lambda)f(y)$$
for all $x, y \in \mathbb{R}$, $x \neq y$, and all $0 < \lambda < 1$. Prove that one cannot find four points on the function's graph that are the vertices of a parallelogram.

Solution.

(a) No matter what x is, there is a unique value $\lambda = \frac{b-x}{b-a}$ such that $x = \lambda a + (1 - \lambda)b$. Also, $0 < \frac{b-x}{b-a} < 1 \iff a < x < b$, which proves the claim.

(b) The condition given says that f is strictly convex. Stated geometrically, whenever $x < t < y$ the point $(t, f(t))$ lies strictly below the line joining $(x, f(x))$ and $(y, f(y))$. Suppose there were a parallelogram on the graph of f whose vertices, from left to right, have x-coordinates a, b, d, c. Then either $(b, f(d))$ or $(d, f(d))$ must lie on or above the line joining $(a, f(a))$ and $(c, f(c))$, a contradiction.

Problem 10.1 Find all real numbers x and y satisfying
$$\frac{1}{4^x} + \frac{1}{27^y} = \frac{5}{6}$$
$$\log_{27} y - \log_4 x \geq \frac{1}{6}$$
$$27^y - 4^x \leq 1.$$

Solution. First, for the second equation to make sense we must have $x, y > 0$ and thus $27^y > 1$. Now from the third equation we have
$$\frac{1}{27^y} \geq \frac{1}{4^x + 1},$$
which combined with the first equation gives
$$\frac{1}{4^x} + \frac{1}{4^x + 1} \leq \frac{5}{6},$$
whence $x \geq \frac{1}{2}$. Similarly, the first and third equations also give
$$\frac{5}{6} \leq \frac{1}{27^y - 1} + \frac{1}{27^y},$$

whence $y \leq \frac{1}{3}$. If either $x > \frac{1}{2}$ or $y < \frac{1}{3}$, we would have $\log_{27} y - \log_4 x < \frac{1}{6}$, contradicting the second given equation. Thus, the only solution is $(x, y) = \left(\frac{1}{2}, \frac{1}{3}\right)$, which indeed satisfies all three equations.

Problem 10.2 A plane intersects edges AB, BC, CD, DA of the regular tetrahedron $ABCD$ at points M, N, P, Q, respectively. Prove that

$$MN \cdot NP \cdot PQ \cdot QM \geq AM \cdot BN \cdot CP \cdot DQ.$$

Solution. By the law of cosines in triangle MBN, we have

$$MN^2 = MB^2 + BN^2 - MB \cdot BN \geq MB \cdot BN.$$

Similarly, $NP^2 \geq CN \cdot CP$, $PN^2 \geq DP \cdot DQ$, and $MQ^2 \geq AQ \cdot AM$. Multiplying these inequalities yields

$$(MN \cdot NP \cdot PQ \cdot MQ)^2 \geq$$
$$(BM \cdot CN \cdot DP \cdot AQ) \cdot (AM \cdot BN \cdot CP \cdot DQ).$$

Now the given plane is different from plane (ABC) and (ADC). Thus if it intersects line AC at some point T, then points M, N, T must be collinear — because otherwise, the only plane containing M, N, T would be plane (ABC). Therefore it intersects line AC at most one point T, and by Menelaus' Theorem applied to triangle ABC and line MNT we have

$$\frac{AM \cdot BN \cdot CT}{MB \cdot NC \cdot TA} = 1.$$

Similarly, P, Q, T are collinear and

$$\frac{AQ \cdot DP \cdot CT}{QD \cdot PC \cdot TA} = 1.$$

Equating these two fractions and cross-multiplying, we find that

$$AM \cdot BN \cdot CP \cdot DQ = BM \cdot CN \cdot DP \cdot AQ.$$

This is true even if the plane does not actually intersect line AC: in this case, we must have $MN \parallel AC$ and $PQ \parallel AC$, in which case ratios of similar triangles show that $AM \cdot BN = BM \cdot CN$ and $CP \cdot DQ = DP \cdot AQ$.

Combining this last equality with the inequality from the first paragraph, we find that

$$(MN \cdot NP \cdot PQ \cdot QM)^2 \geq (AM \cdot BN \cdot CP \cdot DQ)^2,$$

which implies the desired result.

Problem 10.3 Let a, b, c ($a \neq 0$) be complex numbers. Let z_1 and z_2 be the roots of the equation $az^2 + bz + c = 0$, and let w_1 and w_2 be the roots of the equation

$$(a + \bar{c})z^2 + (b + \bar{b})z + (\bar{a} + c) = 0.$$

Prove that if $|z_1|, |z_2| < 1$, then $|w_1| = |w_2| = 1$.

Solution. We begin by proving that $\mathrm{Re}\,(b)^2 \leq |a + \bar{c}|^2$. If $z_1 = z_2 = 0$, then $b = 0$ and the claim is obvious. Otherwise, write $a = m + ni$ and $c = r + si$, and write $z_1 = x + yi$ where $t = |z_1| = \sqrt{x^2 + y^2} < 1$. Also note that

$$r^2 + s^2 = |c|^2 = |az_1 z_2|^2 < |a|^2 |z_1|^2 = (m^2 + n^2)t^2. \tag{1}$$

Assume without loss of generality that $z_1 \neq 0$. Then $|\mathrm{Re}\,(b)| = |\mathrm{Re}\,(-b)| = |\mathrm{Re}\,(az_1 + c/z_1)| = |\mathrm{Re}\,(az_1) + \mathrm{Re}\,(c/z_1)|$; that is,

$$|\mathrm{Re}\,(b)| = |(mx - ny) + (rx + sy)/t^2|$$
$$= |x(m + r/t^2) + y(s/t^2 - n)|$$
$$\leq \sqrt{x^2 + y^2}\sqrt{(m + r/t^2)^2 + (s/t^2 - n)^2}$$
$$= t\sqrt{(m + r/t^2)^2 + (s/t^2 - n)^2},$$

where the inequality follows from the Cauchy-Schwarz inequality. Proving our claim then reduces to showing that

$$t^2 \left((m + r/t^2)^2 + (s/t^2 - n)^2\right) \leq (m + r)^2 + (n - s)^2$$
$$\iff (mt^2 + r)^2 + (st^2 - n)^2 \leq t^2 \left((m + r)^2 + (n - s)^2\right)$$
$$\iff (r^2 + s^2)(1 - t^2) < (m^2 + n^2)(t^4 - t^2)$$
$$\iff (1 - t^2)\left((m^2 + n^2)t^2 - (r^2 + s^2)\right).$$

$1 - t^2 > 0$ by assumption, and $(m^2 + n^2)t^2 - (r^2 + s^2) > 0$ from (1). Therefore our claim is true.

Now because $|c/a| = |z_1 z_2| < 1$, we have $|c| < |a|$ and $a + \bar{c} \neq 0$. Then by the quadratic equation, the roots to $(a + \bar{c})z^2 + (b + \bar{b})z + (\bar{a} + c) = 0$ are given by

$$\frac{-(b + \bar{b}) \pm \sqrt{(b + \bar{b})^2 - 4(a + \bar{c})(\bar{a} + c)}}{2(a + \bar{c})},$$

or (dividing the numerator and denominator by 2)

$$\frac{-\operatorname{Re}(b) \pm \sqrt{\operatorname{Re}(b)^2 - |a+\overline{c}|^2}}{a+\overline{c}} = \frac{-\operatorname{Re}(b) \pm i\sqrt{|a+\overline{c}|^2 - \operatorname{Re}(b)^2}}{a+\overline{c}}.$$

When evaluating either root, the absolute value of the numerator is $\sqrt{\operatorname{Re}(b)^2 + (|a+\overline{c}|^2 - \operatorname{Re}(b)^2)} = |a+\overline{c}|$, and the absolute value of the denominator is clearly $|a+\overline{c}|$ as well. Therefore indeed $|w_1| = |w_2| = 1$, as desired.

Problem 10.4

(a) Let $x_1, y_1, x_2, y_2, \ldots, x_n, y_n$ be positive real numbers such that

  (i) $x_1 y_1 < x_2 y_2 < \cdots < x_n y_n$;

  (ii) $x_1 + x_2 + \cdots + x_k \geq y_1 + y_2 + \cdots + y_k$ for all $k = 1, 2, \ldots, n$.

Prove that

$$\frac{1}{x_1} + \frac{1}{x_2} + \cdots + \frac{1}{x_n} \leq \frac{1}{y_1} + \frac{1}{y_2} + \cdots + \frac{1}{y_n}.$$

(b) Let $A = \{a_1, a_2, \ldots, a_n\} \subseteq \mathbb{N}$ be a set such that for all distinct subsets $B, C \subseteq A$, $\sum_{x \in B} x \neq \sum_{x \in C} x$. Prove that

$$\frac{1}{a_1} + \frac{1}{a_2} + \cdots + \frac{1}{a_n} < 2.$$

Solution. (a) Let $\pi_i = \frac{1}{x_i y_i}$, $\delta_i = x_i - y_i$ for all $1 \leq i \leq n$. We are given that $\pi_1 > \pi_2 > \cdots > \pi_n > 0$ and that $\sum_{i=1}^{k} \delta_i \geq 0$ for all $1 \leq k \leq n$. Note that

$$\sum_{k=1}^{n} \left(\frac{1}{y_k} - \frac{1}{x_k} \right) = \sum_{k=1}^{n} \pi_k \delta_k$$

$$= \pi_n \sum_{i=1}^{n} \delta_i + \sum_{k=1}^{n-1} (\pi_k - \pi_{k+1})(\delta_1 + \delta_2 + \cdots + \delta_k) \geq 0,$$

as desired.

(b) Assume without loss of generality that $a_1 < a_2 < \cdots < a_n$, and let $y_i = 2^{i-1}$ for all i. Clearly,

$$a_1 y_1 < a_2 y_2 < \cdots < a_n y_n.$$

For any k, the $2^k - 1$ sums made by choosing at least one of the numbers $a_1, a_2, \ldots, a_k$ are all distinct. Hence the largest of them, $\sum_{i=1}^{k} a_i$, must

be at least $2^k - 1$. Thus for all $k = 1, 2, \ldots, n$ we have
$$a_1 + a_2 + \cdots + a_k \geq 2^k - 1 = y_1 + y_2 + \cdots + y_k.$$
Then by part (a), we must have
$$\frac{1}{a_1} + \frac{1}{a_2} + \cdots + \frac{1}{a_n} < \frac{1}{y_1} + \frac{1}{y_2} + \cdots + \frac{1}{y_n} = 2 - \frac{1}{2^{n-1}} < 2,$$
as desired.

IMO Selection Tests
Problem 1
(a) Show that out of any 39 consecutive positive integers, it is possible to choose one number with the sum of its digits divisible by 11.

(b) Find the first 38 consecutive positive integers, none with the sum of its digits divisible by 11.

Solution. Call an integer *deadly* if its sum of digits is divisible by 11, and let $d(n)$ equal the sum of the digits of a positive integer n.

If n ends in a 0, then the numbers $n, n+1, \ldots, n+9$ differ only in their units digits, which range from 0 to 9. Hence $d(n), d(n+1), \ldots, d(n+9)$ is an arithmetic progression with common difference 1. Thus if $d(n) \not\equiv 1 \pmod{11}$, then one of these numbers is deadly.

Next suppose that if n ends in $k \geq 0$ nines. Then $d(n+1) = d(n) + 1 - 9k$: the last k digits of $n+1$ are 0's instead of 9's, and the next digit to the left is 1 greater than the corresponding digit in n.

Finally, suppose that n ends in a 0 and that $d(n) \equiv d(n+10) \equiv 1 \pmod{11}$. Because $d(n) \equiv 1 \pmod{11}$, we must have $d(n+9) \equiv 10 \pmod{11}$. If $n+9$ ends in k 9's, then we have $2 \equiv d(n+10) - d(n+9) \equiv 1 - 9k \Longrightarrow k \equiv 6 \pmod{11}$.

(a) Suppose we had 39 consecutive integers, none of them deadly. One of the first ten must end in a 0: call it n. Because none of $n, n+1, \ldots, n+9$ are deadly, we must have $d(n) \equiv 1 \pmod{11}$. Similarly, $d(n+10) \equiv 1 \pmod{11}$ and $d(n+20) \equiv 1 \pmod{11}$. From our third observation above, this implies that both $n+9$ and $n+19$ must end in at least six 9's. This is impossible, because $n+10$ and $n+20$ can't both be multiples of one million!

(b) Suppose we have 38 consecutive numbers $N, N+1, \ldots, N+37$, none of which is deadly. By an analysis similar to that in part (a), none of the first nine can end in a 0. Hence, $N + 9$ must end in a 0, as must $N + 19$ and $N + 29$. Then we must have $d(N + 9) \equiv d(N + 19) \equiv 1 \pmod{11}$. Therefore $d(N + 18) \equiv 10 \pmod{11}$. Furthermore, if $N + 18$ ends in k 9's we must have $k \equiv 6 \pmod{11}$.

The smallest possible such number is 999999, yielding the 38 consecutive numbers 999981, 999982, ..., 1000018. Indeed, none of these numbers is deadly: their sums of digits are congruent to 1, 2, ..., 10, 1, 2, ..., 10, 1, 2, ..., 10, 2, 3, ..., 9, and 10 (mod 11), respectively.

Problem 2 Let ABC be an acute triangle with angle bisectors $\overline{BL}$ and $\overline{CM}$. Prove that $\angle A = 60°$ if and only if there exists a point K on $\overline{BC}$ ($K \neq B, C$) such that triangle KLM is equilateral.

Solution. Let I be the intersection of lines BL and CM. Then $\angle BIC = 180° - \angle ICB - \angle CBI = 180° - \frac{1}{2}(\angle C + \angle B) = 180° - \frac{1}{2}(180° - \angle A) = 90° + \angle A$, and thus $\angle BIC = 120°$ if and only if $\angle A = 60°$.

For the "only if" direction, suppose that $\angle A = 60°$. Let K be the intersection of $\overline{BC}$ and the internal angle bisector of $\angle BIC$. We claim that triangle KLM is equilateral. Because $\angle BIC = 120°$, we know that $\angle MIB = \angle KIB = 60°$. In addition, $\angle IBM = \angle IBK$ and $IB = IB$. Hence by ASA congruency we have $\triangle IBM \cong \triangle IBK$, and $IM = IK$. Similarly, $IL = IK = IM$. Combined with the relation $\angle KIL = \angle LIM = \angle MIK = 120°$, this equality implies that triangle KLM is equilateral.

For the "if" direction, suppose that K is on $\overline{BC}$ and triangle KLM is equilateral. Consider triangles BLK and BLM: $BL = BL$, $LM = LK$, and $\angle MBL = \angle KBL$. There is no SSA congruency, but we do then know that either $\angle LKB + \angle BML = 180°$ or $\angle LKB = \angle BML$. Because $\angle KBM < 90°$ and $\angle MLK = 60°$, we know that $\angle LKB + \angle BML > 210°$. Thus $\angle LKB = \angle BML$, whence $\triangle BLK \cong \triangle BLM$, and $BK = BM$. It follows that $IK = IM$. Similarly, $IL = IK$, and I is the circumcenter of triangle KLM. Thus $\angle LIM = 2\angle LKM = 120°$, giving $\angle BIC = \angle LIM = 120°$ and $\angle A = 60°$.

Problem 3 Show that for any positive integer n, the number

$$S_n = \binom{2n+1}{0} \cdot 2^{2n} + \binom{2n+1}{2} \cdot 2^{2n-2} \cdot 3 + \cdots + \binom{2n+1}{2n} \cdot 3^n$$

is the sum of two consecutive perfect squares.

Solution. Let $\alpha = 1+\sqrt{3}$, $\beta = 1-\sqrt{3}$, and $T_n = \frac{1}{2}\left(\alpha^{2n+1}+\beta^{2n+1}\right)$. Note that $\alpha\beta = -2$, $\frac{\alpha^2}{2} = 2+\sqrt{3}$, and $\frac{\beta^2}{2} = 2-\sqrt{3}$. Also, applying the binomial expansion to $(1+\sqrt{3})^n$ and $(1-\sqrt{3})^n$, we find that $T_n = \sum_{k=0}^{n} \binom{2n+1}{2k} 3^k$ — which is an integer for all n.

Applying the binomial expansion to $(2+\sqrt{3})^{2n+1}$ and $(2-\sqrt{3})^{2n+1}$ instead, we find that

$$S_n = \frac{\left(\frac{\alpha^2}{2}\right)^{2n+1} + \left(\frac{\beta^2}{2}\right)^{2n+1}}{4}$$

$$= \frac{\alpha^{4n+2} + \beta^{4n+2}}{2^{2n+3}}$$

$$= \frac{\alpha^{4n+2} + 2(\alpha\beta)^{2n+1} + \beta^{4n+2}}{2^{2n+3}} + \frac{1}{2}$$

$$= \frac{\left(\alpha^{2n+1} + \beta^{2n+1}\right)^2}{2^{2n+3}} + \frac{1}{2}$$

$$= \frac{T_n^2}{2^{2n+1}} + \frac{1}{2}.$$

Thus $2^{2n+1} S_n = T_n^2 + 2^{2n}$. Then $2^{2n} \mid T_n^2$ but $2^{2n+1} \nmid T_n^2$, and hence $T_n \equiv 2^n \pmod{2^{n+1}}$. Therefore

$$S_n = \frac{T_n^2}{2^{2n+1}} + \frac{1}{2} = \left(\frac{T_n - 2^n}{2^{n+1}}\right)^2 + \left(\frac{T_n + 2^n}{2^{n+1}}\right)^2$$

is indeed the sum of two consecutive perfect squares.

Problem 4 Show that for all positive real numbers $x_1, x_2, \cdots, x_n$ such that

$$x_1 x_2 \cdots x_n = 1,$$

the following inequality holds:

$$\frac{1}{n-1+x_1} + \frac{1}{n-1+x_2} + \cdots + \frac{1}{n-1+x_n} \leq 1.$$

First Solution. Let $a_1 = \sqrt[n]{x_1}$, $a_2 = \sqrt[n]{x_2}$, ..., $a_n = \sqrt[n]{x_n}$. Then $a_1 a_2 \cdots a_n = 1$ and

$$\frac{1}{n-1+x_k} = \frac{1}{n-1+a_k^n} = \frac{1}{n-1+\frac{a_k^{n-1}}{a_1 \cdots a_{k-1} a_{k+1} \cdots a_n}}$$

$$\leq \frac{1}{n-1+\frac{(n-1)a_k^{n-1}}{a_1^{n-1}+\cdots+a_{k-1}^{n-1}+a_{k+1}^{n-1}+\cdots+a_n^{n-1}}}$$

by the AM-GM inequality. It follows that
$$\frac{1}{n-1+x_k} \le \frac{a_1^{n-1} + \cdots + a_{k-1}^{n-1} + a_{k+1}^{n-1} + \cdots + a_n^{n-1}}{(n-1)(a_1^{n-1} + a_2^{n-1} + \cdots + a_n^{n-1})}.$$

Summing up yields $\sum_{k=1}^{n} \frac{1}{n-1+x_k} \le 1$, as desired.

Second Solution. Let $f(x) = \frac{1}{n-1-x}$. We wish to prove that $\sum_{i=1}^{n} f(x_i) \le 1$. Note that
$$f(y) + f(z) = \frac{2(n-1) + y + z}{(n-1)^2 + yz + (y+z)(n-1)}.$$

Suppose that any of our x_i does not equal 1, so that $x_j < 1 < x_k$ for some j, k. If $f(x_j) + f(x_k) \le \frac{1}{n-1}$, then all the other $f(x_i)$ are less than $\frac{1}{n-1}$. Then $\sum_{i=1}^{n} f(x_i) < 1$ and we are done.

Otherwise, $f(x_j) + f(x_k) > \frac{1}{n-1}$. Now set $x'_j = 1$ and $x'_k = x_j x_k$. We have $x'_j x'_k = x_j x_k$, while $x_j < 1 < x_k \Rightarrow (1-x_j)(x_k-1) > 0 \Rightarrow x_j + x_k > x'_j + x'_k$. Let $a = 2(n-1)$, $b = (n-1)^2 + x_j x_k = (n-1)^2 + x'_j x'_k$, and $c = \frac{1}{n-1}$. Also let $m = x_j + x_k$ and $m' = x'_j + x'_k$. Then we have
$$f(x_j) + f(x_k) = \frac{a + cm}{b + m} \quad \text{and} \quad f(x'_j) + f(x'_k) = \frac{a + cm'}{b + m'}.$$

Now $\frac{a+cm}{b+m} > c \Rightarrow a + cm > (b+m)c \Rightarrow \frac{a}{b} > c$, and
$$(a - bc)(m - m') > 0 \Rightarrow \frac{a + cm'}{b + m'}$$
$$> \frac{a + cm}{b + m} \Rightarrow f(x'_j) + f(x'_k) = f(x_j) + f(x_k).$$

Hence as long as no pair $f(x_j) + f(x_k) \le \frac{1}{n-1}$ and the x_i do not all equal 1, we can continually replace pairs x_j and x_k (neither equal to 1) by 1 and $x_j x_k$. This keeps the product $x_1 x_2 \cdots x_n$ equal to 1 while increasing $\sum_{i=1}^{n} f(x_i)$. Then eventually our new $\sum_{i=1}^{n} f(x_i) \le 1$, which implies that our original $\sum_{i=1}^{n} f(x_i)$ was also at most 1. This completes the proof.

Third Solution. Suppose, for the sake of contradiction, that $\frac{1}{n-1+x_1} + \frac{1}{n-1+x_2} + \cdots + \frac{1}{n-1+x_n} > 1$. Letting $y_i = x_i/(n-1)$ for $i = 1, 2, \ldots, n$, we have
$$\frac{1}{1+y_1} + \frac{1}{1+y_2} + \cdots + \frac{1}{1+y_n} > n - 1$$
and hence
$$\frac{1}{1+y_1} > \left(1 - \frac{1}{1+y_2}\right) + \left(1 - \frac{1}{1+y_3}\right) + \cdots + \left(1 - \frac{1}{1+y_n}\right)$$

$$= \frac{y_2}{1+y_2} + \frac{y_3}{1+y_3} + \cdots + \frac{y_n}{1+y_n}$$
$$> (n-1) \sqrt[n-1]{\frac{y_2 y_3 \cdots y_n}{(1+y_2)(1+y_3)\cdots(1+y_n)}}.$$

We write analogous inequalities with $\frac{1}{1+y_2}, \frac{1}{1+y_n}, \ldots, \frac{1}{1+y_n}$ on the left hand side. Multiplying these n inequalities together gives

$$\prod_{k=1}^{n} \frac{1}{1+y_k} > (n-1)^n \frac{y_1 y_2 \cdots y_n}{(1+y_1)(1+y_2)\cdots(1+y_n)}$$

or

$$1 > ((n-1)y_1)((n-1)y_2)\cdots((n-1)y_n) = x_1 x_2 \cdots x_n,$$

a contradiction.

Problem 5 Let $x_1, x_2, \ldots, x_n$ be distinct positive integers. Prove that

$$x_1^2 + x_2^2 + \cdots + x_n^2 \geq \frac{(2n+1)(x_1 + x_2 + \cdots + x_n)}{3}.$$

Solution. Assume without loss of generality that $x_1 < x_2 < \cdots < x_n$. We will prove that $3x_k^2 \geq 2(x_1 + x_2 + \cdots + x_{k-1}) + (2k+1)x_k$. Then, summing this inequality over $k = 1, 2, \ldots, n$, we will have the desired inequality.

First, $x_1 + x_2 + \cdots + x_{k-1} \leq (x_k - (k-1)) + (x_k - (k-2)) + \cdots + (x_k - 1) = (k-1)x_k - \frac{k(k-1)}{2}$. Thus,

$$2(x_1 + x_2 + \cdots + x_{k-1}) + (2k+1)x_k \leq (4k-1)x_k - k(k-1).$$

Now

$$3x_k^2 - [(4k-1)x_k - k(k-1)] = x_k(3x_k - 4k + 1) + k(k-1),$$

which is minimized at $x_k = \frac{2}{3}k$. Because $x_k \geq k$,

$$x_k(3x_k - 4k + 1) + k(k-1) \geq k(3k - 4k + 1) + k(k-1) = 0$$

so

$$3x_k^2 \geq (4k-1)x_k - k(k-1) \geq 2(x_1 + x_2 + \cdots + x_{k-1}) + (2k+1)x_k,$$

and we have finished.

Problem 6 Prove that for any integer n, $n \geq 3$, there exist n positive integers $a_1, a_2, \ldots, a_n$ in arithmetic progression, and n positive integers $b_1, b_2, \ldots, b_n$ in geometric progression, such that

$$b_1 < a_1 < b_2 < a_2 < \cdots < b_n < a_n.$$

Give an example of such progressions $a_1, a_2, \ldots, a_n$ and $b_1, b_2, \ldots, b_n$ each having at least 5 terms.

Solution. Our strategy is to find progressions where $b_n = a_{n-1} + 1$ and $b_{n-1} = a_{n-2} + 1$. Write $d = a_{n-1} - a_{n-2}$. Then for all $2 \leq i, j \leq n-1$ we have $b_{i+1} - b_i \leq b_n - b_{n-1} = d$, so that $b_j = b_n + \sum_{i=j}^{n-1}(b_i - b_{i+1}) > a_{n-1} + (n-j)d = a_{j-1}$.

If we simply ensure that $b_1 < a_1$, then $b_j = b_1 + \sum_{i=1}^{j-1}(b_{i+1} - b_i) \leq a_1 + (j-1)d = a_j$ for all j. Thus the chain of inequalities would be satisfied.

Let $b_1, b_2, \ldots, b_n$ equal $k^{n-1}, k^{n-2}(k+1), \ldots, k^0(k+1)^{n-1}$, where k is a value to be determined later. Also set $a_{n-1} = b_n - 1$ and $a_{n-2} = b_{n-1} - 1$, and define the other a_i accordingly. Then $d = a_n - a_{n-1} = b_n - b_{n-1} = (k+1)^{n-2}$, and $a_1 = (k+1)^{n-2}(k+3-n) - 1$. Thus, we need only pick k such that

$$(k+1)^{n-2}(k+3-n) - 1 - k^{n-1} > 0.$$

Viewing the left hand side as a polynomial in k, the coefficient of k^{n-1} is zero but the coefficient of k^{n-2} is 1. Therefore, it is positive for sufficiently large k and we can indeed find satisfactory sequences $a_1, a_2, \ldots, a_n$ and $b_1, b_2, \ldots, b_n$.

For $n = 5$, we seek k such that

$$(k+1)^3(k-2) - 1 - k^4 > 0.$$

Computation shows that $k = 5$ works, yielding

$$625 < 647 < 750 < 863 < 900 < 1079 < 1080 < 1295 < 1296 < 1511.$$

Problem 7 Let a be a positive real number and $\{x_n\}$ $(n \geq 1)$ be a sequence of real numbers such that $x_1 = a$ and

$$x_{n+1} \geq (n+2)x_n - \sum_{k=1}^{n-1} kx_k,$$

for all $n \geq 1$. Show that there exists a positive integer n such that $x_n > 1999!$.

Solution. We will prove by induction on $n \geq 1$ that

$$x_{n+1} > \sum_{k=1}^{n} kx_k > a \cdot n!.$$

For $n = 1$, we have $x_2 \geq 3x_1 > x_1 = a$.

Now suppose that the claim holds for all values up through n. Then

$$x_{n+2} \geq (n+3)x_{n+1} - \sum_{k=1}^{n} kx_k$$

$$= (n+1)x_{n+1} + 2x_{n+1} - \sum_{k=1}^{n} kx_k$$

$$> (n+1)x_{n+1} + 2\sum_{k=1}^{n} kx_k - \sum_{k=1}^{n} kx_k = \sum_{k=1}^{n+1} kx_k,$$

as desired. Furthermore, $x_1 > 0$ by definition and $x_2, x_3, \ldots, x_n$ are also positive by the induction hypothesis. Thus $x_{n+2} > (n+1)x_{n+1} > (n+1)(a \cdot n!) = a \cdot (n+1)!$. This completes the inductive step.

Therefore for sufficiently large n, we have $x_{n+1} > n! \cdot a > 1999!$.

Problem 8 Let O, A, B, C be variable points in the plane such that $OA = 4$, $OB = 2\sqrt{3}$, and $OC = \sqrt{22}$. Find the maximum possible area of triangle ABC.

Solution. We first look for a tetrahedron $MNPQ$ with the following properties: (i) if H is the foot of the perpendicular from M to plane (NPQ), then $HN = 4$, $HP = 2\sqrt{3}$, and $HQ = \sqrt{22}$; and (ii) lines MN, MP, MQ are pairwise perpendicular.

If such a tetrahedron exists, then let $O = H$ and draw triangle ABC in plane (NPQ). We have $MA = \sqrt{MO^2 + OA^2} = \sqrt{MH^2 + HN^2} = MN$, and similarly $MB = MP$ and $MC = MQ$. Hence

$$[ABCM] \leq \frac{1}{3}[ABM] \cdot MC \leq \frac{1}{3} \cdot \left(\frac{1}{2}MA \cdot MB\right) \cdot MC$$

$$= \frac{1}{3} \cdot \left(\frac{1}{2}MN \cdot MP\right) \cdot MQ = [MNPQ],$$

and therefore the maximum possible area of triangle $[ABC]$ is $[NPQ]$.

It remains to find tetrahedron $MNPQ$. Let $x = MH$, so that $MN = \sqrt{x^2 + 16}$, $MP = \sqrt{x^2 + 12}$, and $MQ = \sqrt{x^2 + 22}$. By the Pythagorean Theorem on triangle MHN, we have $NH = 4$. Next let lines NH and PQ intersect at R. In similar right triangles MHN and MRN, we have $MR = MH \cdot \frac{MN}{NH} = \frac{1}{4}(x^2 + 16)$.

Because $MN \perp (MPQ)$ we have $MN \perp PQ$. Because $MH \perp (NPQ)$, we have $MH \perp PQ$ as well. Hence $PQ \perp (MNHR)$, so that $\overline{MR}$ is an altitude in the right triangle MPQ. Therefore $MR \cdot PQ =$

$2[MPQ] = MP \cdot MQ$, or (after squaring both sides)
$$\sqrt{\frac{(x^2+16)^2}{16} - (x^2+16)}\sqrt{x^2+12+x^2+22} = \sqrt{x^2+12}\sqrt{x^2+22}.$$
Setting $4y = x^2 + 16$ and squaring both sides, we obtain
$$(y^2 - 4y)(8y + 2) = (4y - 4)(4y + 6)$$
$$(y - 6)(4y^2 + y - 2) = 0.$$
Because $y = \frac{1}{4}(x^2 + 16) > 4$, the only solution is $y = 6 \implies x = \sqrt{8}$. Then by taking $MN = \sqrt{24}, MP = \sqrt{20}, MQ = \sqrt{30}$, we get the required tetrahedron.

Then $[MNPQ]$ equals both $\frac{1}{3}MH \cdot [MPQ]$ and $\frac{1}{6}MN \cdot MP \cdot MQ$. Setting these two expressions equal, we find that the maximum area of $[ABC]$ is
$$[NPQ] = \frac{MN \cdot MP \cdot MQ}{2 \cdot MH} = 15\sqrt{2}.$$

Problem 9 Let a, n be integers and let p be a prime such that $p > |a| + 1$. Prove that the polynomial $f(x) = x^n + ax + p$ cannot be represented as a product of two nonconstant polynomials with integer coefficients.

Solution. Let z be a complex root of the polynomial. We shall prove that $|z| > 1$. Suppose $|z| \leq 1$. Then, $z^n + az = -p$, we deduce that
$$p = |z^n + az| = |z||z^{n-1} + a| \leq |z^{n-1}| + |a| \leq 1 + |a|,$$
which contradicts the hypothesis.

Now, suppose $f = gh$ is a decomposition of f into nonconstant polynomials with integer coefficients. Then $p = f(0) = g(0)h(0)$, and either $|g(0)| = 1$ or $|h(0)| = 1$. Assume without loss of generality that $|g(0)| = 1$. If $z_1, z_2, \ldots, z_k$ are the roots of g then they are also roots of f. Therefore
$$1 = |g(0)| = |z_1 z_2 \cdots z_k| = |z_1||z_2|\cdots|z_k| > 1,$$
a contradiction.

Problem 10 Two circles meet at A and B. Line ℓ passes through A and meets the circles again at C and D respectively. Let M and N be the midpoints of arcs $\widehat{BC}$ and $\widehat{BD}$, which do not contain A, and let K be the midpoint of $\overline{CD}$. Prove that $\angle MKN = 90°$.

Solution. All angles are directed modulo $180°$. Let M' be the reflection of M across K. Then triangles MKC and $M'KD$ are congruent in that order, and $M'D = MC$. Because M is the midpoint of $\widehat{BC}$, we have $M'D = MC = MB$. Similarly, because N is the midpoint of $\widehat{BD}$ we have $BN = DN$. Next,

$$\angle MBN = (180° - \angle ABM) + (180° - \angle NBA)$$
$$= \angle MCA + \angle ADN = \angle M'DA + \angle ADN = \angle M'DN.$$

Hence $\triangle M'DN \cong \triangle MBN$, and $MN = M'N$. Therefore $\overline{NK}$ is the median to the base of isosceles triangle MNM', so it is also an altitude and $NK \perp MK$.

Problem 11 Let $n \geq 3$ and $A_1, A_2, \ldots, A_n$ be points on a circle. Find the greatest number of acute triangles having vertices in these points.

Solution. Without loss of generality assume the points $A_1, A_2, \ldots, A_n$ are ordered in that order counterclockwise. Also, take indices modulo n so that $A_{n+1} = A_1, A_{n+2} = A_2$, and so on. Denote by $A_i A_j$ the arc of the circle starting from A_i and ending in A_j in the counterclockwise direction; let $m(A_i A_j)$ denote the angle measure of the arc; and call an arc $A_i A_j$ obtuse if $m(A_i A_j) \geq 180°$. Obviously, $m(A_i A_j) + m(A_j A_i) = 360°$, and thus at least one of the arcs $A_i A_j$ and $A_j A_i$ is obtuse. Let x_s be the number of obtuse arcs each having exactly $s - 1$ points along their interiors. If $s \neq \frac{n}{2}$, then for each i at least one of the arcs $A_i A_{i+s}$ or $A_{i+s} A_i$ is obtuse. Summing over all i, we deduce that

$$x_s + x_{n-s} \geq n \tag{1}$$

for every $s \neq \frac{n}{2}$. Similar reasoning shows that this inequality also holds even when $s = \frac{n}{2}$. For all s, equality holds if and only if there are no diametrically opposite points A_i, A_{i+s}.

The number of non-acute triangles $A_i A_j A_k$ equals the number of non-acute angles $\angle A_i A_j A_k$. For each obtuse arc $A_i A_k$ containing $s - 1$ points in its interior, there are $n - s - 1$ such angles $A_i A_j A_k$: namely, with those A_j in the interior of arc $A_k A_i$. It follows that the number N of non-acute triangles is

$$N = x_1(n-2) + x_2(n-3) + \ldots + x_{n-3} \cdot 2 + x_{n-2} \cdot 1 + x_{n-1} \cdot 0.$$

By regrouping terms and using (1) we obtain

$$N \geq \sum_{s=1}^{\frac{n-1}{2}} (s-1) \cdot (x_{n-s} + x_s)$$

$$\geq n \left(1 + 2 + \cdots + \frac{n-3}{2}\right) = \frac{n(n-1)(n-3)}{8}$$

if n is odd, and

$$N \geq \sum_{s=1}^{\frac{n-2}{2}} (s-1) \cdot (x_{n-s} + x_s) + \frac{n-2}{2} x_{n/2}$$

$$\geq n \left(1 + 2 + \cdots + \frac{n-4}{2}\right) + \frac{n-2}{2} \cdot \frac{n}{2} = \frac{n(n-2)^2}{8}$$

if n is even.

Equality is obtained when there are no diametrically opposite points, and when $x_k = 0$ for $k < \frac{n}{2}$. When n is odd, for instance, this happens when the points form a regular n-gon. When n is even, equality occurs when $m(A_1 A_2) = m(A_2 A_3) = \cdots = m(A_{n-1} A_n) = \frac{360°}{n} + \epsilon$ where $0 < \epsilon < \frac{360°}{n^2}$.

Finally, note that the total number of triangles having vertices in the n points is $\binom{n}{3} = \frac{n(n-1)(n-2)}{6}$. Subtracting the minimum values of N found above, we find that the maximum number of acute angles is $\frac{(n-1)n(n+1)}{24}$ if n is odd, and $\frac{(n-2)n(n+2)}{24}$ if n is even.

Problem 12 The scientists at an international conference are either *native* or *foreign*. Each native scientist sends exactly one message to a foreign scientist and each foreign scientist sends exactly one message to a native scientist, although at least one native scientist does not receive a message. Prove that there exists a set S of native scientists and a set T of foreign scientists such that the following conditions hold: (i) the scientists in S sent messages to exactly those foreign scientists who were not in T (that is, every foreign scientist not in T received at least one message from somebody in S, but none of the scientists in T received any messages from scientists in S); and (ii) the scientists in T sent messages to exactly those native scientists not in S.

Solution. Let A be the set of native scientists and B be the set of foreign scientists. Let $f : A \to B$ and $g : B \to A$ be the functions defined as follows: $f(a)$ is the foreign scientist receiving a message from a, and $g(b)$ is the native scientist receiving a message from b. If such subsets S, T

exist we must have $T = B - f(S)$. Hence we prove that there must exist a subset $S \subseteq A$ such that $A - S = g(B - f(S))$.

For each subset $X \subseteq A$, let $h(X) = A - g(B - f(X))$. If $X \subseteq Y$, then

$$f(X) \subseteq f(Y) \implies B - f(Y) \subseteq B - f(X) \implies g(B - f(Y))$$
$$\subseteq g(B - f(X)) \implies A - g(B - f(X))$$
$$\subseteq A - g(B - f(Y)) \implies h(X) \subseteq h(Y).$$

Let $M = \{X \subseteq A \mid h(X) \subseteq X\}$. The set M is nonempty because $A \in M$. Furthermore, it is given that g is not surjective, so that some native scientist a_0 is never in $g(B - f(X))$ and thus always in $h(X)$ for all $X \subseteq A$. Thus every subset in M contains a_0, so that $S = \bigcap_{X \in M} X$ is nonempty.

From the definition of S we have $h(S) \subseteq S$. From the monotony of h it follows that $h(h(S)) \subseteq h(S)$. Thus, $h(S) \in M$ and $S \subseteq h(S)$. Combining these results, we have $S = h(S)$, as desired.

Problem 13 A polyhedron P is given in space. Determine whether there must exist three edges of P that can be the sides of a triangle.

Solution. The answer is "yes." Assume, for the purpose of contradiction, that there exists a polyhedron P in which no three edges can form the sides of a triangle. Let the edges of P be $E_1, E_2, E_3, \ldots, E_n$, in non-increasing order of length, and let e_i be the length of E_i. Consider the two faces that share E_1: for each of those faces, the sum of the lengths of all its edges except E_1 is greater than e_1. Therefore,

$$e_2 + e_3 + \cdots + e_n > 2e_1.$$

Because we are assuming that no three edges of P can form the sides of a triangle, we have $e_{i+1} + e_{i+2} \leq e_i$ for $i = 1, 2, \ldots, n - 2$. Hence,

$$2(e_2 + e_3 + \cdots + e_n)$$
$$= e_2 + (e_2 + e_3) + (e_3 + e_4) + \cdots + (e_{n-1} + e_n) + e_n$$
$$\leq e_2 + (e_1) + (e_2) + \cdots + (e_{n-2}) + e_n,$$

so

$$e_2 + e_3 + \cdots + e_n \leq e_1 + e_2 - e_{n-1} < e_1 + e_1 + 0 = 2e_1,$$

a contradiction. Thus, our assumption was incorrect and some three edges *can* be the sides of a triangle.

1999 National Contests: Problems and Solutions   137

17 Russia

Fourth round

Problem 8.1 A father wishes to take his two sons to visit their grandmother, who lives 33 kilometers away. He owns a motorcycle whose maximum speed is 25 km/h. With one passenger, its maximum speed drops to 20 km/h. (He cannot carry two passengers.) Each brother walks at a speed of 5 km/h. Show that all three of them can reach the grandmother's house in 3 hours.

Solution. Have the father drive his first son 24 kilometers, which takes $\frac{6}{5}$ hours; then drive back to meet his second son 9 kilometers from home, which takes $\frac{3}{5}$ hours; and finally drive his second son $\frac{6}{5}$ more hours. Each son spends $\frac{6}{5}$ hours riding 24 kilometers, and $\frac{9}{5}$ hours walking 9 kilometers. Thus, they reach their grandmother's house in exactly 3 hours — as does the father, who arrives at the same time as his second son.

Problem 8.2 The natural number A has the following property: the sum of the integers from 1 to A, inclusive, has decimal expansion equal to that of A followed by three digits. Find A.

Solution. We know that
$$k = (1 + 2 + \cdots + A) - 1000A$$
$$= \frac{A(A+1)}{2} - 1000A = A\left(\frac{A+1}{2} - 1000\right)$$
is between 0 and 999, inclusive. If $A < 1999$ then k is negative. If $A \geq 2000$ then $\frac{A+1}{2} - 1000 \geq \frac{1}{2}$ and $k \geq 1000$. Therefore $A = 1999$, and indeed $1 + 2 + \cdots + 1999 = 1999000$.

Problem 8.3 On sides BC, CA, AB of triangle ABC lie points A_1, B_1, C_1 such that the medians A_1A_2, B_1B_2, C_1C_2 of triangle $A_1B_1C_1$ are parallel to AB, BC, CA, respectively. Determine in what ratios the points A_1, B_1, C_1 divide the sides of ABC.

First Solution. A_1, B_1, C_1 divide sides BC, CA, AB in $1:2$ ratios (so that $\frac{BA_1}{A_1C} = \frac{1}{2}$, and so on).

---

[1] Problems are numbered as they appeared in the contests. Problems that appeared more than once in the contests are only printed once in this book.

Lemma. *In any triangle XYZ, the medians can be translated to form a triangle. Furthermore, the medians of this new triangle are parallel to the sides of triangle XYZ.*

Proof: Let x, y, z denote the vectors $\overrightarrow{YZ}$, $\overrightarrow{ZX}$, $\overrightarrow{XY}$ respectively. Observe that $x + y + z = \vec{0}$. Also, the vectors representing the medians of triangle XYZ are
$$m_x = z + \frac{x}{2}, \quad m_y = x + \frac{y}{2}, \quad m_z = y + \frac{z}{2}.$$
These vectors add up to $\frac{3}{2}(x + y + z) = \vec{0}$, so the medians indeed form a triangle.

Furthermore, the vectors representing the medians of the new triangle are
$$m_x + \frac{m_y}{2} = x + y + z - \frac{3}{4}y = -\frac{3}{4}y,$$
and similarly $-\frac{3}{4}z$ and $-\frac{3}{4}x$. Therefore, these medians are parallel to XZ, YX, and ZY. ∎

Let D, E, F be the midpoints of sides BC, CA, AB, and let l_1, l_2, l_3 be the segments A_1A_2, B_1B_2, C_1C_2.

Because l_1, l_2, l_3 are parallel to AB, BC, CA, the medians of the triangle formed by l_1, l_2, l_3 are parallel to CF, AD, BE. From the lemma, they are also parallel to B_1C_1, C_1A_1, A_1B_1.

Therefore, $BE \parallel A_1B_1$, and hence $\triangle BCE \sim \triangle A_1CB_1$. Then
$$\frac{B_1C}{AC} = \frac{1}{2} \cdot \frac{B_1C}{EC} = \frac{1}{2} \cdot \frac{A_1C}{BC} = \frac{1}{2}\left(1 - \frac{A_1B}{CB}\right).$$
Similarly
$$\frac{C_1A}{BA} = \frac{1}{2}\left(1 - \frac{B_1C}{AC}\right)$$
$$\frac{A_1B}{CB} = \frac{1}{2}\left(1 - \frac{C_1A}{BA}\right).$$
Solving these three equations gives
$$\frac{B_1C}{AC} = \frac{C_1A}{BA} = \frac{A_1B}{CB} = \frac{1}{3},$$
as claimed. It is straightforward to verify with the above equations that these ratio indeed work.

Second Solution. As above, we know that $A_1B_1 \parallel BE$, $B_1C_1 \parallel AD$, $C_1A_1 \parallel CF$.

Let A', B', C' be the points dividing the sides BC, CA, AB in $1:2$ ratios. Because $\frac{CA'}{CB} = \frac{CB'}{\frac{1}{2}CA}$, we know $A'B' \parallel BE \parallel A_1B_1$, and so on.

Suppose, by way of contradiction, that A_1 were closer to B than A'. Because $A_1B_1 \parallel A'B'$, B_1 is farther from C than B'. Similarly, C_1 is closer to A than C', and A_1 is *farther* from B than A' — a contradiction.

Likewise, A_1 cannot be farther from B than A'. Thus $A_1 = A'$, $B_1 = B'$, and $C_1 = C'$.

Problem 8.4 We are given 40 balloons, the air pressure inside each of which is unknown and may differ from balloon to balloon. It is permitted to choose up to k of the balloons and equalize the pressure in them (to the arithmetic mean of their respective original pressures). What is the smallest k for which it is always possible to equalize the pressures in all of the balloons?

Solution. $k = 5$ is the smallest such value.

First suppose that $k = 5$. Note that we can equalize the pressure in any 8 balloons $A, B, \ldots, H$: first equalize the pressure in the groups $\{A, B, C, D\}$ and $\{E, F, G, H\}$, and then equalize the pressure in $\{A, B, E, F\}$ and $\{C, D, G, H\}$.

Divide the 40 balloons into eight 5-*groups* of five and equalize the pressure in each group. Then form five new groups of eight — containing one balloon from each 5-group — and equalize the pressure in each of these new groups.

Now suppose that $k \leq 4$. Let $b_1, b_2, \ldots, b_{40}$ denote the original air pressures inside the balloons. It is simple to verify that the pressure in each balloon can always be written as a linear combination $a_1 b_1 + \cdots + a_{40} b_{40}$, where the a_i are rational with denominators not divisible by any primes except 2 and 3. Thus if we the b_j are linearly independent over the rationals (for instance, if $b_j = e^j$), we can never obtain

$$\frac{1}{40}b_1 + \frac{1}{40}b_2 + \cdots + \frac{1}{40}b_{40}$$

in a balloon. In this case, we can never equalize the pressures in all 40 balloons.

Problem 8.5 Show that the numbers from 1 to 15 cannot be divided into a group A of 2 numbers and a group B of 13 numbers in such a way that the sum of the numbers in B is equal to the product of the numbers in A.

Solution. Suppose, by way of contradiction, this were possible, and let a and b be the two numbers in A. Then we have

$$(1 + 2 + \cdots + 15) - a - b = ab$$
$$120 = ab + a + b$$
$$121 = (a+1)(b+1),$$

Because a and b are integers between 1 and 15, the only possible solution to this equation is $(a, b) = (10, 10)$. However, a and b must be distinct, a contradiction.

Problem 8.6 Given an acute triangle ABC, let A_1 be the reflection of A across the line BC, and let C_1 be the reflection of C across the line AB. Show that if A_1, B, C_1 lie on a line and $C_1B = 2A_1B$, then $\angle CA_1B$ is a right angle.

Solution. By the given reflections, we have $\triangle ABC \cong \triangle ABC_1 \cong \triangle A_1BC$.

Because $\angle B$ is acute, C_1 and A lie on the same side of BC. Thus C_1 and A_1 lie on opposite sides of BC as well.

Because C_1, B, A_1 lie on a line we have

$$180° = \angle C_1BA + \angle ABC + \angle CBA_1$$
$$= \angle ABC + \angle ABC + \angle ABC,$$

so that $\angle ABC = 60°$. We also know that

$$C_1B = 2A_1B \implies CB = 2AB,$$

implying that triangle ABC is a 30°-60°-90° triangle and $\angle CA_1B = \angle BAC = 90°$.

Problem 8.7 In a box lies a complete set of 1×2 dominoes. (That is, for each pair of integers i, j with $0 \leq i \leq j \leq n$, there is one domino with i on one square and j on the other.) Two players take turns selecting one domino from the box and adding it to one end of an open (straight) chain on the table, so that adjacent dominoes have the same numbers on their adjacent squares. (The first player's move may be any domino.) The first player unable to move loses. Which player wins with correct play?

Solution. The first player has a winning strategy. If $n = 0$, this is clear. Otherwise, have the first player play the domino $(0, 0)$. If on his first move the second player plays $(0, a)$, have the first player play (a, a).

At this point, the second player faces a chain whose ends are either 0 or a. Also, the domino $(0, k)$ is on the table if and only if the domino (a, k) is on the table. In such a *good* situation, if the second player plays $(0, k)$, then the first player can play (k, a) next to it. On the other hand, if the second player plays (a, k), then the first player can play $(k, 0)$. In both cases, the same conditions for a good situation occur.

Therefore the first player can always play a domino with this strategy, forcing the second player to lose.

Problem 8.8 An open chain of 54 squares of side length 1 is made so that each pair of consecutive squares is joined at a single vertex, and each square is joined to its two neighbors at opposite vertices. Is it possible to cover the surface of a $3 \times 3 \times 3$ cube with this chain?

Solution. It is not possible; suppose, by way of contradiction, it were.

Create axes so that the cube has corners at $(3i, 3j, 3k)$ for $i, j, k \in \{0, 1\}$, and place the chain onto the cube. Imagine that every two adjacent squares in the chain are connected by pivots, and also let the start and end vertices of the chain be "pivots."

Consider some pivot P at (x, y, z). The next pivot Q in the chain is either at $(x, y \pm 1, z \pm 1)$, $(x \pm 1, y, z \pm 1)$, or $(x \pm 1, y \pm 1, z)$. In any case, the sum of the coordinates of P has the same parity as the sum of the coordinates of Q — and hence *all* the pivots' sums of coordinates have the same parity. Suppose without loss of generality the sums are even.

Form a graph whose vertices are the lattice points on the cube with even sums of coordinates, and join two vertices with an edge if the two lattice points are opposite corners of a unit square. Every square in our chain contains one of these edges — but because there are exactly 54 such edges (one across each unit square on the cube's surface), and 54 squares in our chain, every edge is used exactly once. Then as we travel from pivot to pivot along our chain, we create an Eulerian path visiting all the edges. However, four vertices — at $(0, 0, 0)$, $(0, 1, 1)$, $(1, 0, 1)$, and $(1, 1, 0)$ — have odd degree 3, so this is impossible.

Problem 9.1 Around a circle are written all of the positive integers from 1 to N, $N \geq 2$, in such a way that any two adjacent integers have at least one common digit in their decimal expansions. Find the smallest N for which this is possible.

Solution. $N = 29$. Because 1 must be adjacent to two numbers, we must have $N \geq 11$. Then 9 must be adjacent to two numbers, and the next smallest numbers containing 9 as a digit are 19 and 29. Therefore $N \geq 29$, and indeed $N = 29$ suffices:

$$19, 9, 29, 28, 8, 18, 17, 7, 27, \ldots, 13, 3, 23, 2, 22, 21, 20, 12, 11, 10, 1.$$

Problem 9.2 In triangle ABC, points D and E are chosen on side CA such that $AB = AD$ and $BE = EC$ (E lying between A and D). Let F be the midpoint of the arc BC of the circumcircle of ABC. Show that B, E, D, F lie on a circle.

Solution. Let I be the incenter of triangle ABC. Because $AD = AB$, $\angle ADB = 90° - \frac{\angle CAB}{2}$ and

$$\angle BDC = 180° - \angle ADB = 90° + \frac{1}{2}\angle CAB$$
$$= 180° - \angle ICB - \angle CBI = \angle BIC.$$

Therefore, $BIDC$ is cyclic.

Note that because $\angle BAF = \angle FAC$, F lies on line AI. Some angle-chasing then shows that F is the center of the circle passing through B, I, and C. From above, this circle passes through D. Thus, $FD = FC$ and $\angle FDC = \angle DCF = \angle BCA + \frac{1}{2}\angle CAB$.

Because $BE = EC$ and $BF = FC$, line EF is the perpendicular bisector of $\overline{BC}$. Therefore,

$$\angle BEF = 90° - \angle BCA = 180° - \angle ADB - \angle FDC = \angle BDF,$$

and $BEDF$ is indeed cyclic.

Problem 9.3 The product of the positive real numbers x, y, z is 1. Show that if

$$\frac{1}{x} + \frac{1}{y} + \frac{1}{z} \geq x + y + z,$$

then

$$\frac{1}{x^k} + \frac{1}{y^k} + \frac{1}{z^k} \geq x^k + y^k + z^k$$

for all positive integers k.

Solution. Observe that

$$(x-1)(y-1)(z-1) = xyz - xy - yz - zx + x + y + z - 1$$
$$= 1 - \frac{1}{z} - \frac{1}{x} - \frac{1}{y} + x + y + z - 1$$
$$= x + y + z - \frac{1}{x} - \frac{1}{y} - \frac{1}{z} \leq 0.$$

For any positive integer k, $x^k > 1 \iff x > 1$ and $x^k < 1 \iff x < 1$. Similar relations hold for y^k and z^k. Therefore,

$$(x-1)(y-1)(z-1)$$
$$\leq 0 \Rightarrow (x^k-1)(y^k-1)(z^k-1)$$
$$\leq 0 \Rightarrow x^k y^k z^k - x^k y^k - y^k z^k - z^k x^k + x^k + y^k + z^k - 1$$
$$\leq 0 \Rightarrow x^k + y^k + z^k$$
$$\geq \frac{1}{x^k} + \frac{1}{y^k} + \frac{1}{z^k},$$

as desired.

Note. At first glance, it may be unclear why we examined the expression $(x-1)(y-1)(z-1)$. We might have instead written $x = \frac{a}{b}$, $y = \frac{b}{c}$, $z = \frac{c}{a}$ for some positive numbers a, b, c. (For example, we could take $a = 1$, $b = \frac{1}{x}$, $c = \frac{1}{xy}$.) The first given inequality then becomes

$$\frac{b}{a} + \frac{c}{b} + \frac{a}{c} \geq \frac{a}{b} + \frac{b}{c} + \frac{c}{a} \iff a^2 b + b^2 c + c^2 a$$
$$\geq ab^2 + bc^2 + ca^2 \iff 0 \geq (a-b)(b-c)(c-a)$$
$$= abc\left(\frac{a}{b}-1\right)\left(\frac{b}{c}-1\right)\left(\frac{c}{a}-1\right) \iff 0$$
$$\geq (x-1)(y-1)(z-1).$$

Problem 9.4 A maze consists of an 8×8 grid, in each 1×1 cell of which is drawn an arrow pointing up, down, left or right. The top edge of the top right square is the exit from the maze. A token is placed on the bottom left square, and then is moved in a sequence of turns. On each turn, the token is moved one square in the direction of the arrow. Then the arrow in the square the token moved from is rotated 90° clockwise. If the arrow points off of the board (and not through the exit), the token stays put and the arrow is rotated 90° clockwise. Prove that sooner or later the token will leave the maze.

Solution. Suppose, by way of contradiction, that the token could stay in the maze indefinitely. Because there are only finitely many squares, the token visits at least one square infinitely many times. The arrow on this square must make infinitely many 90° rotations, so the token must visit all the squares adjacent to it infinitely many times. It follows that the token visits *all* the squares on the board infinitely many times. Specifically, the token visits the upper-right square at least four times. At some point, then, this square's arrow will point out of the maze. When the token next lands on this square after this point, it will exit — a contradiction.

Problem 9.5 Each square of an infinite grid is colored in one of 5 colors, in such a way that every 5-square (Greek) cross contains one square of each color. Show that every 1×5 rectangle also contains one square of each color.

Solution. Label the centers of the grid squares with coordinates, and suppose that square $(0,0)$ is colored maroon. The Greek cross centered at $(1,1)$ must contain a maroon-colored square. However, the squares $(0,1)$, $(1,0)$, and $(1,1)$ cannot be maroon because each of these squares is in a Greek cross with $(0,0)$. Thus either $(1,2)$ or $(2,1)$ is maroon — without loss of generality, say $(1,2)$.

Then by a similar analysis on square $(1,2)$ and the Greek cross centered at $(2,1)$, one of the squares $(2,0)$ and $(3,1)$ must be maroon. $(2,0)$ is in a Greek cross with $(0,0)$ though, so $(3,1)$ is maroon.

Repeating the analysis on square $(2,0)$ shows that $(2,-1)$ is maroon. Spreading outward, every square of the form $(i+2j, 2i-j)$ is maroon. Because these squares are the centers of Greek crosses that tile the plane, no other squares can be maroon; because no two of these squares are in the same 1×5 rectangle, no two maroon squares can be in the same 1×5 rectangle.

The same argument applies to all the other colors — lavender, tickle-me-pink, green, neon orange. Therefore the five squares in each 1×5 rectangle have distinct colors, as desired.

Problem 9.7 Show that each natural number can be written as the difference of two natural numbers having the same number of prime factors.

Solution. If n is even, then we can write it as $(2n) - (n)$. If n is odd, let d be the smallest odd prime that does not divide n. Then write $n = (dn) - ((d-1)n)$. The number dn contains exactly one more prime factor than n. As for $(d-1)n$, it is divisible by 2 because $d-1$ is even. Its odd factors are less than d so they all divide n. Therefore $(d-1)n$ also contains exactly one more prime factor than n, and dn and $(d-1)n$ have the same number of prime factors.

Problem 9.8 In triangle ABC, with $AB > BC$, points K and M are the midpoints of sides AB and CA, and I is the incenter. Let P be the intersection of the lines KM and CI, and Q be the point such that $QP \perp KM$ and $QM \parallel BI$. Prove that $QI \perp AC$.

Solution. Draw point S on ray CB such that $CS = CA$. Let P' be the midpoint of AS. Because triangle ACS is isosceles, P' lies on line CI. Also, P' and M are midpoints of AS and AC, implying that have $P'M \parallel SC$. It follows that $P = P'$.

Let the incircle touch BC, CA, AB at D, E, F respectively. Writing $a = BC$, $b = CA$, $c = AB$, and $s = \frac{1}{2}(a+b+c)$, we have

$$SD = SC - DC = b - (s-c) = \frac{1}{2}(b+c-a) = FA,$$

$$BF = s - b = DB,$$

$$AP = PS.$$

Therefore

$$\frac{SD}{DB} \frac{BF}{FA} \frac{AP}{PS} = 1,$$

and by Menelaus' Theorem applied to triangle ABS, P lies on line DF.

Then triangle PDE is isosceles, and $\angle DEP = \angle PDE = \angle FEA = 90° - \frac{\angle A}{2}$ while $\angle CED = 90° - \frac{\angle C}{2}$. Therefore

$$\angle PEA = 180° - \angle DEP - \angle CED = 90° - \frac{\angle B}{2}.$$

Now let Q' be the point such that $Q'I \perp AC$ and $Q'M \parallel BI$. Then $\angle Q'EP = 90° - \angle PEA = \frac{\angle B}{2}$.

We also know that $\angle Q'MP = \angle IBC$ (from parallel lines $BC \parallel MP$ and $IB \parallel Q'M$), and $\angle IBC = \frac{\angle B}{2}$ as well. Therefore $\angle Q'MP = \angle Q'EP$, quadrilateral $Q'EMP$ is cyclic, and $\angle Q'PM = \angle Q'EM = 90°$. Therefore $Q = Q'$, and QI is indeed perpendicular to AC.

Problem 10.2 In the plane is given a circle ω, a point A inside ω, and a point B not equal to A. Consider all possible triangles BXY such that X and Y lie on ω and A lies on the chord XY. Show that the circumcenters of these triangles all lie on a line.

Solution. We use directed distances. Let O be the circumcenter and R be the circumradius of triangle BXY. Drop the perpendicular OO' to line AB.

The power of A with respect to circle BXY equals both $AX \cdot AY$ and $AO^2 - R^2$. Therefore

$$BO' - O'A = \frac{BO'^2 - O'A^2}{BO' + O'A}$$
$$= \frac{(BO^2 - O'O^2) - (OA^2 - OO'^2)}{AB}$$
$$= \frac{XA \cdot AY}{AB}.$$

This is constant because $AX \cdot AY$ also equals the power of A with respect to ω.

Because $BO' - O'A$ and $BO' + O'A = AB$ are constant, BO' and $O'A$ are constant as well. Thus O' is fixed regardless of the choice of X and Y. Therefore O lies on the line through O' perpendicular to AB, as desired.

Problem 10.3 In space are given n points in general position (no three points are collinear and no four are coplanar). Through any three of them is drawn a plane. Show that for any $n - 3$ points in space, there exists one of the drawn planes not passing through any of these points.

Solution. Call the given n points *given* and the $n - 3$ points *random*, and call all these points *level-0*. Because there are more given points than random points, one of the given points is not random: say, A. Draw a plane $\mathcal{P}$ not passing through A, and for each of the other points X (whether given or random) let X' be the intersection of line AX with $\mathcal{P}$. Call these points X' *level-1* points, and call X' given or random if and only if X is given or random, respectively.

Because no four level-0 given points are coplanar, no three level-1 given points are collinear; because no three level-0 given points are collinear, no two level-1 given points coincide. Thus we have $n-1$ level-1 given points and at most $n-3$ level-1 random points.

Now perform a similar construction — one of the level-1 given points, say B', is different from all the level-1 random points. Draw a line ℓ in $\mathcal{P}$ not passing through B' not parallel to $B'X'$ for any level-1 point $X' \neq B'$. Then for each other level-2 point (whether given or random), let X'' be the intersection of $B'X'$ with this line. Call these points X'' *level-2* points, and again call X'' given or random if and only if X is given or random, respectively.

Because no three level-1 given points were collinear, all of the level-2 given points are distinct. Thus we have $n - 2$ level-2 given points but at most $n-3$ level-2 random points. Therefore one of these given points C'' is not random.

Consider the drawn plane (ABC). If it contained some level-0 random point — say, Q — then Q' would be collinear with B' and C', and thus Q'' would equal C''', a contradiction. Therefore plane (ABC) does not pass through any of the level-0 random points, as desired.

Problem 10.5 Do there exist 10 distinct integers, the sum of any 9 of which is a perfect square?

Solution. Yes, there do exist 10 such integers. Write $S = a_1 + a_2 + \cdots + a_{10}$, and consider the linear system of equations

$$S - a_1 = 9 \cdot 1^2$$
$$S - a_2 = 9 \cdot 2^2$$
$$\vdots$$
$$S - a_{10} = 9 \cdot 10^2.$$

Adding all these gives

$$9S = 9 \cdot (1^2 + 2^2 + \cdots + 10^2)$$

so that

$$a_k = S - 9k^2 = 1^2 + 2^2 + \cdots + 10^2 - 9k^2.$$

Then all the a_k's are distinct integers, and any nine of them add up to a perfect square.

Problem 10.6 The incircle of triangle ABC touches sides BC, CA, AB at A_1, B_1, C_1, respectively. Let K be the point on the circle diametrically opposite C_1, and D the intersection of the lines $B_1 C_1$ and $A_1 K$. Prove that $CD = CB_1$.

Solution. Draw D' on $B_1 C_1$ such that $CD' \parallel AB$. Then $\angle D'CB_1 = \angle C_1 AB_1$ and $\angle CD'B_1 = \angle AC_1 B_1$, implying that $\triangle AB_1 C_1 \sim \triangle CB_1 D'$.

Thus, triangle $CB_1 D'$ is isosceles and $CD' = CB_1$. Recall that $CB_1 = CA_1$, and hence triangle $CA_1 D'$ is also isosceles. Also, $\angle D'CA_1 = 180° - \angle B$, implying that $\angle CA_1 D' = \frac{\angle B}{2}$.

Note that $\angle CA_1 K = \angle A_1 C_1 K = 90° - \angle C_1 K A_1 = 90° - \angle C_1 A_1 B = \frac{1}{2}\angle B$. Therefore, D' lies on $A_1 K$, and by definition it lies on $B_1 C_1$. Hence $D' = D$.

Thus, $CD = CD' = CB_1$, as desired.

Problem 10.7 Each voter in an election marks on a ballot the names of n candidates. Each ballot is placed into one of $n+1$ boxes. After the election, it is observed that each box contains at least one ballot, and that for any $n+1$ ballots, one in each box, there exists a name which is marked on all of these ballots. Show that for at least one box, there exists a name which is marked on all ballots in the box.

Solution. Suppose, by way of contradiction, that in every box, no name is marked on all the ballots. Label the boxes $1, 2, \ldots, n$, and look at an arbitrary ballot from the first box.

Suppose it has n "chosen" names Al, Bob, ..., Zed. By assumption, some ballot in the second box does not have the name Al on it; some ballot in the third box does not have the name Bob on it; and so on, so some ballot in the $(i+1)$-th box does not have the i-th chosen name on it. Then on these $n+1$ ballots, one from each box, there is no name marked on all the ballots — a contradiction.

Problem 10.8 A set of natural numbers is chosen so that among any 1999 consecutive natural numbers, there is a chosen number. Show that there exist two chosen numbers, one of which divides the other.

Solution. Construct a table with 1999 columns and 2000 rows. In the first row write $1, 2, \ldots, 1999$.

Define the entries in future rows recursively as follows: suppose the entries in row i are $k+1, k+2, \ldots, k+1999$, and that their product is M. Then fill row $i+1$ with $M+k+1, M+k+2, \ldots, M+k+1999$. All the entries in row $i+1$ are bigger than the entries in row i. Furthermore, every entry divides the entry immediately below it (and therefore *all* the entries directly below it).

In each row there are 1999 consecutive numbers, and hence each row contains a chosen number. Because we have 2000 rows, there are two chosen numbers in the same column — and one of them divides another, as desired.

Problem 11.1 The function $f(x)$ is defined on all real numbers. It is known that for all $a > 1$, the function $f(x) + f(ax)$ is continuous. Show that $f(x)$ is continuous.

Solution. We know that for $a > 1$, the functions
$$P(x) = f(x) + f(ax),$$

$$Q(x) = f(x) + f(a^2x),$$
$$P(ax) = f(ax) + f(a^2x)$$

are all continuous. Thus the function

$$\frac{1}{2}\left(P(x) + Q(x) - P(ax)\right) = f(x)$$

is continuous as well.

Problem 11.3 In a class, each boy is friends with at least one girl. Show that there exists a group of at least half of the students, such that each boy in the group is friends with an odd number of the girls in the group.

Solution. We perform strong induction on the total number of students. The base case of zero students is obvious.

Now suppose that we know the claim is true for any number of students less than n (where $n > 0$), and we wish to prove it for n. There must be at least one girl; pick any girl from the n students. We now partition the class into three subsets: the subset A consisting of this girl, the subset B consisting of this girl's male friends, and the subset C consisting of the remaining students.

Because we are using strong induction, the induction hypothesis states that there must be a subset C' of C, with at least $\frac{|C|}{2}$ students, such that any boy in C' is friends with an odd number of girls in C'.

Let B_O be the set of boys in B who are friends with an odd number of girls in C', and let B_E be the set of boys in B who are friends with an even number of girls in C'. Then there are two possible cases:

(i) $|B_O| \geq \frac{|A \cup B|}{2}$.

The set $S = B_O \cup C'$ will realize the claim, i.e., S will have at least $\frac{n}{2}$ elements, and each boy in S will be friends with an odd number of girls in S.

(ii) $|A \cup B_E| \geq \frac{|A \cup B|}{2}$.

Let $T = A \cup B_E \cup C'$. Each boy in C' will be friends with an odd number of girls in C' but not the girl in A. Each boy in B_E will be friends with an even number of girls in C' and the girl in A for a total odd number of girls in T. Finally, T has at least $\frac{n}{2}$ elements, and hence T realizes the claim.

Thus the induction is complete.

Note. With a similar proof, it is possible to prove a slightly stronger result: suppose each boy in a class is friends with at least one girl, and that every boy has a parity, either "even" or "odd." Then there is a group of at least half the students, such that each boy in the group is friends with the same parity of girls as his own parity. (By letting all the boys' parity be "odd," we have the original result.)

Problem 11.4 A polyhedron is circumscribed about a sphere. We call a face big if the projection of the sphere onto the plane of the face lies entirely within the face. Show that there are at most 6 big faces.

Solution.

Lemma. *Given a sphere of radius R, let a "slice" of the sphere be a portion cut off by two parallel planes. The surface area of the sphere contained in this slice is $2\pi RW$, where W is the distance between the planes.*

Proof: Orient the sphere so that the slice is horizontal. Take an infinitesimal horizontal piece of this slice, shaped like a frustum (a small sliver from the bottom of a radially symmetric cone). Let its width be w, its radius be r, and its slant height be ℓ. Its lateral surface area (for infinitesimal w) is then $2\pi r\ell$. If the side of the cone makes an angle θ with the horizontal, then $\ell \sin\theta = w$ and $R\sin\theta = r$ so that the surface area also equals $2\pi Rw$. Adding over all infinitesimal pieces, the complete slice has lateral surface area $2\pi RW$, as desired. ∎

Let the inscribed sphere have radius R and center O. For each big face F in the polyhedron, project the sphere onto F to form a circle k. Next connect k with O to form a cone. Because the interiors of the cones are pairwise disjoint, the cones intersect the sphere's surface in several non-overlapping circular regions.

Each circular region is a slice of the sphere with width $R(1 - \frac{1}{2}\sqrt{2})$ and bounds an region of area $2\pi R^2(1 - \frac{1}{2}\sqrt{2}) > \frac{1}{7}(4\pi R^2)$ of the sphere's surface. Thus each circular region takes up more than $\frac{1}{7}$ of the surface area of the sphere, implying there must be less than 7 such regions and at most six big faces.

Problem 11.5 Do there exist real numbers a, b, c such that for all real numbers x, y,

$$|x + a| + |x + y + b| + |y + c| > |x| + |x + y| + |y|?$$

Solution. No such numbers exist. Suppose they did. Let $y = -b - x$. Then for all real x we have

$$|x + a| + |-b - x + c| > |x| + |-b| + |-b - x|.$$

If we pick x negative with sufficiently large absolute value, this gives

$$(-x - a) + (-b - x + c) > (-x) + |b| + (-b - x) \Rightarrow -a + c$$
$$> |b| \geq 0,$$

so $c > a$. On the other hand, if we pick x positive and sufficiently large, this gives

$$(x + a) + (b + x - c) > (x) + |b| + (b + x) \Rightarrow a - c$$
$$> |b| \geq 0,$$

so $c < a$ as well — a contradiction.

Problem 11.6 Each cell of a 50×50 square is colored in one of four colors. Show that there exists a cell which has cells of the same color directly above, directly below, directly to the left, and directly to the right of it (though not necessarily adjacent to it).

Solution. By the pigeonhole principle, at least one-quarter of the squares (625) are the same color: say, red.

Of these red squares, at most 50 are the topmost red squares of their columns, and at most 50 are the bottommost red squares of their columns. Similarly, at most 50 are the leftmost red squares in their rows and at most 50 are the rightmost red squares in their rows. This gives at most 200 squares. The remaining 425 or more red squares then have red squares directly above, directly below, directly to the left, and directly to the right of them.

Problem 11.8 A polynomial with integer coefficients has the property that there exist infinitely many integers which are the value of the polynomial evaluated at more than one integer. Prove that there exists at most one integer which is the value of the polynomial at exactly one integer.

Solution. First observe that the polynomial cannot be constant. Now let $P(x) = c_n x^n + c_{n-1} x^{n-1} + \cdots + c_0$ be the polynomial with $c_n \neq 0$. The problem conditions imply that n is even and at least 2. We may assume without loss of generality that $c_n > 0$.

Because $P(x)$ has positive leading coefficient and it is not constant, there exists a value N such that $P(x)$ is decreasing for all $x < N$. Consider the pairs of integers (s,t) with $s < t$ and $P(s) = P(t)$. Because there are infinitely many pairs, there must be infinitely such pairs many with $s < N$.

For any integer k, look at the polynomial $P(x) - P(k-x)$. Some algebra shows that the coefficient of x^n is zero and that the coefficient of x^{n-1} is $f(k) = 2c_{n-1} + c_n(nk)$.

Let K be the largest integer such that $f(K) < 0$ (such an integer exists because from assumptions made above, $c_n \cdot n > 0$). Then for sufficiently large t we have

$$P(t) < P(K-t) < P(K-1-t) < \cdots$$

and

$$P(t) \geq P(K+1-t) > P(K+2-t) > P(K+3-t) > \cdots > P(N).$$

Therefore we must have $s = K+1-t$ and $P(t) - P(K+1-t) = 0$ for infinitely many values of t. However, P has finite degree, implying that $P(x) - P(K+1-x)$ is *identically* zero.

Then if $P(a) = b$ for some integers a, b, we also have $P(K+1-a) = b$. Therefore there is at most one value b that could possibly be the value of $P(x)$ at exactly one integer x — specifically, $b = P(\frac{K+1}{2})$.

Fifth round

Problem 9.1 In the decimal expansion of A, the digits occur in increasing order from left to right. What is the sum of the digits of $9A$?

Solution. Write $A = a_1 a_2 \ldots a_k$. By performing the subtraction

$$\begin{array}{ccccccc} & a_1 & a_2 & a_3 & \cdots & a_k & 0 \\ - & & a_1 & a_2 & \cdots & a_{k-1} & a_k \end{array}$$

we find that the digits of $9A = 10A - A$ are

$$a_1,\ a_2 - a_1,\ a_3 - a_2,\ \ldots,\ a_{k-1} - a_{k-2},\ a_k - a_{k-1} - 1,\ 10 - a_k.$$

These digits sum to $10 - 1 = 9$.

Problem 9.3 Let S be the circumcircle of triangle ABC. Let A_0 be the midpoint of the arc BC of S not containing A, and C_0 the midpoint of the arc AB of S not containing C. Let S_1 be the circle with center A_0

tangent to $\overline{BC}$, and let S_2 be the circle with center C_0 tangent to $\overline{AB}$. Show that the incenter I of ABC lies on a common external tangent to S_1 and S_2.

Solution. We prove a stronger result: I lies on a common external tangent to S_1 and S_2 which is parallel to line AC.

Let ℓ be the line tangent to S at B; we claim that it is a common external tangent to S_1 and S_2.

We are given that $\overline{BC}$ is tangent to S_1. Draw the other tangent from B to S_1, intersecting S_1 at P. Because rays BP and BC are reflections of each other across line BA_0, $\angle CBP = 2\angle CBA_0 = 2\angle CAA_0 = \angle CAB$. Therefore, line BP is tangent to S, and it must be the same line as ℓ.

Similarly, ℓ is tangent to S_2. Also, because ℓ does not pass through the interior of S, it does not intersect $\overline{A_0B_0}$. Therefore, it is a common external tangent to S_1 and S_2.

The reflection ℓ' of ℓ across line A_0B_0 is the other common external tangent to S_1 and S_2. Because $\angle BCA_0 = \angle A_0C_0C = \angle A_0IC$ and $\angle C_0A_0B = \angle AA_0C_0 = \angle IA_0C_0$, I and B are reflections of each other across line A_0B_0. Therefore, I lies on ℓ', as desired.

Furthermore, the angle between line CB and ℓ is $\angle CBP = \angle CAB$, and the angle between line CB and line C_0A_0 is $\frac{1}{2}(\widehat{CA_0} + \widehat{BA_0}) = \frac{1}{2}(\angle CAB + \angle BCA)$. Hence, the angle between line CB and ℓ' is $2 \cdot \frac{1}{2}(\angle CAB + \angle BCA) - \angle CAB = \angle BCA$, implying that ℓ' is indeed parallel to line AC.

Problem 9.4 The numbers from 1 to 1000000 can be colored black or white. A permissible move consists of selecting a number from 1 to 1000000 and changing the color of that number and each number not relatively prime to it. Initially all of the numbers are black. Is it possible to make a sequence of moves after which all of the numbers are colored white?

First Solution. It is possible. We begin by proving the following lemma:

Lemma. *Given a set S of positive integers, there is a subset $T \subseteq S$ such that every element of S divides an odd number of elements in T.*

Proof: We prove the claim by induction on $|S|$, the number of elements in S. If $|S| = 1$ then let $T = S$.

If $|S| > 1$, then let a be the smallest element of S. Consider the set $S' = S \setminus \{a\}$ — the set of the largest $|S| - 1$ elements in S. By induction there is a subset $T' \subseteq S'$ such that every element in S' divides an odd number of elements in T'.

If a also divides an odd number of elements in T', then the set $T = T'$ suffices. Otherwise, consider the set $T = T' \cup \{a\}$. a divides an odd number of elements in T. Every other element in T is bigger than a and can't divide it, but divides an odd number of elements in $T' = T \setminus \{a\}$. Hence T suffices, completing the induction and the proof of the lemma. ∎

Now, write each number $n > 1$ in its prime factorization

$$n = p_1^{a_1} p_2^{a_2} \cdots p_k^{a_k},$$

where the p_i are distinct primes and the a_i are positive integers. Notice that the color of n will always be the same as the color of $P(n) = p_1 p_2 \cdots p_k$.

Apply the lemma to the set $S = \bigcup_{i=2}^{1000000} P(i)$ to find a subset $T \subseteq S$ such that every element of S divides an odd number of elements in T. For each $q \in S$, let $t(q)$ equal the number of elements in T that q divides, and let $u(q)$ equal the number of primes dividing q.

Select all the numbers in T, and consider how the color of a number $n > 1$ changes. By the inclusion-exclusion principle, the number of elements in T not relatively prime to n equals

$$\sum_{q \mid P(n),\, q > 1} (-1)^{u(q)+1} t(q).$$

In particular, if $q \mid P(n)$ is divisible by exactly $m > 0$ primes, then it is counted $\binom{m}{1} - \binom{m}{2} + \binom{m}{3} - \cdots = 1$ time in the sum. (For example, if $n = 6$ then the number of elements in T divisible by 2 or 3 equals $t(2) + t(3) - t(6)$.)

By the definition of T, each of the values $t(q)$ is odd. Because there are $2^k - 1$ divisors $q > 1$ of $P(n)$, the above quantity is the sum of $2^k - 1$ odd numbers and is odd itself. Therefore after selecting T, every number $n > 1$ will switch color an odd number of times and will turn white.

Finally, select 1 to turn 1 white to complete the process.

Note. In fact, a slight modification of the above proof shows that T is unique. With some work, this stronger result implies that there is in essence *exactly* one way to make all the numbers white up to trivial manipulations.

Second Solution. Yes, it is possible. We prove a more general statement, where we replace 1000000 in the problem by some arbitrary positive integer

m. We also focus on the numbers divisible by just a few primes instead of all the primes.

Lemma. *For a finite set of distinct primes $S = \{p_1, p_2, \ldots, p_n\}$, let $Q_m(S)$ be the set of numbers between 2 and m divisible only by primes in S. The elements of $Q_m(S)$ can be colored black or white. A permissible move consists of selecting a number in $Q_m(S)$ and changing the color of that number and each number not relatively prime to it. Then it is possible to reverse the coloring of $Q_m(S)$ by selecting several numbers in a subset $R_m(S) \subseteq Q_m(S)$.*

Proof: We prove the lemma by induction on n. If $n = 1$, then selecting p_1 suffices. Now suppose $n > 1$, and assume without loss of generality that the numbers are all black to start with.

Let $T = \{p_1, p_2, \ldots, p_{n-1}\}$, and define t to be the largest integer such that $tp_n \leq m$. We can assume $t \geq 1$ because otherwise we could ignore p_n and just use the smaller set T, and we'd be done by our induction hypothesis.

Now select the numbers in $R_m(T)$, $R_t(T)$, and $p_n R_t(T) = \{p_n x \mid x \in R_t(T)\}$, and consider the effect of this action on a number y:

- y is not a multiple of p_n. Selecting the numbers in $R_m(T)$ makes y white. If selecting $x \in R_t(T)$ changes y's color, selecting xp_n will change it back so that y will become white.

- y is a power of p_n. Selecting the numbers in $R_m(T)$ and $R_t(T)$ has no effect on y, but each of the $|R_t(T)|$ numbers in $xR_t(T)$ changes y's color.

- $p_n \mid y$ but y is not a power of p_n. Selecting the numbers in $R_m(T)$ makes y white. Because $y \neq p_n^i$, it is divisible by some prime in T so selecting the numbers in $R_t(T)$ makes y black again. Finally, each of the $|R_t(T)|$ numbers in $xR_t(T)$ changes y's color.

Therefore, all the multiples of p_n are the same color (black if $|R_t(T)|$ is even, white if $|R_t(T)|$ is odd), while all the other numbers in $Q_m(S)$ are white. If the multiples of p_n are still black, we can select p_n to make them white, and we are done. ∎

We now return to the original problem. Set $m = 1000000$, and let S be the set of all primes under 1000000. From the lemma, we can select numbers between 2 and 1000000 so that all the numbers $2, 3, \ldots, 1000000$ are white. Finally, complete the process by selecting 1.

Problem 9.5 An equilateral triangle of side length n is drawn with sides along a triangular grid of side length 1. What is the maximum number of grid segments on or inside the triangle that can be marked so that no three marked segments form a triangle?

Solution. The grid is made up of $\frac{n(n+1)}{2}$ small equilateral triangles of side length 1. In each of these triangles, at most 2 segments can be marked so we can mark at most $\frac{2}{3} \cdot \frac{3n(n+1)}{2} = n(n+1)$ segments in all. Every segment points in one of three directions, so we can achieve the maximum $n(n+1)$ by marking all the segments pointing in two of the directions.

Problem 9.6 Let $\{x\} = x - \lfloor x \rfloor$ denote the fractional part of x. Prove that for every natural number n,

$$\sum_{k=1}^{n^2} \{\sqrt{k}\} \leq \frac{n^2 - 1}{2}.$$

Solution. We prove the claim by induction on n. For $n = 1$, we have $0 \leq 0$. Now supposing that the claim is true for n, we prove it is true for $n + 1$.

Each of the numbers $\sqrt{n^2 + 1}, \sqrt{n^2 + 2}, \ldots, \sqrt{n^2 + 2n}$ is between n and $n + 1$. Thus

$$\{\sqrt{n^2 + i}\} = \sqrt{n^2 + i} - n$$
$$< \sqrt{n^2 + i + \frac{i^2}{4n^2}} - n$$
$$= \frac{i}{2n}.$$

Therefore we have

$$\sum_{k=1}^{(n+1)^2} \{\sqrt{k}\} = \sum_{k=1}^{n^2} \{\sqrt{k}\} + \sum_{k=n^2+1}^{(n+1)^2} \{\sqrt{k}\}$$
$$< \frac{n^2 - 1}{2} + \frac{1}{2n} \sum_{i=1}^{2n} i + 0$$
$$= \frac{n^2 - 1}{2} + \frac{2n + 1}{2}$$
$$= \frac{(n+1)^2 - 1}{2},$$

completing the inductive step and the proof.

Problem 9.7 A circle passing through vertices A and B of triangle ABC intersects side BC again at D. A circle passing through vertices B and C intersects side AB again at E, and intersects the first circle again at F. Suppose that the points A, E, D, C lie on a circle centered at O. Show that $\angle BFO$ is a right angle.

Solution. Because $AEDC$ is cyclic with O as its center,
$$\angle COA = 2\angle CDA = \angle CDA + \angle CEA$$
$$= (180° - \angle ADB) + (180° - \angle BEC).$$
Because $BDFA$ and $BEFC$ are cyclic, $\angle ADB = \angle AFB$ and $\angle BEC = \angle BFC$. Hence
$$\angle COA = 360° - \angle AFB - \angle BFC = \angle CFA,$$
so $AFOC$ is cyclic. Therefore
$$\angle OFA = 180° - \angle ACO = 180° - \frac{180° - \angle COA}{2} = 90° + \angle CDA.$$
Because $ABDF$ is cyclic,
$$\angle OFA + \angle AFB = 90° + \angle CDA + \angle ADB = 270°.$$
Hence $\angle BFO = 90°$, as desired.

Problem 9.8 A circuit board has 2000 contacts, any two of which are connected by a lead. The hooligans Vasya and Petya take turns cutting leads: Vasya (who goes first) always cuts one lead, while Petya cuts either one or three leads. The first person to cut the last lead from some contact loses. Who wins with correct play?

Solution. Petya wins with correct play. Arrange the contacts in a circle and label them $1, 2, \ldots, 2000$, and let (x, y) denote the lead between contacts x and y (where labels are taken modulo 2000).

If Vasya disconnects $(a, 1000+a)$, Petya can disconnect $(500+a, 1500+a)$. Otherwise, if Vasya disconnects (a, b), Petya can disconnect the three leads $(a+500, b+500)$, $(a+1000, b+1000)$, and $(a+1500, b+1500)$. Notice that in each case, Petya and Vasya tamper with different contacts.

Using this strategy, after each of Petya's turns the circuit board is symmetrical under $90°$, $180°$, and $270°$ rotations, ensuring that he can always make the above moves — for example, if $(a+1500, b+1500)$ were already disconnected during Petya's turn, then (a, b) must have been as well before Vasya's turn.

Also, Petya can never lose, because if he disconnected the last lead (x, y) from some contact x, then Vasya must have already disconnected the last lead $(x - 1500, y - 1500)$, $(x - 1000, y - 1000)$, or $(x - 500, y - 500)$ from some other contact, a contradiction.

Problem 10.1 Three empty bowls are placed on a table. Three players A, B, C, whose order of play is determined randomly, take turns putting one token into a bowl. A can place a token in the first or second bowl, B in the second or third bowl, and C in the third or first bowl. The first player to put the 1999th token into a bowl loses. Show that players A and B can work together to ensure that C will lose.

Solution. Suppose A plays only in the first bowl until it contains 1998 tokens, then always plays in the second bowl. Also suppose B plays only in the third bowl until it contains 1998 tokens, then always plays in the second bowl as well.

Suppose, by way of contradiction, that C doesn't lose. Without loss of generality, suppose that the first bowl fills up to 1998 tokens before the third bowl does. Call this point in time the *critical point*.

First suppose that the third bowl never contains 1998 tokens. At most 999 rounds pass after the critical point because during each round, the third bowl gains 2 tokens (one from B, one from C). Thus A plays at most 999 tokens into the second bowl and doesn't lose. Hence *nobody* loses, a contradiction.

Thus the third bowl contains 1998 tokens some $k \leq 999$ more rounds after the critical point. After this k-th round A has played at most k tokens into the second bowl, and B has possibly played at most one token into the second bowl during the k-th round. Thus the second bowl has at most 1000 tokens. However, the first and third bowls each have 1998 tokens. Hence, during the next round, C will lose.

Problem 10.2 Find all infinite bounded sequences $a_1, a_2, \ldots$ of positive integers such that for all $n > 2$,

$$a_n = \frac{a_{n-1} + a_{n-2}}{\gcd(a_{n-1}, a_{n-2})}.$$

Solution. The only such sequence is $2, 2, 2, \ldots$, which clearly satisfies the given condition.

Let $g_n = \gcd(a_n, a_{n+1})$. Then g_{n+1} divides both a_{n+1} and a_{n+2}, so it divides $g_n a_{n+2} - a_{n+1} = a_n$ as well. Thus g_{n+1} divides both a_n and a_{n+1}, and it divides their greatest common divisor g_n.

Therefore, the g_i form a nonincreasing sequence of positive integers and eventually equal some positive constant g. At this point, the a_i satisfy the recursion
$$ga_n = a_{n-1} + a_{n-2}.$$
If $g = 1$, then $a_n = a_{n-1} + a_{n-2} > a_{n-1}$ so the sequence is increasing and unbounded.

If $g \geq 3$, then
$$a_n = \frac{a_{n-1} + a_{n-2}}{g} < \frac{a_{n-1} + a_{n-2}}{2} \leq \max\{a_{n-1}, a_{n-2}\}.$$
Similarly, $a_{n+1} < \max\{a_{n-1}, a_n\} \leq \max\{a_{n-2}, a_{n-1}\}$, so that
$$\max\{a_n, a_{n+1}\} < \max\{a_{n-2}, a_{n-1}\}.$$
Therefore the maximum values of successive pairs of terms form an infinite decreasing sequence of positive integers, a contradiction.

Thus $g = 2$ and eventually we have $2a_n = a_{n-1} + a_{n-2}$ or $a_n - a_{n-1} = -\frac{1}{2}(a_{n-1} - a_{n-2})$. This implies that $a_i - a_{i-1}$ converges to 0 and that the a_i are eventually constant as well. From $2a_n = a_{n-1} + a_{n-2}$, this constant must be 2.

Now if $a_n = a_{n+1} = 2$ for $n > 1$, then $\gcd(a_{n-1}, a_n) = \gcd(a_{n-1}, 2)$ either equals 1 or 2. Now
$$2 = a_{n+1} = \frac{a_{n-1} + a_n}{\gcd(a_{n-1}, 2)},$$
implying either that $a_{n-1} = 0$ — which is impossible — or that $a_{n-1} = 2$. Therefore all the a_i equal 2, as claimed.

Problem 10.3 The incircle of triangle ABC touches sides AB, BC, CA at K, L, M, respectively. For each two of the incircles of AMK, BKL, CLM is drawn the common external tangent not lying along a side of ABC. Show that these three tangents pass through a single point.

Solution. Let D, E, F be the midpoints of minor arcs MK, KL, LM of the incircle, respectively, and let S_1, S_2, S_3 be the incircles of triangles AMK, BKL, and CLM, respectively.

Because AK is tangent to the incircle, $\angle AKD = \angle KLD = \angle KMD = \angle DKM$. Similarly, $\angle AMD = \angle DMK$. Thus, D is the incenter of AMK and the center of S_1.

Likewise, E is center of S_2 and F is the center of S_3. By the result proved in Problem 9.3, the incenter I of triangle KLM lies on a common external tangent to S_1 and S_2. Because I does not lie on line AB, it must lie on the other external tangent. Similarly, the common external tangent to S_2 and S_3 (not lying on line BC) passes through I, as does the common external tangent to S_3 and S_1 (not lying on line CA). Therefore the three tangents all pass through I, as desired.

Problem 10.4 An $n \times n$ square is drawn on an infinite checkerboard. Each of the n^2 cells contained in the square initially contains a token. A move consists of jumping a token over an adjacent token (horizontally or vertically) into an empty square; the token jumped over is removed. A sequence of moves is carried out in such a way that at the end, no further moves are possible. Show that at least $\frac{n^2}{3}$ moves have been made.

Solution. At the end of the game no two adjacent squares contain tokens. Otherwise (because no more jumps are possible) they would have to be in an infinitely long line of tokens, which is not allowed. During the game, each time a token on square A jumps over another token on square B, imagine putting a 1×2 domino over squares A and B. At the end, every tokenless square on the checkerboard is covered by a tile, so no two uncovered squares are adjacent. We now prove there must be at least $\frac{n^2}{3}$ dominoes, implying that at least $\frac{n^2}{3}$ moves have been made:

Lemma. *If an $n \times n$ square board is covered with 1×2 rectangular dominoes (possibly overlapping, and possibly with one square off the board) in such a way that no two uncovered squares are adjacent, then at least $\frac{n^2}{3}$ tiles are on the board.*

Proof: Call a pair of adjacent squares on the checkerboard a *tile*. If a tile contains two squares on the border of the checkerboard, call it an *outer tile*. Otherwise, call it an *inner tile*.

Now for each domino D, consider any tile it partly covers. If this tile is partly covered by exactly m dominoes, we say that D *destroys* $\frac{1}{m}$ of that tile. Summing over all the tiles that D lies on, we find the total quantity a of outer tiles destroyed by D, and the total quantity b of inner tiles destroyed by D. We then say that D scores $1.5a + b$ points.

Consider the vertical domino D consisting of the upper-left square in the chessboard and the square immediately below it. It partly destroys two horizontal tiles. One of the two squares immediately to D's right must be covered, so if D destroys all of one horizontal tile, it can only destroy at most half of the other.

Armed with this type of analysis, some quick checking shows that any domino scores at most 6 points. Also, it can be verified that any domino scoring 6 points (i) lies completely on the board; (ii) does not contain a corner square of the chessboard; (iii) does not overlap any other dominoes; and (iv) does not have either length-1 edge border any other domino.

In a valid arrangement of dominoes, every tile is destroyed completely. Because there are $4(n-1)$ outer tiles and $2(n-1)(n-2)$ inner tiles, this means that a total of $1.5 \cdot 4(n-1) + 2(n-1)(n-2) = 2(n^2-1)$ points are scored. Therefore, there must be at least

$$\lceil \frac{2(n^2-1)}{6} \rceil = \lceil \frac{n^2-1}{3} \rceil$$

dominoes.

Suppose, by way of contradiction, that we have *exactly* $\frac{n^2-1}{3}$ dominoes. For this to be an integer, n must not be divisible by 3. In addition, the restrictions described earlier must hold for every domino.

Suppose we have any horizontal domino not at the bottom of the chessboard. One of the two squares directly below it must be covered. To satisfy our restrictions, this square must be covered by a horizontal domino (not a vertical one). Thus we can find a chain of horizontal dominoes stretching to the bottom of the board. Similarly, we can follow this chain to the top of the board.

Likewise, if there is any vertical domino then some chain of vertical dominoes stretches across the board. However, we cannot have both a horizontal *and* a vertical chain that do not overlap, so all the dominoes must have the same orientation: without loss of generality, suppose they are all horizontal.

To cover the tiles in any given row while satisfying the restrictions, we must alternate between blank squares and horizontal dominoes. In the top row, because no dominoes contain corner squares we must start and end with blank squares. Thus, $n \equiv 1 \pmod{3}$. Then in the second row, we must start with a horizontal domino (to cover the top-left vertical tiles). After alternating between dominoes and blank squares, the end of the row will contain two blank squares — a contradiction. Thus it is impossible to cover the chessboard with exactly $\frac{n^2-1}{3}$ dominoes, and indeed at least $\frac{n^2}{3}$ dominoes are needed. ∎

Note. When n is even, there is a simpler proof of the main result: split the n^2 squares of the board into 2×2 mini-boards, each containing four (overlapping) 1×2 tiles. At the end of the game, none of these n^2 tiles can contain two checkers (because no two checkers can be adjacent at the end of the game). Any jump removes a checker from at most three full tiles, implying that at least $\frac{n^2}{3}$ moves must take plac.e

A similar approach for odd n yields a lower bound of only $\frac{n^2-n-1}{3}$ moves. However, for sufficiently large n we can count the number of tokens that end up completely *outside* the $(n+2) \times (n+2)$ area around the checkerboard. Each made a jump that freed at most two full tiles, and from here we can show that $\frac{n^2}{3}$ moves are necessary.

Problem 10.5 The sum of the decimal digits of the natural number n is 100, and that of $44n$ is 800. What is the sum of the digits of $3n$?

Solution. The sum of the digits of $3n$ is 300.

Let $S(x)$ denote the sum of the digits of x. Then $S(a+b)$ equals $S(a)+S(b)$, minus nine times the number of carries in the addition $a+b$. Therefore, $S(a+b) \leq S(a) + S(b)$. Applying this inequality repeatedly, we have $S(a_1 + \cdots + a_k) \leq S(a_1) + \cdots + S(a_k)$.

Suppose that d is a digit between 0 and 9, inclusive. If $d \leq 2$ then $S(44d) = 8d$, and if $d = 3$ then $S(8d) = 6 < 8d$. If $d \geq 4$, then $44d \leq 44(9)$ has at most 3 digits so that $S(44d) \leq 27 < 8d$.

Now write $n = \sum n_i \cdot 10^i$, so that the n_i are the digits of n in base 10. Then

$$\sum 8n_i = S(44n) \leq \sum S(44n_i \cdot 10^i)$$
$$= \sum S(44n_i) \leq \sum 8n_i,$$

so equality must occur in the second inequality — that is, each of the n_i must equal 0, 1, or 2. Then each digit of $3n$ is simply three times the corresponding digit of n, and $S(3n) = 3S(n) = 300$, as claimed.

Problem 10.7 The positive real numbers x and y satisfy

$$x^2 + y^3 \geq x^3 + y^4.$$

Show that $x^3 + y^3 \leq 2$.

Solution. Equivalently we can prove that if $x^3 + y^3 > 2$, then

$$x^2 + y^3 < x^3 + y^4.$$

First notice that

$$\sqrt{\frac{x^2+y^2}{2}} \leq \sqrt[3]{\frac{x^3+y^3}{2}}$$

by the Power-Mean Inequality. This implies that
$$\begin{aligned} x^2 + y^2 &\le (x^3 + y^3)^{2/3} \cdot 2^{1/3} \\ &< (x^3 + y^3)^{2/3}(x^3 + y^3)^{1/3} \\ &= x^3 + y^3, \end{aligned}$$
or $x^2 - x^3 < y^3 - y^2$. Now $0 \le y^2(y-1)^2 \Rightarrow y^3 - y^2 \le y^4 - y^3$, so that
$$\begin{aligned} x^2 - x^3 < y^4 - y^3 &\Rightarrow x^2 + y^3 \\ &< x^3 + y^4, \end{aligned}$$
as desired.

Problem 10.8 In a group of 12 people, among every 9 people one can find 5 people, any two of whom know each other. Show that there exist 6 people in the group, any two of whom know each other.

Solution. Suppose, by way of contradiction, that no 6 people know each other. Draw a complete graph with twelve vertices corresponding to the people, labelling the people (and their corresponding vertices) $A, B, \ldots, L$. Color the edge between two people red if they know each other, and blue otherwise. Then among every nine vertices there is at least one red K_5, and among any six vertices there is at least one blue edge.

We prove that there are no blue cycles of odd length in this graph. Suppose, for sake of contradiction, that there is a blue cycle of length (i) 3 or 5, (ii) 7, (iii) 9, or (iv) 11.

(i) First suppose, for sake of contradiction, there is a blue 3-cycle (without loss of generality, ABC) or a blue 5-cycle (without loss of generality, $ABCDE$). In the first case, there is a blue edge among $DEFGHI$ (say, DE). Any red K_5 contains at most one vertex from $\{A, B, C\}$ and at most one vertex from $\{D, E\}$. In the second case, any K_5 still contains at most two vertices from $\{A, B, C, D, E\}$.

Now, $FGHIJK$ contains some other blue edge. Without loss of generality, suppose FG is blue. Now for each edge V_1V_2 in $HIJKL$, there must be a red K_5 among $ABCDEFGV_1V_2$. As before, this K_5 can contain at most two vertices from $\{A, B, C, D, E\}$, and it contains at most one vertex from each of $\{F, G\}$, $\{V_1\}$, and $\{V_2\}$. Therefore V_1 and V_2 must be connected by a red edge, so $HIJKL$ is a red K_5. Now $FHIJKL$ cannot be a red K_6, so without loss of generality suppose FH is blue. Similarly, $GHIJKL$ cannot be a red K_6, so without loss of generality either GH or GI is blue. In either case, $ABCDEFGHI$ must contain some red K_5. If GH is blue then this K_5 contains at most four vertices, two from $\{A, B, C, D, E\}$ and one from each of $\{F, G, H\}$ and $\{I\}$. If GI is blue then this K_5 still contains at most four vertices, two from $\{A, B, C, D, E\}$ and one from each of $\{F, H\}$ and $\{G, I\}$. Either possibility yields a contradiction.

(ii) If there is some blue 7-cycle, assume without loss of generality that it is $ABCDEFG$. As before, any K_5 contains at most three vertices from $\{A, B, \ldots, G\}$, so $HIJKL$ must be a red K_5. Now for each of the $\binom{5}{2} = 10$ choices of pairs $\{V_1, V_2\} \subseteq \{H, I, J, K, L\}$, there must be a red K_5 among $ABCDEFGV_1V_2$. Hence, for each edge in $HIJKL$, some red triangle in $ABCDEFG$ forms a red K_5 with that edge. $ABCDEFG$ contains at most 7 red triangles: ACE, BDF, ..., and GBD. Thus some triangle corresponds to two edges. Without loss of generality, either ACE corresponds to both HI and HJ, or else ACE corresponds to both HI and JK. In either case, $ACEHIJ$ is a red K_6, a contradiction.

(iii) Next suppose that there is some blue 9-cycle. Among these nine vertices there can be no red K_5, a contradiction.

(iv) Finally, suppose that there is some blue 11-cycle. Without loss of generality, suppose it is $ABCDEFGHIJK$. There is a red K_5 among $\{A, B, C, D, E, F, G, H, I\}$, which must be $ACEGI$. Likewise, $DFHJA$ must be a red K_5, so AC, AD, ..., AH are all red. Similarly, *every* edge in $ABCDEFGHI$ is red except for those in the blue 11-cycle.

Now among $\{A, B, C, D, E, F, G, H, L\}$ there is some red K_5, either $ACEGL$ or $BDFHL$. Without loss of generality, assume the former. Because $ACEGLI$ and $ACEGLJ$ cannot be red 6-cycles, AI and AJ must be blue. Then AIJ is a blue 3-cycle, a contradiction. ∎

Thus there are indeed no blue cycles of odd length, so the blue edges form a bipartite graph — that is, the twelve vertices can be partitioned into two sets S_1 and S_2 containing no blue edges. One of these sets, say S_1, has at least 6 vertices. But then S_1 is a red K_6, a contradiction. Therefore our original assumption was false. Hence there *is* some red K_6, so some six people do indeed know each other.

Problem 11.1 Do there exist 19 distinct natural numbers which add to 1999 and which have the same sum of digits?

Solution. No such integers exist. Suppose, by way of contradiction, that such integers did exist.

The average of the numbers is $\frac{1999}{19} < 106$, so one number is at most 105 and has digit sum at most 18.

Every number is congruent to its digit sum modulo 9, so all the numbers and their digit sums are congruent modulo 9, say congruent to k. Then $19k \equiv 1999 \Rightarrow k \equiv 1 \pmod{9}$, so the common digit sum is either 1 or 10.

If it is 1 then all the numbers equal 1, 10, 100, or 1000 so that some two are equal. This is not allowed. Thus the common digit sum is 10. Note that the twenty smallest numbers with digit sum 10 are:

$$19, 28, 37, \ldots, 91, 109, 118, 127, \ldots, 190, 208.$$

The sum of the first nine numbers is

$$(10 + 20 + \cdots + 90) + (9 + 8 + \cdots + 1) = 450 + 45 = 495,$$

while the sum of the next nine numbers is

$$(900) + (10 + 20 + \cdots + 80) + (9 + 8 + 7 + \cdots + 1) = 900 + 360 + 45 = 1305.$$

Hence the first eighteen numbers add up to 1800.

Because $1800 + 190 \neq 1999$, the largest number among the nineteen must be at least 208. Hence the smallest eighteen numbers add up to at least 1800, giving a total sum of at least $2028 > 1999$, a contradiction.

Problem 11.2 At each rational point on the real line is written an integer. Show that there exists a segment with rational endpoints, such that the sum of the numbers at the endpoints does not exceed twice the number at the midpoint.

Solution. Let $f : \mathbb{Q} \to \mathbb{Z}$ be the function that maps each rational point to the integer written at that point. Suppose, by way of contradiction, that for all $q, r \in \mathbb{Q}$,

$$f(q) + f(r) > 2f\left(\frac{q+r}{2}\right).$$

For $i \geq 0$, let $a_i = \frac{1}{2^i}$ and $b_i = -\frac{1}{2^i}$. We shall prove that for some k, $f(a_k)$ and $f(b_k)$ are both less than $f(0)$. Suppose that for some i, $f(a_i) \geq f(0)$. Now we apply the condition:

$$f(a_{i+1}) < \frac{f(a_i) + f(0)}{2} \leq f(a_i).$$

Because the range of f is the integers, $f(a_{i+1}) \leq f(a_i) - 1$ as long as $f(a_i) \geq f(0)$. Therefore, there exists some m such that $f(a_m) < f(0)$.

Then
$$f(a_{m+1}) < \frac{f(a_m) + f(0)}{2} < \frac{2f(0)}{2},$$
so $f(a_i) < f(0)$ for $i \geq m$.

Similarly, there exists n such that $f(b_i) < f(0)$ for $i \geq n$. Now if we take $k = \max\{m, n\}$, we have a contradiction,
$$f(a_k) + f(b_k) < 2f(0).$$

Problem 11.3 A circle inscribed in quadrilateral $ABCD$ touches sides DA, AB, BC, CD at K, L, M, N, respectively. Let S_1, S_2, S_3, S_4 be the incircles of triangles AKL, BLM, CMN, DNK, respectively. The common external tangents to S_1 and S_2, to S_2 and S_3, to S_3 and S_4, and to S_4 and S_1, not lying on the sides of $ABCD$, are drawn. Show that the quadrilateral formed by these tangents is a rhombus.

Solution. Let P be the intersection of the two common external tangents involving S_1, and let Q, R, S be the intersections of the pairs of tangents involving S_2, S_3, S_4, respectively.

As in problem 10.3, the centers of S_1, S_2, S_3, S_4 are the midpoints of arcs KL, LM, MN, NK, respectively. Line AB does not pass through the incenter I of triangle KLM. Therefore, by the results proved in problem 9.3, the other external tangent $\overline{PQ}$ must pass through I, and be parallel to KM. Likewise, $RS \parallel KM$ so we have $PQ \parallel RS$.

Similarly, $QR \parallel LN \parallel SP$, so $PQRS$ is a parallelogram.

Let $\langle X \mid \omega \rangle$ denote the length of the tangent from point X to circle ω, and let $\langle \omega_1 \mid \omega_2 \rangle$ denote the length of the external tangent to circles ω_1 and ω_2. Then we also know
$$AB = \langle A \mid S_1 \rangle + \langle S_1 \mid S_2 \rangle + \langle S_2 \mid B \rangle$$
$$= \langle A \mid S_1 \rangle + \langle S_1 \mid P \rangle + PQ + \langle Q \mid S_2 \rangle + \langle S_2 \mid B \rangle,$$
with analogous expressions holding for BC, CD, and DA. Substituting these into $AB + CD = BC + DA$, which is true because $ABCD$ is circumscribed about a circle, we find that $PQ + RS = QR + SP$.

Because $PQRS$ is a parallelogram, $PQ = RS$ and $QR = SP$, implying that $PQ = QR = RS = SP$ and that $PQRS$ is a rhombus.

Problem 11.5 Four natural numbers have the property that the square of the sum of any two of the numbers is divisible by the product of the other two. Show that at least three of the four numbers are equal.

Solution. Suppose, by way of contradiction, four such numbers did exist with no three of them equal. Select such numbers a, b, c, d so that $a+b+c+d$ is minimal. If some prime p divided both a and b, then from $a \mid (b+c)^2$ and $a \mid (b+d)^2$ we know that p divides c and d as well. Then $\frac{a}{p}, \frac{b}{p}, \frac{c}{p}, \frac{d}{p}$ are a counterexample with smaller sum. Therefore, the four numbers are pairwise relatively prime.

Suppose that some prime $p > 2$ divided a. Because a divides each of $(b+c)^2$, $(c+d)^2$, $(d+b)^2$, we know that p divides $b+c$, $c+d$, $d+b$. Hence p divides $(b+c) + (c+d) + (d+b)$ and thus $p \mid b+c+d$. Therefore $p \mid (b+c+d) - (b+c) = d$, and similarly $p \mid c$ and $p \mid b$, giving a contradiction.

Thus each of a, b, c, d are powers of 2. Because they are pairwise relatively prime, three of them must equal 1 — a contradiction. Therefore our original assumption was false, and no such counterexample exists.

Problem 11.6 We are given three convex polygons $\mathcal{P}_1$, $\mathcal{P}_2$, and $\mathcal{P}_3$ in the plane. Show that the following two conditions are equivalent:

(i) No line ℓ intersects all three polygons.

(ii) For $i = 1, 2, 3$, there exists a line ℓ_i not intersecting any of the polygons, such that $\mathcal{P}_i$ lies on the opposite side of ℓ_i from the other two polygons.

Solution. In this proof, "polygon" refers to both the border and interior of a polygon — the problem statement is not affected by this assumption, because a line intersecting the interior of a polygon must intersect its border as well.

Suppose that some line ℓ intersects all three polygons. Orient the figure to make ℓ horizontal, and suppose it intersects each polygon $\mathcal{P}_i$ at some point A_i, where A_1, A_2, and A_3 lie on ℓ from left to right in that order. Any line m not intersecting any of the polygons is either parallel to ℓ; intersects ℓ to the left of A_2; or intersects ℓ to the right of A_2. In all of these cases, m does not separate A_2 from both A_1 and A_3, so m cannot separate $\mathcal{P}_2$ from the other polygons. (In the first two cases A_2 and A_3 are not separated, and in the first and third cases A_1 and A_2 are not separated.)

To prove the other direction, we begin by proving an intuitively obvious but nontrivial lemma:

Lemma. *Given two non-intersecting convex polygons, there is a line that separates them.*

Let V be the convex hull of the two polygons. If all its vertices are in one polygon, then this polygon contains the other — a contradiction. Also, for any four vertices A, B, C, D in that order on V (not necessarily adjacent), because AC and BD intersect we cannot have A and C in one polygon and B and D in the other. Thus one run of adjacent vertices $V_1, \ldots, V_m$ is in one polygon $\mathcal{Q}$, and the remaining vertices $W_1, \ldots, W_n$ are in the other polygon $\mathcal{R}$.

Then $V_1 V_m$ is contained in polygon $\mathcal{Q}$, so line $V_1 V_m$ does not intersect $\mathcal{R}$. Therefore we can simply choose a line extremely close to $V_1 V_m$ that doesn't intersect $\mathcal{Q}$, and separates $\mathcal{Q}$ and $\mathcal{R}$. ∎

Now suppose that no line intersects all three polygons. Then every two polygons are disjoint. For instance, if M were in $\mathcal{P}_1 \cap \mathcal{P}_2$ and $N \neq M$ were in $\mathcal{P}_3$, then the line MN would intersect all three polygons.

Triangulate the convex hull H of $\mathcal{P}_1$ and $\mathcal{P}_2$ (that is, divide it into triangles whose vertices are vertices of H). If $\mathcal{P}_3$ intersects H at some point M, then M is on or inside of these triangles, XYZ. Without loss of generality suppose $X \in \mathcal{P}_1$ and $Y, Z \in \mathcal{P}_2$ (otherwise both triangle XYZ and M are inside either $\mathcal{P}_1$ or $\mathcal{P}_2$, so this polygon intersects $\mathcal{P}_3$). Then line XM intersects both $\mathcal{P}_1$ and $\mathcal{P}_3$. Because line XM intersects YZ, it intersects $\mathcal{P}_2$ as well — a contradiction.

Thus H is disjoint from $\mathcal{P}_3$, and from the lemma we can draw a line separating the two, and thus separating $\mathcal{P}_1$ and $\mathcal{P}_2$ from $\mathcal{P}_3$, as desired. We can repeat this construction for $\mathcal{P}_1$ and $\mathcal{P}_2$, so we are done.

Problem 11.7 Through vertex A of tetrahedron $ABCD$ passes a plane tangent to the circumscribed sphere of the tetrahedron. Show that the lines of intersection of the plane with the planes ABC, ACD, ABD form six equal angles if and only if

$$AB \cdot CD = AC \cdot BD = AD \cdot BC.$$

Solution. Perform an inversion about A with arbitrary radius r. Because the given plane P is tangent to the circumscribed sphere of $ABCD$, the sphere maps to a plane parallel to P containing B', C', D', the images of B, C, D under inversion. Planes P, ABC, ACD, and ABD stay fixed under the inversion because they all contain A.

Now, because $C'D'$ is in a plane parallel to P, plane $ACD = AC'D'$ intersects P in a line parallel to $C'D'$. For a more rigorous argument, complete parallelogram $C'D'AX$. Then X is both in plane $AC'D' =$

ACD and in plane P (because $PX \parallel C'D'$), so the intersection of ACD and P is the line PX, parallel to $C'D'$.

Similarly, plane ADB intersects P in a line parallel to $D'B'$, and plane ABC intersects P in a line parallel to $B'C'$. These lines form six equal angles if and only if $C'D'$, $D'B'$, $B'C'$ form equal angles — that is, if triangle $C'D'B'$ is equilateral and $C'D' = D'B' = B'C'$. Under the inversion distance formula, this is true if and only if

$$\frac{CD \cdot r^2}{AC \cdot AD} = \frac{DB \cdot r^2}{AD \cdot AB} = \frac{BC \cdot r^2}{AB \cdot AC},$$

which is equivalent to desired result.

18 Slovenia

Problem 1 The sequence of real numbers $a_1, a_2, a_3, \ldots$ satisfies the initial conditions $a_1 = 2, a_2 = 500, a_3 = 2000$ as well as the relation
$$\frac{a_{n+2} + a_{n+1}}{a_{n+1} + a_{n-1}} = \frac{a_{n+1}}{a_{n-1}}$$
for $n = 2, 3, 4, \ldots$. Prove that all the terms of this sequence are positive integers and that 2^{2000} divides the number a_{2000}.

Solution. From the recursive relation it follows that $a_{n+2} a_{n-1} = a_{n+1}^2$ for $n = 2, 3, \ldots$. No term of our sequence can equal 0, and hence it is possible to write
$$\frac{a_{n+2}}{a_{n+1} a_n} = \frac{a_{n+1}}{a_n a_{n-1}}$$
for $n = 2, 3, \ldots$. It follows by induction that the value of the expression
$$\frac{a_{n+1}}{a_n a_{n-1}}$$
is constant, namely equal to $\frac{a_3}{a_2 a_1} = 2$. Thus $a_{n+2} = 2 a_n a_{n+1}$ and all terms of the sequence are positive integers.

From this new relation, we also know that $\frac{a_{n+1}}{a_n}$ is an even integer for all positive integers n. Write
$$a_{2000} = \frac{a_{2000}}{a_{1999}} \frac{a_{1999}}{a_{1998}} \cdots \frac{a_2}{a_1} \cdot a_1.$$
In this product each of the 1999 fractions is divisible by 2, and $a_1 = 2$ is even as well. Thus a_{2000} is indeed divisible by 2^{2000}.

Problem 2 Find all functions $f : \mathbb{R} \to \mathbb{R}$ that satisfy the condition
$$f(x - f(y)) = 1 - x - y$$
for all $x, y \in \mathbb{R}$.

Solution. For $x = 0, y = 1$ we get $f(-f(1)) = 0$. For $y = -f(1)$ it follows that $f(x) = 1 + f(1) - x$. Writing $a = 1 + f(1)$ and $f(x) = a - x$, we have
$$1 - x - y = f(x - f(y)) = a - x + f(y) = 2a - x - y$$
so that $a = \frac{1}{2}$. Indeed, the function $f(x) = \frac{1}{2} - x$ satisfies the functional equation.

Problem 3 Let E be the intersection of the diagonals in cyclic quadrilateral $ABCD$, and let F and G be the midpoints of sides AB and CD, respectively. Prove that the three lines through G, F, E perpendicular to $\overline{AC}, \overline{BD}, \overline{AD}$, respectively, intersect at one point.

Solution. All angles are directed modulo $180°$. Drop perpendicular $\overline{GP}$ to diagonal AC and perpendicular $\overline{FQ}$ to diagonal BD. Let R be the intersection of lines PG and FQ, and let H be the foot of the perpendicular from E to side AD. We wish to prove that H, E, R are collinear.

Because F and G are midpoints of corresponding sides in similar triangles DEC and ABE (with opposite orientations), triangles DPE and AQE are similar with opposite orientations as well. Thus $\angle DPE = \angle EQA$ and therefore $AQPD$ is a cyclic quadrilateral. Because $\angle EQR = 90° = \angle EPR$, the quadrilateral $EQRP$ is cyclic, too. So

$$\angle ADQ = \angle APQ = \angle EPQ = \angle ERQ.$$

It follows that $\angle DEH = 90° - \angle ADQ = 90° - \angle ERQ = \angle QER$. Because D, E, Q are collinear, H, E, R must be collinear as well.

Problem 4 Three boxes with at least one marble in each are given. In a *step* we choose two of the boxes, doubling the number of marbles in one of the boxes by taking the required number of marbles from the other box. Is it always possible to empty one of the boxes after a finite number of steps?

Solution. Without loss of generality suppose that the number of marbles in the boxes are a, b, and c with $a \leq b \leq c$. Write $b = qa + r$ where $0 \leq r < a$ and $q \geq 1$. Then express q in binary:

$$q = m_0 + 2m_1 + \cdots + 2^k m_k,$$

where each $m_i \in \{0, 1\}$ and $m_k = 1$. Now for each $i = 0, 1, \ldots, k$, add $2^i a$ marbles to the first box: if $m_i = 1$ take these marbles from the second box; otherwise take them from the third box. In this way we take at most $(2^k - 1)a < qa \leq b \leq c$ marbles from the third box and exactly qa marbles from the second box altogether.

In the second box there are now $r < a$ marbles left. Thus the box with the least number of marbles now contains less than a marbles. Then by repeating the described procedure, we will eventually empty one of the boxes.

19 Taiwan

Problem 1 Determine all solutions (x, y, z) of positive integers such that
$$(x+1)^{y+1} + 1 = (x+2)^{z+1}.$$

Solution. Let $a = x+1$, $b = y+1$, $c = z+1$. Then $a, b, c \geq 2$ and
$$a^b + 1 = (a+1)^c$$
$$((a+1) - 1)^b + 1 = (a+1)^c.$$

Taking the equations mod $(a+1)$ yields $(-1)^b + 1 \equiv 0$, so b is odd.

Taking the second equation mod $(a+1)^2$ after applying the binomial expansion yields
$$\binom{b}{1}(a+1)(-1)^{b-1} + (-1)^b + 1 \equiv 0 \pmod{(a+1)^2}$$
so $(a+1) \mid b$ and a is even. On the other hand, taking the first equation mod a^2 after applying the binomial expansion yields
$$1 \equiv \binom{c}{1}a + 1 \pmod{a^2},$$
so c is divisible by a and is even as well. Write $a = 2a_1$ and $c = 2c_1$. Then
$$2^b a_1^b = a^b = (a+1)^c - 1 = ((a+1)^{c_1} - 1)((a+1)^{c_1} + 1).$$
It follows that $\gcd((a+1)^{c_1} - 1, (a+1)^{c_1} + 1) = 2$. Therefore, using the fact that $2a_1$ is a divisor of $(a+1)^{c_1} - 1$, we may conclude that
$$(a+1)^{c_1} - 1 = 2a_1^b$$
$$(a+1)^{c_1} + 1 = 2^{b-1}.$$

We must have $2^{b-1} > 2a_1^b \Rightarrow a_1 = 1$. Then these equations give $c_1 = 1$ and $b = 3$. Therefore the only solution is $(x, y, z) = (1, 2, 1)$.

Problem 2 There are 1999 people participating in an exhibition. Out of any 50 people, at least 2 do not know each other. Prove that we can find at least 41 people who each know at most 1958 other people.

Solution. Let Y be the set of people who know at least 1959 other people, and let $N(p)$ denote the set of people whom p knows. Assume by way of contradiction that less than 41 people each know at most 1958 people, so

$|Y| \geq 1959$. We now show that some 50 people all know each other, a contradiction.

Pick a person $y_1 \in Y$ and write $B_1 = N(y_1)$ with $|B_1| \geq 1959$. Then $|B_1| + |Y| > 1999$, and there is a person $y_2 \in B_1 \cap Y$.

Now write $B_2 = N(y_1) \cap N(y_2)$ with

$$|B_2| = |B_1| + |N(y_2)| - |B_1 \cup N(y_2)| \geq 1959 + 1959 - 1999$$
$$= 1999 - 40 \cdot 2.$$

Then $|B_2| + |Y| > 1999$, and there is a person $y_3 \in B_2 \cap Y$.

Now continue in a similar way: suppose we have $j \leq 48$ different people $y_1, y_2, \ldots, y_j$ in Y who all know each other, and suppose that $B_j = N(y_1) \cap N(y_2) \cap \cdots \cap N(y_j)$ has at least $1999 - 40j \geq 79 > 40$ elements. Then $|B_j| + |Y| > 1999$, and there is a person $y_{j+1} \in B_j \cap Y$. Thus $B_{j+1} = B_j \cap N(y_{j+1})$ has at least

$$|B_j| + |N(y_{j+1})| - |B_j \cup N(y_{j+1})| \geq (1999 - 40j) + 1959 - 1999$$
$$= 1959 - 40(j+1) > 0$$

elements, and we can continue onward.

Thus we can find 49 people $y_1, y_2, \ldots, y_{49}$ such that $B_{49} = N(y_1) \cap N(y_2) \cap \cdots \cap N(y_{49})$ is nonempty. Thus there is a person $y_{50} \in B_{49}$. But then any two people from $y_1, y_2, \ldots, y_{50}$ know each other, a contradiction.

Problem 3 Let P^* denote all the odd primes less than 10000, and suppose $p \in P^*$. For each subset $S = \{p_1, p_2, \cdots, p_k\}$ of P^*, with $k \geq 2$ and not including p, there exists a $q \in P^* \setminus S$ such that

$$(q+1) \mid (p_1+1)(p_2+1)\cdots(p_k+1).$$

Find all such possible values of p.

Solution. A *Mersenne prime* is a prime that has the form $2^n - 1$ for some integer $n > 0$. Notice that if $2^n - 1$ is prime then $n > 1$ and n is prime because otherwise either (if n were even) $n = 2m$ and $2^n - 1 = (2^m - 1)(2^m + 1)$, or (if n were odd) $n = ab$ for odd a, b and $2^n - 1 = (2^a - 1)(2^{(b-1)a} + 2^{(b-2)a} + \cdots + 2^a + 1)$. Direct calculation shows that the set T of Mersenne primes less than 10000 is

$$\{M_2, M_3, M_5, M_7, M_{13}\} = \{3, 7, 31, 127, 8191\},$$

where $M_p = 2^p - 1$. ($2^{11} - 1$ is not prime: it equals $23 \cdot 89$.) We claim this is the set of all possible values of p.

If some prime p is *not* in T, then look at the set $S = T$. Then there must be some prime $q \notin S$ less than 10000 such that

$$(q+1) \mid (M_2+1)(M_3+1)(M_5+1)(M_7+1)(M_{13}+1) = 2^{30}.$$

Thus, $q+1$ is a power of 2 and q is a Mersenne prime less than 10000 — and therefore $q \in T = S$, a contradiction.

On the other hand, suppose p is in T. Suppose we have a set $S = \{p_1, p_2, \ldots, p_k\} \subseteq P^*$ not including p, with $k \geq 2$ and $p_1 < p_2 < \cdots < p_k$. Suppose, by way of contradiction, that for all $q \in P^*$ such that $(q+1) \mid (p_1+1)\cdots(p_k+1)$, we have $q \in S$. Then

$$4 \mid (p_1+1)(p_2+1) \Longrightarrow M_2 \in S$$
$$8 \mid (M_2+1)(p_2+1) \Longrightarrow M_3 \in S$$
$$32 \mid (M_2+1)(M_3+1) \Longrightarrow M_5 \in S$$
$$128 \mid (M_2+1)(M_5+1) \Longrightarrow M_7 \in S$$
$$8192 \mid (M_3+1)(M_5+1)(M_7+1) \Longrightarrow M_{13} \in S.$$

Then p, a Mersenne prime under 10000, must be in S — a contradiction. Therefore there *is* some prime $q < 10000$ not in S with $q+1 \mid (p_1+1) \cdots (p_k+1)$, as desired. This completes the proof.

Problem 4 The altitudes through the vertices A, B, C of an acute-angled triangle ABC meet the opposite sides at D, E, F, respectively, and $AB > AC$. The line EF meets BC at P, and the line through D parallel to EF meets the lines AC and AB at Q and R, respectively. Let N be a point on the side BC such that $\angle NQP + \angle NRP < 180°$. Prove that $BN > CN$.

Solution. Let M be the midpoint of BC. We claim that P, Q, M, R are concyclic. Given this, we would have

$$\angle MQP + \angle MRP = 180° > \angle NQP + \angle NRP.$$

This can only be true if N is between M and C. Hence $BN > CN$.

Because $\angle BEC = \angle BFC = 90°$, we observe that the points B, C, E, F are concyclic and thus $PB \cdot PC = PE \cdot PF$. Also, the points D, E, F, M lie on the nine-point circle of triangle ABC so that $PE \cdot PF = PD \cdot PM$. (Alternatively, it's easy to show that $DEFM$ is cyclic with some angle-chasing). These two equations yield

$$PB \cdot PC = PD \cdot PM. \tag{1}$$

On the other hand, because $\triangle AEF \sim \triangle ABC$ and $QR \parallel EF$, we have $\angle RBC = \angle AEF = \angle CQR$. Thus $CQBR$ is cyclic and

$$DQ \cdot DR = DB \cdot DC. \tag{2}$$

Now let $MB = MC = a$, $MD = d$, $MP = p$. Then we have $PB = p + a$, $DB = a + d$, $PC = p - a$, $CD = a - d$, $DP = p - d$. Then equation (1) implies $(p+a)(p-a) = (p-d)p$. Therefore $a^2 = dp$ and $(a+d)(a-d) = (p-d)d$, or equivalently

$$DB \cdot DC = DP \cdot DM. \tag{3}$$

Combining (2) and (3) yields $DQ \cdot DR = DP \cdot DM$, so that the points P, Q, M, R are concyclic, as claimed.

Problem 5 There are 8 different symbols designed on n different T-shirts, where $n \geq 2$. It is known that each shirt contains at least one symbol, and for any two shirts, the symbols on them are not all the same. Also, for any k symbols, $1 \leq k \leq 7$, the number of shirts containing at least one of the k symbols is even. Find the value of n.

Solution. Let X be the set of 8 different symbols. Call a subset S of X *stylish* if some shirt contains exactly those symbols in S. Look at a stylish set A with the minimal number of symbols $|A| \geq 1$. Because $n \geq 2$, we must have $|A| \leq 7$. All the other $n - 1$ stylish sets contain at least one of the $k = 8 - |A|$ symbols in $X \setminus A$, so $n - 1$ is even and n is odd.

Observe that any nonempty subset $S \subseteq X$ contains an odd number of stylish subsets: for $S = X$ this number is n; for $|S| \leq 7$, an even number t of stylish sets contain some element of $X \setminus S$, so the remaining *odd* number $n - t$ of stylish sets are contained in S.

Then every nonempty subset of X is stylish. Otherwise, pick a minimal non-stylish subset $S \subseteq X$. Its only stylish subsets are its $2^{|S|} - 2$ proper subsets, which are all stylish by the minimal definition of S. However, this is an *even* number, which is impossible. Thus there must be $2^8 - 1 = 255$ T-shirts. Indeed, given any k symbols ($1 \leq k \leq 7$), an even number $2^8 - 2^{8-k}$ of T-shirts contain at least one of these k symbols.

20 Turkey

Problem 1 Let ABC be an isosceles triangle with $AB = AC$. Let D be a point on $\overline{BC}$ such that $BD = 2DC$, and let P be a point on $\overline{AD}$ such that $\angle BAC = \angle BPD$. Prove that

$$\angle BAC = 2\angle DPC.$$

Solution. Draw X on $\overline{BP}$ such that $BX = AP$. Then $\angle ABX = \angle ABP = \angle DPB - \angle PAB = \angle CAB - \angle PAB = \angle CAP$. Because $AB = CA$ and $BX = AP$, by SAS we have $\triangle ABX \cong \triangle CAP$. Hence $[ABX] = [CAP]$, and $\angle DPC = 180° - \angle CPA = 180° - \angle AXB = \angle PXA$.

Next, because $BD = 2CD$, the distance from B to line AD is twice the distance from C to line AD. Therefore $[ABP] = 2[CAP] \Longrightarrow [ABX] + [AXP] = 2[ABX]$. Hence $[AXP] = [ABX]$ and $XP = BX = AP$. Hence $\angle PXA = \angle XAP$, and $\angle BAC = \angle BPD = \angle PXA + \angle XAP = 2\angle PXA = 2\angle DPC$, as desired.

Problem 2 Prove that

$$(a+3b)(b+4c)(c+2a) \geq 60abc$$

for all real numbers $0 \leq a \leq b \leq c$.

Solution. By the AM-GM inequality we have $a + b + b \geq 3\sqrt[3]{ab^2}$. Multiplying this inequality and the analogous inequalities for $b + 2c$ and $c + 2a$ yields $(a+2b)(b+2c)(c+2a) \geq 27abc$. Then

$$(a+3b)(b+4c)(c+2a)$$
$$\geq \left(a + \frac{1}{3}a + \frac{8}{3}b\right)\left(b + \frac{2}{3}b + \frac{10}{3}c\right)(c+2a)$$
$$= \frac{20}{9}(a+2b)(b+2c)(c+2a) \geq 60abc,$$

as desired.

Problem 3 The points on a circle are colored in three different colors. Prove that there exist infinitely many isosceles triangles with vertices on the circle and of the same color.

First Solution. Partition the points on the circle into infinitely many regular 13-gons. In each 13-gon, by the Pigeonhole Principle, there are at least 5 vertices of the same color: say, red. Later we use some extensive case analysis to show that among these 5 vertices, some three form an isosceles triangle. Then for each 13-gon there is a monochrome isosceles triangle. Thus there are infinitely many monochrome isosceles triangles, as desired.

It suffices now to prove the following claim:

Claim *Suppose 5 vertices of a regular 13-gon are colored red. Then some three red vertices form an isosceles triangle.*

Proof. Suppose none of these 5 vertices did form an isosceles triangle. Label the vertices $P_0, \ldots, P_{12}$ (with indices taken modulo 13).

First we prove that P_i and P_{i+2} cannot both be red. Assume they could be, and say without loss of generality that P_{12} and P_1 were red; then P_{10}, P_0, and P_3 cannot be red. Furthermore, at most one vertex from each pair (P_{11}, P_4), (P_4, P_7), and (P_7, P_8) is red because each of these pairs forms an isosceles triangle with P_1. Similarly, at most one vertex from each pair (P_2, P_9), (P_9, P_6), and (P_6, P_5) is red. Now three vertices from $\{P_{11}, P_4, P_7, P_8\} \cup \{P_2, P_9, P_6, P_5\}$ are red; assume without loss of generality that two vertices from $\{P_{11}, P_4, P_7, P_8\}$ are also red. Vertices P_4 and P_8 can't both be red because they form an isosceles triangle with P_{12}, so vertices P_{11} and P_7 must be red. However, then any remaining vertex forms an isosceles triangle with some two of P_1, P_7, P_{11}, P_{12}. Thus we can't have five red vertices, a contradiction.

Next we prove that P_i and P_{i+1} can't be red. If so, suppose without loss of generality that P_6 and P_7 are red. Then P_4, P_5, P_8, and P_9 cannot be red as seen in the last paragraph. P_0 cannot be red either, because triangle $P_0 P_6 P_7$ is isosceles. Now each pair (P_3, P_{11}) and (P_{11}, P_1) contains at most one red vertex because triangles $P_3 P_7 P_{11}$ and $P_1 P_6 P_{11}$ are isosceles. Also, P_1 and P_3 can't both be red as shown earlier. Thus at most one of $\{P_1, P_3, P_{11}\}$ can be red. Similarly, at most one of $\{P_{12}, P_{10}, P_2\}$ can be red. Then we have at most four red vertices, again a contradiction.

Thus if P_i is red then $P_{i-2}, P_{i-1}, P_{i+1}, P_{i+2}$ cannot be red. Then we can have at most four red vertices, a contradiction. ■

Second Solution. Suppose we have $k \geq 1$ colors and a number $n \geq 3$. Then Van der Warden's theorem states that we can find N such that for any coloring of the numbers $1, 2, \ldots, N$ in the k colors, there are n numbers in arithmetic progression which are colored the same. Apply this theorem with $k = n = 3$ to find such an N, and partition the points on the circle into infinitely many regular N-gons rather than 13-gons. For each

N-gon $P_1P_2\ldots P_N$, there exist i,j,k (between 1 and N) in arithmetic progression such that P_i, P_j, P_k are all the same color. Hence triangle $P_iP_jP_k$ is a monochrome isosceles triangle. It follows that because we have infinitely many such N-gons, there are infinitely many monochrome isosceles triangles.

Problem 4 Let $\angle XOY$ be a given angle, and let M and N be two points on the rays OX and OY, respectively. Determine the locus of the midpoint of $\overline{MN}$ as M and N vary along the rays OX and OY such that $OM + ON$ is constant.

Solution. Let $\hat{x}$ and $\hat{y}$ be the unit vectors in the directions of rays OX and OY. Suppose we want $OM + ON$ to equal the constant k. If $OM = c$, then $ON = k - c$, and thus the midpoint of $\overline{MN}$ is $\frac{1}{2}(c\hat{x} + (k-c)\hat{y})$. As c varies from 0 to k, this midpoint traces out the line segment connecting $\frac{1}{2}k\hat{x}$ with $\frac{1}{2}k\hat{y}$; that is, the segment $\overline{M'N'}$ where $OM' = ON' = \frac{1}{2}k$, $M' \in \overrightarrow{OX}$, and $N' \in \overrightarrow{OY}$.

Problem 5 Some of the vertices of the unit squares of an $n \times n$ chessboard are colored such that any $k \times k$ square formed by these unit squares has a colored point on at least one of its sides. If $l(n)$ denotes the minimum number of colored points required to ensure the above condition, prove that
$$\lim_{n\to\infty} \frac{l(n)}{n^2} = \frac{2}{7}.$$

Solution. For each colored point P, consider any 1×1 square of the board that contains it. If this square contains m colored points including P, we say that P gains $\frac{1}{m}$ points from that square. Summing over all the 1×1 squares that P lies on, we find the total number of points that P accrues.

Any colored point on the edge of the chessboard gains at most 2 points. As for a colored point P on the chessboard's interior, the 2×2 square centered at P must have a colored point Q on its border. Then P and Q both lie on some unit square, which P gains at most half a point from. Thus P accrues at most $\frac{7}{2}$ points.

Therefore any colored point collects at most $\frac{7}{2}$ points, and $l(n)$ colored points collectively accrue at most $\frac{7}{2}l(n)$ points. For the given condition to hold, the total number of points accrued must be n^2. It follows that $\frac{7}{2}l(n) \geq n^2$ and thus $\frac{l(n)}{n^2} \geq \frac{2}{7}$.

Now, given some $n \times n$ board, embed it as the corner of an $n' \times n'$ board where $7 \mid (n'+1)$ and $n \leq n' \leq n+6$. To each 7×7 grid of vertices on the $n' \times n'$ board, color the vertices as below:

```
● ○ ● ○ ○ ○ ○
○ ○ ○ ○ ● ○ ●
○ ● ○ ● ○ ○ ○
● ○ ○ ○ ○ ● ○
○ ○ ● ○ ● ○ ○
○ ● ○ ○ ○ ○ ●
○ ○ ○ ● ○ ● ○
```

Then any $k \times k$ square on the chessboard has a colored point on at least one of its sides. Because we color $\frac{2}{7}(n'+1)^2$ vertices in this coloring, we have
$$l(n) \leq \frac{2}{7}(n'+1)^2 \leq \frac{2}{7}(n+7)^2$$
so that
$$\frac{l(n)}{n^2} \leq \frac{2}{7}\left(\frac{n+7}{n}\right)^2.$$

As $n \to \infty$, the right hand side becomes arbitrarily close to $\frac{2}{7}$. Because $\frac{l(n)}{n^2} \geq \frac{2}{7}$ for all n, we have $\lim_{n \to \infty} \frac{l(n)}{n^2} = \frac{2}{7}$.

Problem 6 Let $ABCD$ be a cyclic quadrilateral, and let L and N be the midpoints of diagonals AC and BD, respectively. Suppose that $\overline{BD}$ bisects $\angle ANC$. Prove that $\overline{AC}$ bisects $\angle BLD$.

Solution. For now, let $ABCD$ be *any* cyclic quadrilateral and let L and N be the midpoints of $\overline{AC}$ and $\overline{BD}$. Perform an inversion with arbitrary radius about B. A, D, C are mapped to collinear points A', D', C', while N is mapped to the point N' such that D' is the midpoint of $\overline{BN'}$. There are only two points X on line $A'D'$ such that $\angle BXN' = \angle BA'N'$: the point A' itself, and the reflection of A' across D'. Then

$$\angle ANB = \angle BNC \iff \angle BA'N' = \angle BC'N' \iff A'D'$$
$$= D'C' \iff \frac{AD}{BA \cdot BD} = \frac{DC}{BD \cdot BC} \iff AD \cdot BC$$
$$= BA \cdot DC.$$

Similarly, $\angle BLA = \angle DLA \iff AD \cdot BC = BA \cdot DC$. Therefore $\angle ANB = \angle BNC \iff \angle BLA = \angle DLA$. In other words, $\overline{BD}$ bisects $\angle ANC$ if and only if $\overline{AC}$ bisects $\angle BLD$. This proves the claim.

Problem 7 Determine all functions $f : \mathbb{R} \to \mathbb{R}$ such that the set

$$\left\{ \frac{f(x)}{x} \mid x \in \mathbb{R} \text{ and } x \neq 0 \right\}$$

is finite and

$$f(x - 1 - f(x)) = f(x) - x - 1$$

for all $x \in \mathbb{R}$.

Solution. First we show that the set $\{x - f(x) \mid x \in \mathbb{R}\}$ is finite. If not, there exist infinitely many $k \neq 1$ such that for some x_k, $x_k - f(x_k) = k$. Then

$$\frac{f(k-1)}{k-1} = \frac{f(x_k - 1 - f(x_k))}{k-1} = \frac{f(x_k) - x_k - 1}{k-1}$$

$$= -1 - \frac{2}{k-1}.$$

Because k takes on infinitely many values, $\frac{f(k-1)}{k-1}$ does as well — a contradiction.

Now choose x_0 so that $|x - f(x)|$ is maximal for $x = x_0$. Then for $y = x_0 - 1 - f(x_0)$ we have

$$y - f(y) = y - (f(x_0) - x_0 - 1) = 2(x_0 - f(x_0)).$$

Because $|x_0 - f(x_0)|$ is maximal, we must have $y - f(y) = x_0 - f(x_0) = 0$. Therefore $f(x) = x$ for all x. This function indeed satisfies the given conditions.

Problem 8 Let the area and the perimeter of a cyclic quadrilateral C be A_C and P_C, respectively. If the area and the perimeter of the quadrilateral which is tangent to the circumcircle of C at the vertices of C are A_T and P_T, respectively, prove that

$$\frac{A_C}{A_T} \geq \left(\frac{P_C}{P_T}\right)^2.$$

Solution. Let the outer quadrilateral be $EFGH$ with angles $\angle E = 2\alpha_1$, $\angle F = 2\alpha_2$, $\angle G = 2\alpha_3$, $\angle H = 2\alpha_4$. Let the circumcircle of C have radius r and center O, and let sides EF, FG, GH, HE be tangent to C at I, J, K, L.

In right triangle EIO, we have $IO = r$ and $\angle OEI = \alpha_1$ so that $EI = r \cot \alpha_1$. After finding similar expressions for $IF, FJ, \ldots, LE$, we find that $P_T = 2r \sum_{i=1}^{4} \cot \alpha_i$. Also, $[EFO] = \frac{1}{2} EF \cdot IO = \frac{1}{2} EF \cdot r$. Finding $[FGO], [GHO], [HEO]$, similarly shows that $A_T = \frac{1}{2} P_T \cdot r$.

Note that
$$IJ = 2r\sin\angle IKJ = 2r\sin\angle FIJ = 2r\sin(90° - \alpha_2)$$
$$= 2r\cos\alpha_2.$$

Similar expressions hold for JK, KL, LI leading to $P_C = 2r\sum_{i=1}^{4}\cos\alpha_i$. Also note that $\angle IOJ = 180° - \angle JFI = 180° - 2\alpha_2$, and hence
$$[IOJ] = \frac{1}{2}OI \cdot OJ\sin\angle IOJ = \frac{1}{2}r^2\sin(2\alpha_2) = r^2\sin\alpha_2\cos\alpha_2.$$

Adding this to the analogous expressions for $[JOK]$, $[KOL]$, $[LOI]$, we find that
$$A_C = r^2\sum_{i=1}^{4}\sin\alpha_i\cos\alpha_i.$$

Therefore the inequality we wish to prove is
$$A_C \cdot P_T^2 \geq A_T \cdot P_C^2 \iff r^2\sum_{i=1}^{4}\sin\alpha_i\cos\alpha_i \cdot P_T^2$$
$$\geq \left(\frac{1}{2}P_T \cdot r\right) \cdot 4r^2\left(\sum_{i=1}^{4}\cos\alpha_i\right)^2 \iff P_T \cdot \sum_{i=1}^{4}\sin\alpha_i\cos\alpha_i$$
$$\geq 2r \cdot \left(\sum_{i=1}^{4}\cos\alpha_i\right)^2 \iff \sum_{i=1}^{4}\cot\alpha_i \cdot \sum_{i=1}^{4}\sin\alpha_i\cos\alpha_i$$
$$\geq \left(\sum_{i=1}^{4}\cos\alpha_i\right)^2.$$

This is true by the Cauchy-Schwarz inequality $\sum a_i^2 \sum b_i^2 \geq (\sum a_ib_i)^2$ applied with each $a_i = \sqrt{\cot\alpha_i}$ and $b_i = \sqrt{\sin\alpha_i\cos\alpha_i}$.

Problem 9 Prove that the plane is not a union of the inner regions of finitely many parabolas. (The outer region of a parabola is the union of the lines on the plane not intersecting the parabola. The inner region of a parabola is the set of points on the plane that do not belong to the outer region of the parabola.)

Solution. Suppose, by way of contradiction, we could cover the plane with the inner regions of finitely many parabolas — say, n of them. Choose some fixed positive acute angle $\theta < \left(\frac{360}{2n}\right)°$.

Take any of the parabolas and (temporarily) choose a coordinate system so that the parabola is the graph of $y = ax^2$ with $a > 0$. (Assume that our coordinates are chosen to scale, so that one unit along the y-axis has the same length as the unit along the x-axis). Draw the tangents to the parabola at $x = \pm\frac{\cot\theta}{2a}$; these lines have slopes $2ax = \pm\cot\theta$. These lines meet on the y-axis at an angle of 2θ, forming a V-shaped region in the plane that contains the inner region of the parabola.

Performing the above procedure with all the parabolas, we obtain n V-shaped regions covering the entire plane. Again choose an x-axis, and let the rays bordering these regions make angles ϕ_j and $\phi_j + 2\theta$ with the positive x-axis (with angles taken modulo 360°). Because $2n\theta < 360°$, there is some angle ϕ' not in any of the intervals $[\phi_j, \phi_j + 2\theta]$. Then consider the ray passing through the origin and making angle of ϕ' with the positive x-axis. For sufficiently large x, the points on this ray cannot lie in any of the V-shaped regions, a contradiction. Thus our original assumption was false, so we *cannot* cover the plane with the inner regions of finitely many parabolas.

21 Ukraine

Problem 1 Let $P(x)$ be a polynomial with integer coefficients. The sequence $\{x_n\}_{n\geq 1}$ satisfies the conditions $x_1 = x_{2000} = 1999$, and $x_{n+1} = P(x_n)$ for $n \geq 1$. Calculate
$$\frac{x_1}{x_2} + \frac{x_2}{x_3} + \cdots + \frac{x_{1999}}{x_{2000}}.$$

Solution. Write $a_i = x_i - x_{i-1}$ for each i, where we take subscripts modulo 1999. Because $c - d$ divides $P(c) - P(d)$ for integers c and d, we have $a_i \mid a_{i+1}$ for all i.

First suppose that all the $a_i \neq 0$. Then $|a_{i+1}| \geq |a_i|$ for all i. Also $|a_1| = |a_{2000}|$, so all the $|a_i|$ must be equal. Let $m > 0$ be this common value. If n of the $a_1, a_2, \ldots, a_{1999}$ equal $m \neq 0$ and the other $1999 - n$ equal $-m$, then their sum $0 = x_{1999} - x_0 = a_1 + a_2 + \cdots + a_{1999}$ equals $m(2n - 1999) \neq 0$, a contradiction.

Thus for some k we have $a_k = 0$. Because a_k divides a_{k+1}, we have $a_{k+1} = 0$. Similarly $a_{k+2} = 0$, and so on. Thus all the x_i are equal and the given expression equals 1999.

Problem 2 For real numbers $0 \leq x_1, x_2, \ldots, x_6 \leq 1$ prove the inequality
$$\frac{x_1^3}{x_2^5 + x_3^5 + x_4^5 + x_5^5 + x_6^5 + 5} + \frac{x_2^3}{x_1^5 + x_3^5 + x_4^5 + x_5^5 + x_6^5 + 5}$$
$$+ \cdots + \frac{x_6^3}{x_1^5 + x_2^5 + x_3^5 + x_4^5 + x_5^5 + 5} \leq \frac{3}{5}.$$

Solution. The condition $0 \leq x_1, x_2, \ldots, x_6 \leq 1$ implies that the left hand side of the inequality is at most
$$\sum_{i=1}^{6} \frac{x_i^3}{x_1^5 + x_2^5 + \cdots + x_6^5 + 4} = \frac{x_1^3 + x_2^3 + \cdots + x_6^3}{x_1^5 + x_2^5 + \cdots + x_6^5 + 4}.$$

For $t \geq 0$ we have
$$\frac{t^5 + t^5 + t^5 + 1 + 1}{5} \geq t^3$$

by the AM-GM inequality. Adding up the six resulting inequalities for $t = x_1, x_2, \ldots, x_6$ and dividing by $(x_1^5 + x_2^5 + \cdots + x_6^5 + 4)$ shows that the above expression is at most $\frac{3}{5}$.

Problem 3 Let $\overline{AA_1}, \overline{BB_1}, \overline{CC_1}$ be the altitudes of an acute triangle ABC, and let O be an arbitrary point inside the triangle $A_1B_1C_1$. Let

M, N, P, Q, R, S be the orthogonal projections of O onto lines AA_1, BC, BB_1, CA, CC_1, AB, respectively. Prove that lines MN, PQ, RS are concurrent.

Solution. Observe that three lines passing through different vertices of a triangle are concurrent if and only if their reflections across the corresponding angle bisectors are also concurrent; this fact is easily proved using the trigonometric form of Ceva's Theorem.

Let A_0, B_0, C_0 be the centers of rectangles OMA_1N, OPB_1Q, OSC_1R, respectively. Under the homothety with center O and ratio $\frac{1}{2}$, triangle $A_1B_1C_1$ maps to triangle $A_0B_0C_0$. Then, because lines AA_1, BB_1, CC_1 are the angle bisectors of triangle $A_1B_1C_1$ (easily proved with angle-chasing), the angle bisectors of triangle $A_0B_0C_0$ are parallel to lines AA_1, BB_1, CC_1.

Because OMA_1N is a rectangle, diagonals OA_1 and MN are reflections of each across the line through A_0 parallel to line AA_1. From above, this line is precisely the angle bisector of $\angle C_0A_0B_0$ in triangle $A_0B_0C_0$. Similarly, lines OB_1 and OC_1 are reflections of lines PQ and RS across the other angle bisectors. Because lines OA_1, OB_1, OC_1 are concurrent at O, from our initial observation lines MN, PQ, RS are concurrent as well.

22 United Kingdom

Problem 1 I have four children. The age in years of each child is a positive integer between 2 and 16 inclusive and all four ages are distinct. A year ago the square of the age of the oldest child was equal to the sum of the squares of the ages of the other three. In one year's time the sum of the squares of the ages of the oldest and the youngest children will be equal to the sum of the squares of the other two children. Decide whether this information is sufficient to determine their ages uniquely, and find all possibilities for their ages.

Solution. Let the children's present ages be $a+1$, $b+1$, $c+1$, and $d+1$. We are given that $1 \leq a < b < c < d \leq 15$. Note that $b \leq 13$ so that $b - a \leq 12$. We are also given

$$d^2 = a^2 + b^2 + c^2 \tag{1}$$

and

$$(d+2)^2 + (a+2)^2 = (b+2)^2 + (c+2)^2. \tag{2}$$

Subtracting (1) from (2) gives $4(a+d) + a^2 = 4(b+c) - a^2$, or

$$a^2 = 2(b + c - a - d). \tag{3}$$

Thus a must be even because its square is even. Furthermore, because $d > c$,

$$a^2 = 2\bigl(b - a + (c - d)\bigr) < 2(b - a) < 24,$$

and hence either $a = 2$ and $a = 4$.

If $a = 4$ then, because $a^2 < 2(b - a)$, we have $2b > a^2 + 2a = 24$ so that $b > 12$. This forces $b = 13$, $c = 14$, and $d = 15$, which contradicts the given conditions.

Thus $a = 2$. Equation (3) gives $b + c - d = 4$. Substituting $a = 2$ and $d = b + c - 4$ into (1) and simplifying yields

$$(b-4)(c-4) = 10 = 1 \cdot 10 = 2 \cdot 5.$$

Therefore we have $(b, c) = (5, 14)$ or $(6, 9)$, in which cases $d = 15$ and $d = 11$ respectively.

Hence the only possible solutions are $(a, b, c, d) = (2, 5, 14, 15)$ or $(2, 6, 9, 11)$, and these indeed satisfy (1) and (2). It follows that there is no unique solution, and it is not possible to determine the children's' ages.

Problem 2 A circle has diameter $\overline{AB}$ and X is a fixed point on the segment AB. A point P, distinct from A and B, lies on the circle. Prove that, for all possible positions of P,
$$\frac{\tan \angle APX}{\tan \angle PAX}$$
is a constant.

Solution. Let Q be the projection of X onto $\overline{AP}$. Note that $\angle APB = 90°$, and thus $\tan \angle PAX = \frac{PB}{PA}$. Also, $XQ \parallel PB$ so $\triangle AQX \sim \triangle APB$. Therefore,
$$\tan \angle APX = \frac{QX}{QP} = \frac{\frac{AX \cdot BP}{AB}}{\frac{BX \cdot AP}{AB}} = \frac{AX \cdot BP}{BX \cdot AP},$$
and
$$\frac{\tan \angle APX}{\tan \angle PAX} = \frac{AX}{BX}$$
is fixed.

Problem 3 Determine a positive constant c such that the equation
$$xy^2 - y^2 - x + y = c$$
has exactly three solutions (x, y) in positive integers.

Solution. When $y = 1$ the left-hand side is 0. Thus we can rewrite our equation as
$$x = \frac{y(y-1) + c}{(y+1)(y-1)}.$$
The numerator is congruent to $-1(-2) + c$ modulo $(y+1)$, and it is also congruent to c modulo $(y-1)$. Hence we must have $c \equiv -2 \pmod{(y+1)}$ and $c \equiv 0 \pmod{(y-1)}$. Because $c = y - 1$ satisfies these congruences, we must have $c \equiv y - 1 \pmod{\text{lcm}(y-1, y+1)}$. When y is even, $\text{lcm}(y-1, y+1) = y^2 - 1$; when y is odd, $\text{lcm}(y-1, y+1) = \frac{1}{2}(y^2 - 1)$.

Then for $y = 2, 3, 11$ we have $c \equiv 1 \pmod 3$, $c \equiv 2 \pmod 4$, $c \equiv 10 \pmod{60}$. Hence, we try setting $c = 10$. For x to be an integer we must have $(y-1) \mid 10 \Rightarrow y = 2, 3, 6$, or 11. These values give $x = 4, 2, \frac{2}{7}$, and 1, respectively. Thus there *are* exactly three solutions in positive integers, namely $(x, y) = (4, 2), (2, 3)$, and $(1, 11)$.

Problem 4 Any positive integer m can be written uniquely in base 3 form as a string of 0's, 1's and 2's (not beginning with a zero). For

example,
$$98 = 81 + 9 + 2 \times 3 + 2 \times 1 = (10122)_3.$$

Let $c(m)$ denote the sum of the cubes of the digits of the base 3 form of m; thus, for instance
$$c(98) = 1^3 + 0^3 + 1^3 + 2^3 + 2^3 = 18.$$

Let n be any fixed positive integer. Define the sequence $\{u_r\}$ as
$$u_1 = n, \text{ and } u_r = c(u_{r-1}) \text{ for } r \geq 2.$$

Show that there is a positive integer r such that $u_r = 1, 2,$ or 17.

Solution. If m has $d \geq 5$ digits then we have $m \geq 3^{d-1} = (80 + 1)^{(d-1)/4} \geq 80 \cdot \frac{d-1}{4} + 1 > 8d$ by Bernoulli's inequality. Thus $m > c(m)$.

If $m > 32$ has 4 digits in base 3, then $c(m) \leq 2^3 + 2^3 + 2^3 + 2^3 = 32 < m$. On the other hand, if $27 \leq m \leq 32$, then m starts with the digits 10 in base 3 and $c(m) < 1^3 + 0^3 + 2^3 + 2^3 = 17 < m$.

Therefore $0 < c(m) < m$ for all $m \geq 27$. Hence, eventually, we have $u_s < 27$. Because u_s has at most three digits, u_{s+1} can only equal $8, 16, 24, 1, 9, 17, 2, 10,$ or 3. If it equals $1, 2,$ or 17 we are already done; if it equals 3 or 9 then $u_{s+2} = 1$. Otherwise a simple check shows that u_r will eventually equal 2:
$$\left. \begin{array}{l} 8 = (22)_3 \\ 24 = (220)_3 \end{array} \right\} \to 16 = (121)_3 \to 10 = (101)_3 \to 2.$$

Problem 5 Consider all functions $f : \mathbb{N} \to \mathbb{N}$ such that

(i) for each positive integer m, there is a unique positive integer n such that $f(n) = m$;

(ii) for each positive integer n, $f(n+1)$ is either $4f(n) - 1$ or $f(n) - 1$.

Find the set of positive integers p such that $f(1999) = p$ for some function f with properties (i) and (ii).

Solution. Imagine hopping along a sidewalk whose blocks are marked from left to right with the positive integers, where at time n we stand on the block marked $f(n)$. Note that if $f(n) - f(n+1) > 0$ then $f(n) - f(n+1) = 1$ — that is, whenever we move to the left we move exactly one block. Likewise, whenever we move to the right from $f(n)$ we must move to $4f(n) - 1$.

Suppose that we are at block $f(a)$ and that $f(a) - 1$ is unvisited. If we move to the right (that is, if $f(a+1) > f(a)$) then at some point we must pass through block $f(a)$ to reach block $f(a) - 1$ again. This is not allowed. Thus we must have $f(a+1) = f(a) - 1$.

Therefore our path is completely determined by the value of $f(1)$: whenever we are at $f(n)$, if $f(n) - 1 > 0$ is unvisited, we must have $f(n+1) = f(n) - 1$. Otherwise, we must have $f(n+1) = 4f(n) - 1$.

If $f(1) = 1$, then consider the function f defined as follows: whenever $2^k \le n < 2^{k+1}$, set $f(n) = (3 \cdot 2^k - 1) - n$. The function f is bijective because for $n = 2^k, 2^k + 1, \ldots, 2^{k+1} - 1$ we have $f(n) = 2^{k+1} - 1, 2^{k+1} - 2, \ldots, 2^k$. A a quick check shows that f satisfies condition (ii) as well. Thus, as proved above, this is the *only* function with $f(1) = 1$, and in this case because $2^{10} \le 1999 < 2^{11}$ we have $f(1999) = (3 \cdot 2^{10} - 1) - 1999 = 1072$.

If $f(1) = 2$, then consider instead the function f defined as follows: for $4^k \le n < 3 \cdot 4^k$, set $f(n) = (4^{k+1} - 1) - n$; for $3 \cdot 4^k \le n < 4^{k+1}$ set $f(n) = (7 \cdot 4^k - 1) - n$. Again, we can check that this function satisfies the conditions and that this is the only function with $f(1) = 2$. In this case because $4^5 \le 1999 < 3 \cdot 4^5$, we have $f(1999) = (4^6 - 1) - 1999 = 2096$.

Finally, suppose that $f(1) \ge 3$. First we must visit $f(1) - 1, f(1) - 2, \ldots, 1$. It follows that $f(n) = 3$ and $f(n+2) = 1$ for some n. Then $f(n+3) = 4 \cdot 1 - 1 = 3 = f(n)$, a contradiction.

Therefore the only possible values of $f(1999)$ are 1072 and 2096.

Problem 6 For each positive integer n, let $S_n = \{1, 2, \ldots, n\}$.

(a) For which values of n is it possible to express S_n as the union of two non-empty disjoint subsets so that the elements in the two subsets have equal sum?

(b) For which values of n is it possible to express S_n as the union of three non-empty disjoint subsets so that the elements in the three subsets have equal sum?

Solution. (a) Let $\sigma(T)$ denote the sum of the elements in a set T. For the condition to hold $\sigma(S_n) = \frac{n(n+1)}{2}$ must be even, and hence we must have $n = 4k - 1$ or $4k$ where $k \in \mathbb{N}$. For such n, let A consist of the second and third elements of each of the sets $\{n, n-1, n-2, n-3\}$, $\{n-4, n-5, n-6, n-7\}, \ldots, \{4, 3, 2, 1\}$ (or if $n = 4k-1$, the last set in this grouping will be $\{3, 2, 1\}$). Also set $B = S_n \setminus A$. Then $\sigma(A) = \sigma(B)$, as desired.

(b) For the condition to hold, $\sigma(S_n) = \frac{n(n+1)}{2}$ must be divisible by 3. Furthermore, the construction is impossible for $n = 3$. Thus n

must be of the form $3k + 2$ or $3k + 3$ where $k \in \mathbb{N}$. We prove all such n work by induction on n. We have $S_5 = \{5\} \cup \{1,4\} \cup \{2,3\}$, $S_6 = \{1,6\} \cup \{2,5\} \cup \{3,4\}$, $S_8 = \{8,4\} \cup \{7,5\} \cup \{1,2,3,6\}$, and $S_9 = \{9,6\} \cup \{8,7\} \cup \{1,2,3,4,5\}$. Now suppose that we can partition S_{n-6} into $A \cup B \cup C$ with $\sigma(A) = \sigma(B) = \sigma(C)$. Then

$$\sigma\bigl(A \cup \{n-5, n\}\bigr) = \sigma\bigl(B \cup \{n-4, n-1\}\bigr) = \sigma\bigl(C \cup \{n-3, n-2\}\bigr),$$

completing the inductive step and the proof of our claim.

Problem 7 Let $ABCDEF$ be a hexagon which circumscribes a circle ω. The circle ω touches sides AB, CD, EF at their respective midpoints P, Q, R. Let ω touch sides BC, DE, FA at X, Y, Z respectively. Prove that lines PY, QZ, RX are concurrent.

Solution. Let O be the center of ω. Because P is the midpoint of $\overline{AB}$, $AP = PB$. By equal tangents, $ZA = AP = PB = BX$. Thus $\angle ZOA = \angle AOP = \angle POB = \angle BOX$. It follows that $\angle ZOP = \angle POX$, and hence $\angle ZYP = \angle PYX$. Therefore line YP is the angle bisector of $\angle XYZ$. Similarly lines XR and ZQ are the angle bisectors of $\angle ZXY$ and $\angle YZX$, and therefore lines PY, QZ, RX meet at the incenter of triangle XYZ.

Problem 8 Some three nonnegative real numbers p, q, r satisfy

$$p + q + r = 1.$$

Prove that

$$7(pq + qr + rp) \leq 2 + 9pqr.$$

Solution. Given a function f of three variables, let $\sum_{\text{cyc}} f(p,q,r)$ denote the cyclic sum $f(p,q,r) + f(q,r,p) + f(r,p,q)$ — for example, $\sum_{\text{cyc}}(pqr + p) = 3pqr + p + q + r$. Because $p + q + r = 1$ the inequality is equivalent to

$$7(pq + qr + rp)(p + q + r)$$
$$\leq 2(p+q+r)^3 + 9pqr \iff 7\sum_{\text{cyc}}\left(p^2 q + pq^2 + pqr\right)$$
$$\leq 9pqr + \sum_{\text{cyc}}\left(2p^3 + 6p^2 q + 6pq^2 + 4pqr\right) \iff \sum_{\text{cyc}} p^2 q + \sum_{\text{cyc}} pq^2$$
$$\leq \sum_{\text{cyc}} 2p^3 = \sum_{\text{cyc}} \frac{2p^3 + q^3}{3} + \sum_{\text{cyc}} \frac{p^3 + 2q^3}{3}.$$

This last inequality is true by the weighted AM-GM inequality.

Problem 9 Consider all numbers of the form $3n^2 + n + 1$, where n is a positive integer.

(a) How small can the sum of the digits (in base 10) of such a number be?

(b) Can such a number have the sum of its digits (in base 10) equal to 1999?

Solution. (a) Let $f(n) = 3n^2 + n + 1$. When $n = 8$, the sum of the digits of $f(8) = 201$ is 3. Suppose that there was some m such that $f(m)$ had a smaller sum of digits. Then the last digit of $f(m)$ must be either 0, 1, or 2. Because $f(n) = n(n+3) + 1 \equiv 1 \pmod{2}$ for all n, $f(m)$ must have units digit 1.

Because $f(n)$ can never equal 1, this means we must have $3m^2 + m + 1 = 10^k + 1$ for some positive integer k, and $m(3m+1) = 10^k$. Because m and $3m+1$ are relatively prime, and $m < 3m+1$, we must either have $(m, 3m+1) = (1, 10^k)$ — which is impossible — or $(m, 3m+1) = (2^k, 5^k)$. For $k = 1$, $5^k \neq 3 \cdot 2^k + 1$; for $k > 1$, we have

$$5^k = 5^{k-2} \cdot 25 > 2^{k-2} \cdot (12 + 1) \geq 3 \cdot 2^k + 1.$$

Therefore, $f(m)$ can't equal $10^k + 1$, and 3 is indeed the minimum value for the sum of digits.

(b) Consider $n = 10^{222} - 1$. $f(n) = 3 \cdot 10^{444} - 6 \cdot 10^{222} + 3 + 10^{222}$. Thus, its decimal expansion is

$$2\underbrace{9\ldots9}_{221}5\underbrace{0\ldots0}_{221}3,$$

and the sum of the digits in $f(10^{222} - 1)$ is 1999.

23 United States of America

Problem 1 Some checkers placed on an $n \times n$ checkerboard satisfy the following conditions:

(i) every square that does not contain a checker shares a side with one that does;

(ii) given any pair of squares that contain checkers, there is a sequence of squares containing checkers, starting and ending with the given squares, such that every two consecutive squares of the sequence share a side.

Prove that at least $\frac{n^2-2}{3}$ checkers have been placed on the board.

Solution. It suffices to show that if m checkers are placed so as to satisfy condition (b), then the number of squares they either cover or are adjacent to is at most $3m + 2$. This is easily seen by induction: it is obvious for $m = 1$, and if m checkers are so placed, some checker can be removed so that the remaining checkers still satisfy (b); they cover at most $3m - 1$ squares, and the new checker allows us to count at most 3 new squares (because the square it occupies was already counted, and one of its neighbors is occupied).

Note. The exact number of checkers required is known for $m \times n$ checkerboards with m small, but only partial results are known in the general case. Contact the authors for more information.

Problem 2 Let $ABCD$ be a convex cyclic quadrilateral. Prove that

$$|AB - CD| + |AD - BC| \geq 2|AC - BD|.$$

First Solution. Let E be the intersection of $\overline{AC}$ and $\overline{BD}$. Then the triangles ABE and DCE are similar, so if we let $x = AE, y = BE, z = AB$, then there exists k such that $kx = DE, ky = CE, kz = CD$. Now

$$|AB - CD| = |k - 1|z$$

and

$$|AC - BD| = |(kx + y) - (ky + x)| = |k - 1| \cdot |x - y|.$$

Because $|x - y| \leq z$ by the triangle inequality, we conclude $|AB - CD| \geq |AC - BD|$, and similarly $|AD - BC| \geq |AC - BD|$. These two inequalities imply the desired result.

Second Solution. Let $2\alpha, 2\beta, 2\gamma, 2\delta$ be the measures of the arcs subtended by AB, BC, CD, DA, respectively, and take the radius of the circumcircle of $ABCD$ to be 1. Assume without loss of generality that $\beta \leq \delta$. Then $\alpha + \beta + \gamma + \delta = 180°$, and (by the Extended Law of Sines)

$$|AB - CD| = 2|\sin\alpha - \sin\gamma| = 4\left|\sin\frac{\alpha-\gamma}{2}\right|\left|\cos\frac{\alpha+\gamma}{2}\right|$$

and

$$|AC - BD| = 2|\sin(\alpha+\beta) - \sin(\beta+\gamma)|$$
$$= 4\left|\sin\frac{\alpha-\gamma}{2}\right|\left|\cos\left(\frac{\alpha+\gamma}{2}+\beta\right)\right|.$$

Because $0 \leq \frac{1}{2}(\alpha+\gamma) \leq \frac{1}{2}(\alpha+\gamma) + \beta \leq 90°$ (by the assumption $\beta \leq \delta$) and the cosine function is nonnegative and decreasing on $[0, 90°]$, we conclude that $|AB-CD| \geq |AC-BD|$, and similarly $|AD-BC| \geq |AC-BD|$.

Problem 3 Let $p > 2$ be a prime and let a, b, c, d be integers not divisible by p, such that

$$\left\{\frac{ra}{p}\right\} + \left\{\frac{rb}{p}\right\} + \left\{\frac{rc}{p}\right\} + \left\{\frac{rd}{p}\right\} = 2$$

for any integer r not divisible by p. Prove that at least two of the numbers $a+b, a+c, a+d, b+c, b+d, c+d$ are divisible by p. Here, for real numbers x, $\{x\} = x - \lfloor x \rfloor$ denotes the fractional part of x.

Solution. For convenience, we write $[x]$ for the unique integer in $\{0, \ldots, p-1\}$ congruent to x modulo p. In this notation, the given condition can be written

$$[ra] + [rb] + [rc] + [rd] = 2p \quad \text{for all } r \text{ not divisible by } p. \quad (1)$$

The conditions of the problem are preserved by replacing a, b, c, d with ma, mb, mc, md for any integer m relatively prime to p. If we choose m so that $ma \equiv 1 \pmod{p}$ and then replace a, b, c, d with $[ma], [mb], [mc], [md]$, respectively, we end up in the case $a = 1$ and $b, c, d \in \{1, \ldots, p-1\}$. Applying (1) with $r = 1$, we see moreover that $a + b + c + d = 2p$.

Now observe that

$$[(r+1)x] - [rx] = \begin{cases} [x] & [rx] < p - [x] \\ -p + [x] & [rx] \geq p - [x]. \end{cases}$$

Comparing (1) applied to two consecutive values of r and using the observation, we see that for each $r = 1, \ldots, p-2$, two of the quantities

$$p - a - [ra], p - b - [rb], p - c - [rc], p - d - [rd]$$

are positive and two are negative. We say that a pair (r, x) is *positive* if $[rx] < p - [x]$ and *negative* otherwise; then for each $r < p-1$, $(r, 1)$ is positive, so exactly one of $(r, b), (r, c), (r, d)$ is also positive.

Lemma. *If $r_1, r_2, x \in \{1, \ldots, p-1\}$ have the property that (r_1, x) and (r_2, x) are negative but (r, x) is positive for all $r_1 < r < r_2$, then*

$$r_2 - r_1 = \left\lfloor \frac{p}{x} \right\rfloor \quad \text{or} \quad r_2 - r_1 = \left\lfloor \frac{p}{x} \right\rfloor + 1.$$

Proof. Note that (r', x) is negative if and only if

$$\{r'x + 1, r'x + 2, \ldots, (r'+1)x\}$$

contains a multiple of p. In particular, exactly one multiple of p lies in $\{r_1 x, r_1 x + 1, \ldots, r_2 x\}$. Because $[r_1 x]$ and $[r_2 x]$ are distinct elements of $\{p - [x], \ldots, p-1\}$, we have

$$p - x + 1 < r_2 x - r_1 x < p + x - 1,$$

from which the lemma follows. ∎

$[rx]$	9	10	0	1	2	3	4	5	6	7	8	9	10	0
is (r,x) + or $-$?	$-$			$+$		$+$		$+$					$-$	
r	**3**			4		5		6					**7**	

(The above diagram illustrates the meanings of *positive* and *negative* in the case $x = 3$ and $p = 11$. Note that the difference between 7 and 3 here is $\lfloor \frac{p}{x} \rfloor + 1$. The next r such that (r, x) is negative is $r = 10$; $10 - 7 = \lfloor \frac{p}{x} \rfloor$.)

Recall that exactly one of $(1, b)$, $(1, c)$, $(1, d)$ is positive; we may as well assume $(1, b)$ is positive, which is to say $b < \frac{p}{2}$ and $c, d > \frac{p}{2}$. Put $s_1 = \lfloor \frac{p}{b} \rfloor$, so that s_1 is the smallest positive integer such that (s_1, b) is negative. Then exactly one of (s_1, c) and (s_1, d) is positive, say the former. Because s_1 is also the smallest positive integer such that (s_1, c) is positive, or equivalently such that $(s_1, p - c)$ is negative, we have $s_1 = \lfloor \frac{p}{p-c} \rfloor$. The lemma states that consecutive values of r for which (r, b) is negative differ by either s_1 or $s_1 + 1$. It also states (when applied with $x = p - c$) that consecutive values of r for which (r, c) is positive differ by either s_1 or $s_1 + 1$. From these observations we will show that (r, d) is always negative.

r	1		s_1	s_1+1		s'	$s'+1$		s	$s+1\stackrel{?}{=}t$
(r,b)	$+$		$-$	$+$		$-$	$+$		$-$	$-?$
(r,c)	$-$	$\ldots$	$+$	$-$	$\ldots$	$+$	$-$	$\ldots$	$-$	$+?$
(r,d)	$-$		$-$	$-$		$-$	$-$		$+$	$-?$

Indeed, if this were not the case, there would exist a smallest positive integer $s > s_1$ such that (s,d) is positive; then (s,b) and (s,c) are both negative. If s' is the last integer before s such that (s',b) is negative (possibly equal to s_1), then (s',d) is negative as well (by the minimal definition of s). Also,

$$s - s' = s_1 \quad \text{or} \quad s - s' = s_1 + 1.$$

Likewise, if t were the next integer after s' such that (t,c) were positive, then

$$t - s' = s_1 \quad \text{or} \quad t - s' = s_1 + 1.$$

From these we deduce that $|t - s| \leq 1$. However, we can't have $t \neq s$ because then both (s,b) and (t,b) would be negative — and any two values of r for which (r,b) is negative differ by at least $s_1 \geq 2$, a contradiction. (The above diagram shows the hypothetical case when $t = s+1$.) Nor can we have $t = s$ because we already assumed that (s,c) is negative. Therefore we *can't* have $|t - s| \leq 1$, contradicting our findings and thus proving that (r,d) is indeed always negative.

Now if $d \neq p-1$, then the unique $s \in \{1,\ldots,p-1\}$ such that $[ds] = 1$ is not equal to $p-1$; and (s,d) is positive, a contradiction. Thus $d = p-1$ and $a+d$ and $b+c$ are divisible by p, as desired.

Problem 4 Let $a_1, a_2, \ldots, a_n$ ($n > 3$) be real numbers such that

$$a_1 + a_2 + \cdots + a_n \geq n \quad \text{and} \quad a_1^2 + a_2^2 + \cdots + a_n^2 \geq n^2.$$

Prove that $\max(a_1, a_2, \ldots, a_n) \geq 2$.

Solution. Let $b_i = 2 - a_i$, and let $S = \sum b_i$ and $T = \sum b_i^2$. Then the given conditions are that

$$(2 - b_1) + \cdots + (2 - b_n) \geq n$$

and

$$(4 - 4b_1 + b_1^2) + \cdots + (4 - 4b_n + b_n^2) \geq n^2,$$

which is to say $S \leq n$ and $T \geq n^2 - 4n + 4S$.

From these inequalities, we obtain

$$T \geq n^2 - 4n + 4S \geq (n-4)S + 4S = nS.$$

On the other hand, if $b_i > 0$ for $i = 1, \ldots, n$, then certainly $b_i < \sum b_i = S \leq n$, and so

$$T = b_1^2 + \cdots + b_n^2 < nb_1 + \cdots + nb_n = nS.$$

Thus we cannot have $b_i > 0$ for $i = 1, \ldots, n$, so $b_i \leq 0$ for some i; then $a_i \geq 2$ for that i, proving the claim.

Note. The statement is false when $n \leq 3$. The example $a_1 = a_2 = \cdots = a_{n-1} = 2$, $a_n = 2 - n$ shows that the bound cannot be improved. Also, an alternate approach is to show that if $a_i \leq 2$ and $\sum a_i \geq n$, then $\sum a_i^2 \leq n^2$ (with the equality case just mentioned), by noticing that replacing a pair a_i, a_j with $2, a_i + a_j - 2$ increases the sum of squares.

Problem 5 The Y2K Game is played on a 1×2000 grid as follows. Two players in turn write either an S or an O in an empty square. The first player who produces three consecutive boxes that spell SOS wins. If all boxes are filled without producing SOS then the game is a draw. Prove that the second player has a winning strategy.

Solution. Call a partially filled board *stable* if there is no SOS and no single move can produce an SOS; otherwise call it *unstable*. For a stable board call an empty square *bad* if either an S or an O played in that square produces an unstable board. Thus a player will lose if the only empty squares available to him are bad, but otherwise he can at least be guaranteed another turn with a correct play.

Claim: A square is bad if and only if it is in a block of 4 consecutive squares of the form S - - S.

Proof. If a square is bad, then an O played there must give an unstable board. Thus the bad square must have an S on one side and an empty square on the other side. An S played there must also give an unstable board, so there must be another S on the other side of the empty square. ∎

From the claim it follows that there are always an even number of bad squares. Thus the second player has the following winning strategy:

(a) If the board is unstable at any time, play the winning move; otherwise continue as below.

(b) On the first move, play an S at least four squares away from either end and at least seven squares from the first player's first move. (The board is long enough that this is possible.)

(c) On the second move, play an S three squares away from the second player's first move, so that the squares in between are empty and so that the board remains stable. (Regardless of the first player's second move, this must be possible on at least one side.) This produces two bad squares; whoever plays in one of them first will lose. Thus the game will not be a draw.

(d) On any subsequent move, play in a square which is not bad — keeping the board stable, of course. Such a square will always exist because if the board is stable, there will be an odd number of empty squares and an even number of bad squares.

Because there exist bad squares after the second player's second move, the game cannot end in a draw; and because the second player can always leave the board stable, the first player cannot win. Therefore eventually the second player will win.

Problem 6 Let $ABCD$ be an isosceles trapezoid with $AB \parallel CD$. The inscribed circle ω of triangle BCD meets CD at E. Let F be a point on the (internal) angle bisector of $\angle DAC$ such that $EF \perp CD$. Let the circumscribed circle of triangle ACF meet line CD at C and G. Prove that the triangle AFG is isosceles.

Solution. We will show that $FA = FG$. Let H be the center of the escribed circle of triangle ACD opposite vertex A. Then H lies on the angle bisector AF. Let K be the point where this escribed circle touches CD. By a standard computation using equal tangents, we see that $CK = \frac{1}{2}(AD + CD - AC)$. By a similar computation in triangle BCD, we see that $CE = \frac{1}{2}(BC + CD - BD) = CK$. Therefore $E = K$ and $F = H$.

Because F is now known to be an excenter, we have that FC is the external angle bisector of $\angle DCA = \angle GCA$. Therefore

$$\angle GAF = \angle GCF = 90° - \frac{1}{2}\angle GCA = 90° - \frac{1}{2}\angle GFA.$$

We conclude that the triangle GAF is isosceles with $FA = FG$, as desired.

24 Vietnam

Problem 1 Solve the system of equations

$$(1 + 4^{2x-y}) \cdot 5^{1-2x+y} = 1 + 2^{2x-y+1}$$
$$y^3 + 4x + 1 + \ln(y^2 + 2x) = 0.$$

Solution. The only solution is $(x, y) = (0, -1)$, which yields $5 = 5$ and $0 = 0$ in the above equations.

We first solve the first equation for $t = 2x - y$. Multiplying the equation by 5^{t-1} yields

$$(1 - 5^{t-1}) + 4(4^{t-1} - 10^{t-1}) = 0.$$

If $t > 1$, then both $1 - 5^{t-1}$ and $4^{t-1} - 10^{t-1}$ are negative; if $t < 1$, then both of these terms are positive. Therefore, $2x - y = 1$.

Substituting $2x = y + 1$ into the second equation, we find that

$$y^3 + 2y + 3 + \ln(y^2 + y + 1) = 0.$$

This has solution $y = -1$. To prove that this is the only solution, it suffices to show that $f(y) = y^3 + 2y + 3 + \ln(y^2 + y + 1)$ is always increasing. The derivative of f is

$$f'(y) = 3y^2 + 2 + \frac{2y+1}{y^2+y+1}.$$

Observe that $2(y^2 + y + 1) + (2y + 1) = 2(y + 1)^2 + 1 > 0$ and that $y^2 + y + 1 > (y + \frac{1}{2})^2 \geq 0$. Thus

$$f'(y) = 3y^2 + \frac{2(y^2+y+1) + (2y+1)}{y^2+y+1} > 3y^2 \geq 0$$

for all y, as desired.

Problem 2 Let A', B', C' be the respective midpoints of the arcs BC, CA, AB, not containing points A, B, C, respectively, of the circumcircle of the triangle ABC. The sides BC, CA, AB meet the pairs of segments

$$\{C'A', A'B'\}, \{A'B', B'C'\}, \{B'C', C'A'\}$$

at the pairs of points

$$\{M, N\}, \{P, Q\}, \{R, S\},$$

respectively. Prove that $MN = PQ = RS$ if and only if the triangle ABC is equilateral.

Solution. If ABC is equilateral then $MN = PQ = RS$ by symmetry.

Now suppose that $MN = PQ = RS$. Observe that $\angle NMA' = \angle BMS = \frac{1}{2}(\widehat{BC'} + \widehat{CA'}) = \frac{1}{2}(\angle C + \angle A)$ and similarly $\angle C'SR = \angle MSB = \frac{1}{2}(\angle A + \angle C)$. Furthermore,

$$\angle A'B'C' = \angle A'B'B + \angle BB'C' = \frac{1}{2}(\angle A + \angle C)$$

as well.

Thus $MB = SB$, and also $\triangle C'RS \sim \triangle C'A'B' \sim \triangle NA'M$. Next, by the law of sines in triangles $C'SB$ and $A'MB$ we have

$$C'S = SB \cdot \frac{\sin \angle C'BS}{\sin \angle SC'B} = SB \cdot \frac{\sin \frac{\angle C}{2}}{\sin \frac{\angle A}{2}}$$

and

$$MA' = MB \cdot \frac{\sin \angle A'BM}{\sin \angle MA'B} = MB \cdot \frac{\sin \frac{\angle A}{2}}{\sin \frac{\angle C}{2}},$$

giving

$$\frac{C'S}{MA'} = \left(\frac{\sin \frac{\angle C}{2}}{\sin \frac{\angle A}{2}}\right)^2.$$

Next, because $\triangle C'RS \sim \triangle C'A'B'$ we have $RS = A'B' \cdot \frac{C'S}{C'B'}$. Because $\triangle NA'M \sim \triangle C'A'B'$ we have $MN = B'C' \cdot \frac{MA'}{B'A'}$. Therefore, $RS = MN$ implies that

$$A'B' \cdot \frac{C'S}{C'B'} = B'C' \cdot \frac{MA'}{B'A'} \implies \frac{C'S}{MA'} = \left(\frac{B'C'}{A'B'}\right)^2$$

$$= \left(\frac{\sin \frac{1}{2}(\angle B + \angle C)}{\sin \frac{1}{2}(\angle B + \angle A)}\right)^2 = \left(\frac{\cos \frac{\angle A}{2}}{\cos \frac{\angle C}{2}}\right)^2 \implies \left(\frac{\sin \frac{\angle C}{2}}{\sin \frac{\angle A}{2}}\right)^2$$

$$= \left(\frac{\cos \frac{\angle A}{2}}{\cos \frac{\angle C}{2}}\right)^2 \implies \left(\sin \frac{\angle C}{2} \cos \frac{\angle C}{2}\right)^2$$

$$= \left(\sin \frac{\angle A}{2} \cos \frac{\angle A}{2}\right)^2 \implies \frac{1}{4}\sin^2 \angle C$$

$$= \frac{1}{4}\sin^2 \angle A \implies \sin \angle C = \sin \angle A.$$

This is only possible when $\angle A = \angle C$ because $\angle A + \angle C < 180°$. Similarly, $\angle A = \angle B$, and therefore triangle ABC is equilateral.

Problem 3 For $n = 0, 1, 2, \ldots$, let $\{x_n\}$ and $\{y_n\}$ be two sequences defined recursively as follows:

$$x_0 = 1, \ x_1 = 4, \ x_{n+2} = 3x_{n+1} - x_n;$$

$$y_0 = 1, \ y_1 = 2, \ y_{n+2} = 3y_{n+1} - y_n.$$

(a) Prove that $x_n^2 - 5y_n^2 + 4 = 0$ for all non-negative integers n.

(b) Suppose that a, b are two positive integers such that $a^2 - 5b^2 + 4 = 0$. Prove that there exists a nonnegative integer k such that $x_k = a$ and $y_k = b$.

Solution. We first prove by induction on n that $(x_{n+1}, y_{n+1}) = \left(\frac{3x_n+5y_n}{2} \text{ and } \frac{x_n+3y_n}{2}\right)$ for $n \geq 0$. For $n = 0$, we have $(4, 2) = \left(\frac{3+5}{2}, \frac{1+3}{2}\right)$, and for $n = 1$, we have $(11, 5) = \left(\frac{12+10}{2}, \frac{4+6}{2}\right)$.

Assume that our formula for (x_{n+1}, y_{n+1}) holds for $n = k$ and $k+1$, where $k \geq 0$. Substituting the expressions for $x_{k+2}, x_{k+1}, y_{k+2}, y_{k+1}$ into $(x_{k+3}, y_{k+3}) = (3x_{k+2} - x_{k+1}, 3y_{k+2} - y_{k+1})$, we find that (x_{k+3}, y_{k+3}) equals

$$\left(\tfrac{3}{2}(3x_{k+1} - x_k) + \tfrac{5}{2}(3y_{k+1} - y_k), \tfrac{1}{2}(3x_{k+1} - x_k) + \tfrac{3}{2}(3y_{k+1} - y_k)\right)$$
$$= \left(\tfrac{1}{2}(3x_{k+2} + 5y_{k+2}), \tfrac{1}{2}(x_{k+2} + 3y_{k+2})\right).$$

This completes the induction step and the proof of our claim.

(a) We prove that $x_n^2 - 5y_n^2 + 4 = 0$ by induction on n. For $n = 0$ we have $1 - 5 + 4 = 0$. Now assume the result is true for n. We prove (with the help of our above observation) that it is true for $n+1$:

$$x_{n+1}^2 - 5y_{n+1}^2 = \left(\frac{3x_n + 5y_n}{2}\right)^2 - 5\left(\frac{x_n + 3y_n}{2}\right)^2$$
$$= \frac{4x_n^2 - 20y_n^2}{4} = x_n^2 - 5y_n^2 = -4,$$

as desired.

(b) Suppose, by way of contradiction, that $a^2 - 5b^2 + 4 = 0$ for integers $a, b > 0$, and that there did *not* exist k such that $(x_k, y_k) = (a, b)$. Choose such a, b so that $a + b$ is minimal.

Let $(a', b') = \left(\frac{3a-5b}{2}, \frac{3b-a}{2}\right)$. We argue that a' and b' are positive integers. This is true if a and b are the same parity, $a < 3b$, and $3a > 5b$. Note that $0 \equiv a^2 - 5b^2 + 4 \equiv a - b \pmod{2}$. Next, $a^2 = 5b^2 - 4 < 9b^2 \Rightarrow a < 3b$. In addition, there are no counterexamples with $a = 1$ or 2. Thus $a^2 > 5$ and $0 = 5a^2 - 25b^2 + 20 < 5a^2 - 25b^2 + 4a^2 \Rightarrow 3a > 5b$.

Using the condition $a^2 - 5b^2 = -4$, some quick algebra shows that $a'^2 - 5b'^2 = -4$ as well. However, $a' + b' = \frac{3a-5b}{2} + \frac{3b-a}{2} = a - b < a + b$. It follows from the minimal definition of (a, b) that there must exist some (a_k, b_k) equal to (a', b'). It is then easy to verify that

$$(a, b) = \left(\frac{3a' + 5b'}{2}, \frac{a' + 3b'}{2}\right) = (a_{k+1}, b_{k+1}),$$

a contradiction. This completes the proof.

Problem 4 Let a, b, c be real numbers such that $abc + a + c = b$. Determine the greatest possible value of the expression

$$P = \frac{2}{a^2 + 1} - \frac{2}{b^2 + 1} + \frac{3}{c^2 + 1}.$$

Solution. The condition is equivalent to $b = \frac{a+c}{1-ac}$, which suggests making the substitutions $A = \operatorname{Tan}^{-1} a$ and $C = \operatorname{Tan}^{-1} c$. Then $b = \tan(A + C)$ and

$$P = \frac{2}{\tan^2 A + 1} - \frac{2}{\tan^2(A+C) + 1} + \frac{3}{\tan^2 C + 1}$$
$$= 2\cos^2 A - 2\cos^2(A + C) + 3\cos^2 C$$
$$= (2\cos^2 A - 1) - (2\cos^2(A + C) - 1) + 3\cos^2 C$$
$$= \cos(2A) - \cos(2A + 2C) + 3\cos^2 C$$
$$= 2\sin(2A + C)\sin C + 3\cos^2 C.$$

Letting $u = |\sin C|$, this expression is at most

$$2u + 3(1 - u^2) = -3u^2 + 2u + 3$$
$$= -3\left(u - \frac{1}{3}\right)^2 + \frac{10}{3} \leq \frac{10}{3}.$$

Equality can be achieved when $\sin(2A + C) = 1$ and $\sin C = \frac{1}{3}$, that is, when $(a, b, c) = (\frac{\sqrt{2}}{2}, \sqrt{2}, \frac{\sqrt{2}}{4})$ or $(-\frac{\sqrt{2}}{2}, -\sqrt{2}, -\frac{\sqrt{2}}{4})$. Thus, the maximum value of P is $\frac{10}{3}$.

Problem 5 In the three-dimensional space let Ox, Oy, Oz, Ot be four nonplanar distinct rays such that the angles between any two of them have the same measure.

(a) Determine this common measure.

(b) Let Or be another ray different from the above four rays. Let $\alpha, \beta, \gamma, \delta$ be the angles formed by Or with Ox, Oy, Oz, Ot, respectively. Put

$$p = \cos\alpha + \cos\beta + \cos\gamma + \cos\delta,$$
$$q = \cos^2\alpha + \cos^2\beta + \cos^2\gamma + \cos^2\delta.$$

Prove that p and q remain constant as Or rotates about the point O.

Solution. Put O at the origin, and let the four rays intersect the unit sphere at A, B, C, D. Then $ABCD$ is a regular tetrahedron, and (letting X also represent the vector $\overrightarrow{OX}$) we have $A + B + C + D = 0$.

(a) Let the common angle be ϕ. Then

$$0 = A \cdot (A + B + C + D) = A \cdot A + A \cdot (B + C + D) = 1 + 3\cos\phi,$$

so $\phi = \cos^{-1}\left(-\frac{1}{3}\right)$.

(b) Without loss of generality let Or intersect the unit sphere at $U = (1, 0, 0)$. Also write $A = (x_1, y_1, z_1)$, and so on. Then

$$p = U \cdot A + U \cdot B + U \cdot C + U \cdot D$$
$$= U \cdot (A + B + C + D)$$
$$= U \cdot \vec{0} = 0,$$

a constant. Also, $(x_1, x_2, x_3, x_4) = (\cos\alpha, \cos\beta, \cos\gamma, \cos\delta)$ and $q = \sum x_i^2$. The following lemma then implies that q will always equal $\frac{4}{3}$:

Lemma. *Suppose we are given a regular tetrahedron T inscribed in the unit sphere and with vertices (x_i, y_i, z_i) for $1 \le i \le 4$. Then we have $\sum x_i^2 = \sum y_i^2 = \sum z_i^2 = \frac{4}{3}$ and $\sum x_i y_i = \sum y_i z_i = \sum z_i x_i = 0$.*

Proof. This is easily verified when the vertices are at

$$A_0 = (0, 0, 1), \quad B_0 = \left(\tfrac{2\sqrt{2}}{3}, 0, -\tfrac{1}{3}\right),$$
$$C_0 = \left(-\tfrac{\sqrt{2}}{3}, \tfrac{\sqrt{6}}{3}, -\tfrac{1}{3}\right), \quad D_0 = \left(-\tfrac{\sqrt{2}}{3}, -\tfrac{\sqrt{6}}{3}, -\tfrac{1}{3}\right).$$

Now assume these equations are true for a tetrahedron $ABCD$, and rotate it about the z-axis through an angle θ. Then each (x_i, y_i, z_i) becomes $(x_i', y_i', z_i') = (x_i \cos\theta - y_i \sin\theta, x_i \sin\theta + y_i \cos\theta, z_i)$, and

$$\sum x_i'^2 = \cos^2\theta \sum x_i^2 - 2\sin\theta\cos\theta \sum x_i y_i + \sin^2\theta \sum y_i^2 = \frac{4}{3}$$

$$\sum y_i'^2 = \sin^2\theta \sum x_i^2 + 2\sin\theta\cos\theta \sum x_i y_i + \cos^2\theta \sum y_i^2 = \frac{4}{3}$$

$$\sum z_i'^2 = \sum z_i^2 = \frac{4}{3}$$

$$\sum x_i' y_i' = \sin\theta \cos\theta \sum (x_i^2 - y_i^2) + (\cos^2\theta - \sin^2\theta) \sum x_i y_i = 0$$

$$\sum y_i' z_i' = \sin\theta \sum x_i z_i + \cos\theta \sum y_i z_i = 0$$

$$\sum z_i' x_i' = \cos\theta \sum z_i x_i - \sin\theta \sum z_i y_i = 0.$$

Similarly, the equations remain true after rotating $ABCD$ about the y- and z-axes.

Now, first rotate our given tetrahedron T about the z-axis until one vertex is in the yz-plane. Next rotate it about the x-axis until this vertex is at $(0,0,1)$. Finally, rotate it about the z-axis again until the tetrahedron corresponds with the initial tetrahedron $A_0 B_0 C_0 D_0$ described above. Because we *know* the above equations are true for $A_0 B_0 C_0 D_0$, if we reverse the rotations to return to T the equations will remain true, as claimed. ∎

Problem 6 Let $\mathcal{S} = \{0, 1, 2, \ldots, 1999\}$ and $\mathcal{T} = \{0, 1, 2, \ldots\}$. Find all functions $f : \mathcal{T} \to \mathcal{S}$ such that

(i) $f(s) = s$ for all $s \in \mathcal{S}$.

(ii) $f(m+n) = f\bigl(f(m) + f(n)\bigr)$ for all $m, n \in \mathcal{T}$.

Solution. Suppose that $f(2000) = 2000 - t$, where $1 \leq t \leq 2000$. We prove by induction on n that $f(n) = f(n-t)$ for all $n \geq 2000$. By assumption it is true for $n = 2000$. Assuming it is true for n, we have

$$f(n+1) = f\bigl(f(n) + f(1)\bigr) = f\bigl(f(n-t) + f(1)\bigr) = f(n-t+1),$$

completing the inductive step. Therefore, the function is completely determined by the value of $f(2000)$, and it follows that there are at most 2000 such functions.

Conversely, given any $c \in S$, we show that there is a valid function f with $f(2000) = c$. Set $t = 2000 - c$, and let f be the function such that $f(s) = s$ for all $s \in \mathcal{S}$ while $f(n) = f(n-t)$ for all $n \geq 2000$. We prove by induction on $m + n$ that condition (ii) holds. If $m + n \leq 2000$, then $m, n \in \mathcal{S}$ and the claim is obvious. Otherwise, $m + n > 2000$. Again, if $m, n \in \mathcal{S}$, then the claim is obvious. Otherwise, assume without loss of generality that $n \geq 2000$. Then

$$f(m+n) = f(m+n-t) = f\bigl(f(m) + f(n-t)\bigr) = f\bigl(f(m) + f(n)\bigr),$$

where the first and third equalities come from our periodic definition of f, and the second equality comes from the induction hypothesis. Therefore there are exactly 2000 valid functions f.

Problem 7 For $n = 1, 2, \ldots$, let $\{u_n\}$ be a sequence defined by

$$u_1 = 1, \quad u_2 = 2, \quad u_{n+2} = 3u_{n+1} - u_n.$$

Prove that

$$u_{n+2} + u_n \geq 2 + \frac{u_{n+1}^2}{u_n}$$

for all n.

Solution. We first prove by induction that $u_n u_{n+2} = u_{n+1}^2 + 1$ for $n \geq 1$. Because $u_3 = 5$, for $n = 1$ we have $1 \cdot 5 = 2^2 + 1$, as desired.

Now assume that our formula is true for $n = k \geq 1$. To prove that it is true for $n = k+1$, it suffices to prove that $u_{k+1} u_{k+3} - u_k u_{k+2} = u_{k+2}^2 - u_{k+1}^2$. This equation is true because it is equivalent to

$$u_{k+1}(u_{k+1} + u_{k+3}) = u_{k+2}(u_k + u_{k+2}),$$

or $u_{k+1} \cdot 3u_{k+2} = u_{k+2} \cdot 3u_{k+1}$. This completes the induction and the proof of our formula.

If u_n and u_{n+1} are positive for $n \geq 1$, then $u_{n+2} = \frac{u_{n+1}^2 + 1}{u_n}$ is positive as well. Because u_1 and u_2 are positive, u_n is positive for all $n \geq 1$. Therefore, for $n \geq 1$, $u_n + \frac{1}{u_n} \geq 2$ by the AM-GM inequality. Hence

$$u_{n+2} + u_n = \frac{u_{n+1}^2 + 1}{u_n} + u_n = \frac{u_{n+1}^2}{u_n} + \left(u_n + \frac{1}{u_n}\right) \geq \frac{u_{n+1}^2}{u_n} + 2,$$

as desired.

Problem 8 Let ABC be a triangle inscribed in circle ω. Construct all points P in the plane (ABC) and not lying on ω, with the property that the lines PA, PB, PC meet ω again at points A', B', C' such that $A'B' \perp A'C'$ and $A'B' = A'C'$.

Solution. All angles are directed modulo $180°$. We solve a more general problem: suppose we have some fixed triangle DEF. We find all points P such that when $A' = PA \cap \omega$, $B' = PB \cap \omega$, $C' = PC \cap \omega$, then triangles $A'B'C'$ and DEF are similar with the same orientations. In other words, we want $\angle B'C'A' = \angle EFD$ and $\angle C'A'B' = \angle FDE$.

Given X, Y on ω, define $\angle \widehat{XY} = \angle XZY$ for any other point Z on ω. Given a point P, we have $\angle BPA = \angle \widehat{BA} + \angle \widehat{B'A'} = \angle BCA + \angle B'C'A'$

and $\angle CPB = \angle\widehat{CB} + \angle\widehat{C'B'} = \angle CAB + \angle C'A'B'$. Thus $\angle B'C'A' = \angle EFD$ if and only if $\angle BPA = \angle BCA + \angle EFD$, while $\angle C'A'B' = \angle FDE$ if and only if $\angle CPB = \angle CAB + \angle FDE$. Therefore our desired point P is the intersection point, different than B, of the two circles $\{P' \mid \angle BP'A = \angle BCA + \angle EFD\}$ and $\{P' \mid \angle CP'B = \angle CAB + \angle FDE\}$.

We now return to the original problem. We wish to find P such that triangle $A'B'C'$ is a 45°-45°-90° triangle with $\angle C'A'B' = 90°$. Because our angles are directed, there are two possible orientations for such a triangle: either $\angle A'B'C' = 45°$, or $\angle A'B'C' = -45°$. Applying the above construction twice with triangle DEF defined appropriately, we find the two desired possible locations of P.

Problem 9 Consider real numbers a, b such that $a \neq 0$, $a \neq b$, and all roots of the equation

$$ax^3 - x^2 + bx - 1 = 0$$

are real and positive. Determine the smallest possible value of the expression

$$P = \frac{5a^2 - 3ab + 2}{a^2(b-a)}.$$

Solution. When the roots of the equation are all $\sqrt{3}$, we have $a = \frac{1}{3\sqrt{3}}$, $b = \sqrt{3}$, and $P = 12\sqrt{3}$. We prove that $12\sqrt{3}$ is minimal.

Let the zeroes of $ax^3 - x^2 + bx - 1$ be $p = \tan A$, $q = \tan B$, and $r = \tan C$ with $0° < A, B, C < 90°$. Then

$$ax^3 - x^2 + bx - 1 = a(x-p)(x-q)(x-r)$$
$$= ax^3 - a(p+q+r)x^2 + a(pq+qr+rp)x - a(pqr).$$

Thus $a = \frac{1}{p+q+r} = \frac{1}{pqr} > 0$ and $p+q+r = pqr$. Then

$$r = \frac{p+q}{pq-1} = -\tan(A+B) = \tan(180° - A - B),$$

implying that $A + B + C = 180°$. Because $\tan x$ is convex along the interval $0° < x < 90°$,

$$\frac{1}{a} = \tan A + \tan B + \tan C \geq 3\tan 60° = 3\sqrt{3}.$$

Hence $a \leq \frac{1}{3\sqrt{3}}$.

Also notice that

$$\frac{b}{a} = pq + qr + rp \geq 3\sqrt[3]{p^2q^2r^2} = 3\sqrt[3]{\frac{1}{a^2}} \geq 9 > 1,$$

so $b > a$. Thus the denominator of P is always positive and is an increasing function of b, while the numerator of P is a decreasing function of b. Therefore, for a constant a, P is a decreasing function of b.

Furthermore,

$$(p-q)^2 + (q-r)^2 + (r-p)^2 \geq 0 \Longrightarrow (p+q+r)^2$$
$$\geq 3(pq+qr+rp) \Longrightarrow \frac{1}{a^2}$$
$$\geq \frac{3b}{a} \Longrightarrow b \leq \frac{1}{3a},$$

and

$$P \geq \frac{5a^2 - 3a(\frac{1}{3a}) + 2}{a^2(\frac{1}{3a} - a)} = \frac{5a^2 + 1}{\frac{a}{3} - a^3}.$$

Thus for $0 < a \leq \frac{1}{3\sqrt{3}}$, it suffices to show that

$$5a^2 + 1 \geq 12\sqrt{3}\left(\frac{a}{3} - a^3\right)$$
$$= 4\sqrt{3}a - 12\sqrt{3}a^3 \iff 12\sqrt{3}a^3 + 5a^2 - 4\sqrt{3}a + 1$$
$$\geq 0 \iff 3\left(a - \frac{1}{3\sqrt{3}}\right)(4\sqrt{3}a^2 + 3a - \sqrt{3})$$
$$\geq 0 \iff 4\sqrt{3}a^2 + 3a - \sqrt{3} \leq 0.$$

The last quadratic has one positive root,

$$\frac{-3 + \sqrt{57}}{8\sqrt{3}} \geq \frac{-3 + 7}{8\sqrt{3}} = \frac{1}{2\sqrt{3}} > \frac{1}{3\sqrt{3}},$$

so it is indeed negative when $0 < a \leq \frac{1}{3\sqrt{3}}$. This completes the proof.

Problem 10 Let $f(x)$ be a continuous function defined on $[0, 1]$ such that

(i) $f(0) = f(1) = 0$;

(ii) $2f(x) + f(y) = 3f\left(\frac{2x+y}{3}\right)$ for all $x, y \in [0, 1]$.

Prove that $f(x) = 0$ for all $x \in [0, 1]$.

Solution. We prove by induction on n that $f\left(\frac{m}{3^n}\right) = 0$ for all integers $n \geq 0$ and all integers $0 \leq m \leq 3^n$. The given conditions show that this claim is true for $n = 0$. Now assuming it is true for $n = k - 1 \geq 0$, we prove it is true for $n = k$.

If $m \equiv 0 \pmod{3}$ then

$$f\left(\frac{m}{3^k}\right) = f\left(\frac{\frac{m}{3}}{3^{k-1}}\right) = 0$$

by the induction hypothesis.

If $m \equiv 1 \pmod{3}$, then $1 \leq m \leq 3^k - 2$ and

$$3f\left(\frac{m}{3^k}\right) = 2f\left(\frac{\frac{m-1}{3}}{3^{k-1}}\right) + f\left(\frac{\frac{m+2}{3}}{3^{k-1}}\right) = 0 + 0 = 0.$$

Thus $f\left(\frac{m}{3^k}\right) = 0$.

Finally, if $m \equiv 2 \pmod{3}$, then $2 \leq m \leq 3^k - 1$ and

$$3f\left(\frac{m}{3^k}\right) = 2f\left(\frac{\frac{m+1}{3}}{3^{k-1}}\right) + f\left(\frac{\frac{m-2}{3}}{3^{k-1}}\right) = 0 + 0 = 0.$$

Hence $f\left(\frac{m}{3^k}\right) = 0$, finishing our induction.

Given any $x \in [0,1]$, we can form a sequence of numbers of the form $\frac{m}{3^k}$ whose limit is x. Because f is continuous, it follows that $f(x) = 0$ for all $x \in [0,1]$, as desired.

Problem 11 The base side and the altitude of a right regular hexagonal prism $ABCDEF - A'B'C'D'E'F'$ are equal to a and h respectively.

(a) Prove that six planes

$$(AB'F'), (CD'B'), (EF'D'), (D'EC), (F'AE), (B'CA)$$

touch the same sphere.

(b) Determine the center and the radius of the sphere.

Solution. (a) Let O be the center of the prism. $(AB'F')$ is tangent to a unique sphere centered at O. The other given planes are simply reflections and rotations of $(AB'F')$ with respect to O. Because the sphere remains fixed under these transformations, it follows that all six planes are tangent to this same sphere.

(b) From part (a), the center of the sphere is the center O of the prism. Let P be the midpoint of $\overline{AE}$, and let P' be the midpoint of $\overline{A'E'}$. Also let Q be the midpoint of $\overline{PF'}$, and let R be the projection of O onto line PF'. Note that P, P', Q, R, O, F' all lie in one plane.

It is straightforward to calculate that $F'P' = \frac{a}{2}$ and $QO = \frac{3a}{4}$. Also, $\angle RQO = \angle PF'P'$ because $QO \parallel F'P'$. Combined with $\angle ORQ = \angle PP'F' = 90°$, this equality implies that $\triangle ORQ \sim \triangle PP'F'$. Hence the radius of the sphere is

$$OR = PP' \cdot \frac{OQ}{PF'} = h \cdot \frac{\frac{3a}{4}}{\sqrt{\left(\frac{a}{2}\right)^2 + h^2}} = \frac{3ah}{2\sqrt{a^2 + 4h^2}}.$$

Problem 12 For $n = 1, 2, \ldots$, two sequences $\{x_n\}$ and $\{y_n\}$ are defined recursively by

$$x_1 = 1, \ y_1 = 2, \ x_{n+1} = 22y_n - 15x_n, \ y_{n+1} = 17y_n - 12x_n.$$

(a) Prove that x_n and y_n are not equal to zero for all $n = 1, 2, \ldots$.

(b) Prove that each sequence contains infinitely many positive terms and infinitely many negative terms.

(c) For $n = 1999^{1945}$, determine whether x_n and y_n are divisible by 7.

Solution. (a) From the recursion for x_{n+1}, we find that $y_n = \frac{1}{22}(15x_n + x_{n+1})$ and $y_{n+1} = \frac{1}{22}(15x_{n+1} + x_{n+2})$ for $n \geq 1$. Substituting these expressions into the recursion for y_{n+1} yields

$$x_{n+2} = 2x_{n+1} - 9x_n.$$

Similar work shows that

$$y_{n+2} = 2y_{n+1} - 9y_n.$$

If x_n is odd, then x_{n+2} is odd as well. Because $x_1 = 1$ and $x_2 = 29$ are odd, all the x_n are odd and hence nonzero. Similarly, if y_n and y_{n+1} are both congruent to 2 modulo 4, then so is y_{n+2}. Because $y_1 = 2$ and $y_2 = 22$ are congruent to 2 modulo 4, then all the y_n are congruent to 2 modulo 4 and hence nonzero.

(b) Note that $x_{n+3} = 2(2x_{n+1} - 9x_n) - 9x_{n+1} = -5x_{n+1} - 18x_n$. Thus if x_n, x_{n+1} are positive (or negative) then x_{n+3} is negative (or positive). Hence, among every four consecutive terms x_n, there is some positive term and some negative term. Therefore there are infinitely many positive terms and infinitely many negative terms in this sequence. A similar proof holds for the y_n.

(c) All congruences are modulo 7 unless stated otherwise. Note that if $x_n \equiv x_{n+1}$ and $x_n \not\equiv 0$, then $(x_{n+2}, x_{n+3}, x_{n+4}, x_{n+5}) \equiv (0, 5x_n, 3x_n, 3x_n)$, where $5x_n \not\equiv 0$ and $3x_n \not\equiv 0$. Because $x_1 \equiv x_2 \equiv 1$, this implies that $x_n \equiv 0$ if and only if $n \equiv 3 \pmod 4$. Because $1999^{1945} \equiv 3^{1945} \equiv 3 \cdot 9^{1944/2} \equiv 3 \pmod 4$, we indeed have $7 \mid x_n$ when $n = 1999^{1945}$.

On the other hand, suppose, by way of contradiction, that $7 \mid y_{n'}$ for some n', and choose the minimal such n'. From the recursion for y_n, we have $y_n \equiv y_{n+1} + 3y_{n+2}$. If $n' \geq 5$, then $y_{n'-2} \equiv y_{n'-1}$, $y_{n'-3} \equiv 4y_{n'-1}$, and $y_{n'-4} \equiv 0$ — contradicting the minimal choice of n'. Thus we have $n' \leq 4$, but $(y_1, y_2, y_3, y_4) \equiv (2, 1, 5, 1)$. Therefore no term y_n is divisible by 7. Specifically, 7 does not divide y_n when $n = 1999^{1945}$.

2

1999 Regional Contests: Problems and Solutions

1 Asian Pacific Mathematical Olympiad

Problem 1 Find the smallest positive integer n with the following property: There does not exist an arithmetic progression of 1999 real numbers containing exactly n integers.

Solution. Look at the 1999-term arithmetic progression with common difference $\frac{1}{q}$ and beginning term $\frac{p}{q}$, where p and q are integers with $1 < q < 1999$; without loss of generality assume that $1 \leq p \leq q$. When $p = 1$, the progression contains the integers $1, 2, \ldots, \left\lfloor \frac{1999}{q} \right\rfloor$. When $p = q$, the progression contains the integers $1, 2, \ldots, 1 + \left\lfloor \frac{1998}{q} \right\rfloor$. Because 1999 is prime, q does not divide 1999 and hence

$$\left\lfloor \frac{1999}{q} \right\rfloor = \left\lfloor \frac{1998}{q} \right\rfloor.$$

Thus the progression contains either

$$\left\lfloor \frac{1999}{q} \right\rfloor \quad \text{or} \quad \left\lfloor \frac{1999}{q} \right\rfloor + 1$$

integers, and any k of this form can be attained. Call such numbers *good*.

Conversely, suppose we have an arithmetic progression containing exactly k integers, where $1 < k < 1999$. Without loss of generality, suppose that its common difference is positive and that it contains 0 as its t-th term. Its common difference cannot be irrational, so it is of the form $\frac{p}{q}$ for some positive, relatively prime integers p, q. Because $1 < k < 1999$, $1 < q < 1999$. Now consider the arithmetic progression with common difference $\frac{1}{q}$ and 0 as its t-th term. It, too, contains exactly k integers, so our previous argument shows that k is good.

Now, for $q \geq 32$ we have

$$1999 < 2q(q+1) \implies \frac{1999}{q} - \frac{1999}{q+1} < 2.$$

Because $\lfloor \frac{1999}{32} \rfloor = 62$ and $\lfloor \frac{1999}{1998} \rfloor = 1$, this implies that every integer k between 1 and 63 is good. Also,

$$\left\lfloor \frac{1999}{31} \right\rfloor = 64, \quad \left\lfloor \frac{1999}{30} \right\rfloor = 66, \quad \left\lfloor \frac{1999}{29} \right\rfloor = 68,$$

$$\left\lfloor \frac{1999}{q} \right\rfloor \leq 62 \text{ when } q \geq 32, \quad \left\lfloor \frac{1999}{q} \right\rfloor \geq 71 \text{ when } q \leq 28.$$

Thus k can take on every value less than 70, but it cannot be equal to 70. Thus the desired value of n is 70.

Problem 2 Let $a_1, a_2, \ldots$ be a sequence of real numbers satisfying

$$a_{i+j} \leq a_i + a_j$$

for all $i, j = 1, 2, \ldots$. Prove that

$$a_1 + \frac{a_2}{2} + \frac{a_3}{3} + \cdots + \frac{a_n}{n} \geq a_n$$

for all positive integers n.

Solution.

Lemma. *If m, n are positive integers with $m \geq n$, then*

$$a_1 + a_2 + \cdots + a_n \geq \frac{n(n+1)}{2m} \cdot a_m.$$

Proof. The result for $m = n$ comes from adding the inequalities $a_1 + a_{n-1} \geq a_n$, $a_2 + a_{n-2} \geq a_n$, \ldots, $a_{n-1} + a_1 \geq a_n$, $2a_n \geq 2a_n$, and then dividing by two. Now for positive integers j, write

$$\beta_j = \frac{a_1 + a_2 + \cdots + a_j}{1 + 2 + \cdots + j}.$$

Then the inequality for $m = n = j = k+1$ is equivalent to both $\beta_j \geq \frac{a_j}{j}$ and $\beta_k \geq \beta_{k+1}$. Thus when $m \geq n$ we have

$$\beta_n \geq \beta_{n+1} \geq \cdots \geq \beta_m \geq \frac{a_m}{m},$$

as desired. ∎

The desired inequality now follows from expressing $a_1 + \frac{a_2}{2} + \cdots + \frac{a_n}{n}$ as an Abed sum and then applying the lemma multiple times:

$$a_1 + \frac{a_2}{2} + \cdots + \frac{a_n}{n} = \frac{1}{n}(a_1 + a_2 + \cdots + a_n)$$
$$+ \sum_{j=1}^{n-1}\left(\frac{1}{j} - \frac{1}{j+1}\right)(a_1 + a_2 + \cdots + a_j)$$
$$\geq \frac{1}{n}\frac{n(n+1)}{2n}a_n + \sum_{j=1}^{n-1}\frac{1}{j(j+1)} \cdot \frac{j(j+1)}{2n}a_n$$
$$= a_n,$$

as desired.

Problem 3 Let ω_1 and ω_2 be two circles intersecting at P and Q. The common tangent, closer to P, of ω_1 and ω_2, touches ω_1 at A and ω_2 at B. The tangent to ω_1 at P meets ω_2 again at C, and the extension of AP meets BC at R. Prove that the circumcircle of triangle PQR is tangent to BP and BR.

Solution. We shall use directed angles. Using tangents and cyclic quadrilaterals, we have $\angle QAR = \angle QAP = \angle QPC = \angle QBC = \angle QBR$, so $QABR$ is cyclic.

Because $\angle BPR$ is an exterior angle to triangle ABP, we have $\angle BPR = \angle BAP + \angle PBA$. Then again using tangents and cyclic quadrilaterals, we have $\angle BAP + \angle PBA = \angle BAR + \angle PCB = \angle BQR + \angle PQB = \angle PQR$. Thus $\angle BPR = \angle PQR$, which implies that line BP is tangent to the circumcircle of triangle PQR.

Next, $\angle PRB$ is an exterior angle to triangle CRP so $\angle PRB = \angle PCR + \angle RPC$. We know that $\angle PCR = \angle PCB = \angle PQB$. Letting T be the intersection of lines CP and AB, we have $\angle RPC = \angle APT = \angle AQP = \angle BAP = \angle BAR = \angle BQR$. Therefore $\angle PRB = \angle PQB + \angle BQR = \angle PQR$. This implies that line BR is tangent to the circumcircle of triangle PQR as well.

Problem 4 Determine all pairs (a, b) of integers for which the numbers $a^2 + 4b$ and $b^2 + 4a$ are both perfect squares.

Solution. If $a = 0$ then b must be a perfect square, and vice versa. Now assume both a and b are nonzero. Also observe that $a^2 + 4b$ and a^2 have the same parity, and similarly $b^2 + 4a$ and b^2 have the same parity.

If b is positive then $a^2 + 4b \geq (|a|+2)^2 = a^2 + 4|a| + 4$ so $|b| \geq |a| + 1$.
If b is negative then $a^2 + 4b \leq (|a|-2)^2 = a^2 - 4|a| + 4$ so $|b| \geq |a| - 1$.
Similarly, $a > 0 \implies |a| \geq |b| + 1$ and $a < 0 \implies |a| \geq |b| - 1$.

Assume without loss of generality that $b > a$. If a and b are positive, then from the last paragraph we have $b \geq a + 1$ and $a \geq b + 1$, a contradiction.

If a and b are negative, then we have either $a = b$ or $a = b - 1$. For $b \geq -5$, only $(a, b) = (-4, -4)$ and $(-6, -5)$ work. Otherwise, we have $(b+4)^2 < b^2 + 4a < (b+2)^2$, a contradiction.

Finally, if a is negative and b is positive, then we have both $|b| \geq |a| + 1$ and $|a| \geq |b| - 1$. Then we must have $|b| = |a| + 1$ and hence $a + b = 1$. Any such pair works, because then $a^2 + 4b = (a-2)^2$ and $b^2 + 4a = (b-2)^2$ are both perfect squares.

Therefore the possible pairs (a, b) are $(-4, -4)$, $(-6, -5)$, $(-5, -6)$, and $(0, n^2)$, $(n^2, 0)$, $(n, 1-n)$ where n is any integer.

Problem 5 Let S be a set of $2n + 1$ points in the plane such that no three are collinear and no four concyclic. A circle will be called *good* if it has 3 points of S on its circumference, $n - 1$ points in its interior, and $n - 1$ in its exterior. Prove that the number of good circles has the same parity as n.

Solution.

Lemma. *Suppose we have $2n \geq 1$ given points in the plane with no three collinear, and one distinguished point A among them. Call a line good if it passes through A and one other given point, and if it separates the remaining $2n - 2$ points: that is, half of them lie on one side of the line, and the other half lie on the other. Then there are an odd number of good lines.*

Proof. We prove the claim by induction. It is trivial for $n = 1$; now assuming it is true for $n - 1$, we prove it is true for n.

Without loss of generality, arrange the $2n - 1$ points different from A on a circle centered at A. From those $2n - 1$ points, choose two points, B and C, that are the greatest distance apart. Then if $P \neq A, B, C$ is a given point, B and C lie on different sides of line AP. Thus line AP is good if and only if it separates the other $2n - 3$ points; by the induction hypothesis, there are an odd number of such lines. Finally, line AB is good if and only if line AC is good — adding either 0 or 2 good lines to our count, so that our total count remains odd. This completes the inductive step. ∎

Suppose we have a pair of points $\{A, B\}$ in $\mathcal{S}$. Perform an inversion about A with arbitrary radius. B and the other $2n - 1$ points $C_1, C_2, \ldots, C_{2n-1}$ map to $2n$ distinct points $B', C_1', C_2', \ldots, C_{2n-1}'$ (no three collinear). Also, the circle passing through A, B, C_k is good if and only if the corresponding line $B'C_k'$ separates the other C_i'. By the lemma, there are an odd number of such lines; so an odd number of good circles pass through any pair of given points.

For each pair of points count the number of good circles passing through these points. Each good circle is counted three times in this manner, so if there are k good circles then our count will be $3k$. There are $\binom{2n+1}{2} = n(2n+1)$ pairs of points, each contributing an odd amount to our count. Therefore $3k \equiv n(2n+1) \implies k \equiv n \pmod{2}$, as desired.

2 Austrian-Polish Mathematics Competition

Problem 1 Let n be a positive integer and $M = \{1, 2, \ldots, n\}$. Find the number of ordered 6-tuples $(A_1, A_2, A_3, A_4, A_5, A_6)$ which satisfy the following two conditions:

(a) $A_1, A_2, A_3, A_4, A_5, A_6$ (not necessarily distinct) are subsets of M;

(b) each element of M belongs to either 0, 3, or 6 of the sets $A_1, A_2, A_3, A_4, A_5, A_6$.

Solution. Given $k \in M$, there are $\binom{6}{0}$ ways to put k into exactly 0 of the 6 sets, $\binom{6}{3}$ ways to put k into exactly 3 of the sets, and $\binom{6}{6}$ ways to put k into all 6 sets. This adds up to $1 + 20 + 1 = 22$ ways to put k into the sets. Every admissible 6-tuple is uniquely determined by the placement of each k. Therefore, there are 22^n possible distributions.

Problem 2 Find the largest real number C_1 and the smallest real number C_2 such that for all positive real numbers a, b, c, d, e the following inequalities hold:

$$C_1 < \frac{a}{a+b} + \frac{b}{b+c} + \frac{c}{c+d} + \frac{d}{d+e} + \frac{e}{e+a} < C_2.$$

Solution. Let $f(a, b, c, d, e) = \frac{a}{a+b} + \cdots + \frac{e}{e+a}$. Note that $f(e, d, c, b, a) = 5 - f(a, b, c, d, e)$. Hence $f(a, b, c, d, e)$ can attain the value x if and only if it can attain the value $5 - x$. Thus if we find C_1, then $C_2 = 5 - C_1$.

If $a = \epsilon^4$, $b = \epsilon^3$, $c = \epsilon^2$, $d = \epsilon$, $e = 1$, then

$$f(a, b, c, d, e) = \frac{4\epsilon}{1+\epsilon} + \frac{1}{1+\epsilon^4},$$

which for small ϵ can become arbitrarily close to 1. We now prove that $f(a, b, c, d, e)$ is *always* bigger than 1. Because f is homogenous, we may assume without loss of generality that $a + b + c + d + e = 1$.

The function $g(x) = \frac{1}{x}$ is convex for all positive x. Thus by Jensen's inequality, $ag(x_1) + bg(x_2) + \cdots + eg(x_5) \geq g(ax_1 + bx_2 + \cdots + ex_5)$. Applying this inequality with $x_1 = a+b$, $x_2 = b+c$, ..., $x_5 = e+a$, we find that

$$f(a, b, c, d, e) \geq \frac{1}{\sum_{\text{cyc}} a(a+b)}$$

$$= \frac{(a+b+c+d+e)^2}{\sum_{\text{cyc}} a(a+b)} > \frac{\sum_{\text{cyc}} a(a+2b)}{\sum_{\text{cyc}} a(a+b)} > 1,$$

as desired. (Here $\sum_{\text{cyc}} h(a,b,c,d,e)$ denotes the sum $h(a,b,c,d,e) + h(b,c,d,e,a) + \cdots + h(e,a,b,c,d)$.) Hence $C_1 = 1$, so from our initial arguments, $C_2 = 4$.

Problem 3 Let $n \geq 2$ be a given integer. Determine all systems of n functions $(f_1, \ldots, f_n)$ where $f_i : \mathbb{R} \to \mathbb{R}$, $i = 1, 2, \ldots, n$, such that for all $x, y \in \mathbb{R}$ the following equalities hold:

$$f_1(x) - f_2(x)f_2(y) + f_1(y) = 0$$
$$f_2(x^2) - f_3(x)f_3(y) + f_2(y^2) = 0$$
$$\vdots$$
$$f_{n-1}(x^{n-1}) - f_n(x)f_n(y) + f_{n-1}(y^{n-1}) = 0$$
$$f_n(x^n) - f_1(x)f_1(y) + f_n(y^n) = 0.$$

Solution. Setting $x = y$ into the k-th equation gives $2f_k(x^k) = f_{k+1}(x)^2$ (where we write $f_{n+1} = f_1$). Thus if $f_k(x) = 0$ for all $x \in \mathbb{R}$, then $f_{k+1}(x)$ is also always zero. Similarly, *all* the f_i's are identically zero. Now assume that no f_k is identically zero.

First consider an odd integer k. Given any real value r, let $x = \sqrt[k]{r}$. Then $2f_k(r) = 2f_k(x^k) = f_{k+1}(x)^2 \geq 0$ for all r.

Next consider an even integer k. For $r \geq 0$, letting $x = \sqrt[k]{r}$ as above we find that $f_k(r) \geq 0$. Furthermore, f_k) is *strictly* positive at some value — because $f_{k+1}(x) \neq 0$ for some x, we have $f_k(x^k) = \frac{1}{2}f_{k+1}(x)^2 > 0$. Also observe that we cannot have $f_k(x) < 0$ and $f_k(y) > 0$ for some x and y because then we'd have $f_{k-1}(x^{k-1}) - f_k(x)f_k(y) + f_{k-1}(y^{k-1}) > 0$, contradicting the $(k-1)$-th equation. It follows that $f_k(x) \geq 0$ for all x.

Hence $f_k(x) \geq 0$ for all k and all x. We now prove by induction on k that f_k is a constant function. For $k = 1$, plugging in $f_2(x) = \sqrt{2f_1(x)}$ and $f_2(y) = \sqrt{2f_1(y)}$ into the first equation gives $f_1(x) - 2\sqrt{f_1(x)f_1(y)} + f_1(y) = 0$ for all $x, y \in \mathbb{R}$. By the AM-GM inequality, this is true only when $f_1(x) = f_1(y)$ for all $x, y \in \mathbb{R}$.

Now assume that $f_k(x) = f_k(y)$ for all x, y. Then $f_{k+1}(x) = \sqrt{2f_k(x^k)} = \sqrt{2f_k(y^k)} = f_{k+1}(y)$ for any $x, y \in \mathbb{R}$, completing the inductive step.

Writing $f_k(x) = c_k$ (where we write $c_{n+1} = c_1$), observe that $c_k = \frac{1}{2}c_{k+1}^2$ for all k. Because $f_k(x) \geq 0$ for all x but f_k is not identically zero, each c_k is positive. If $0 < c_{k+1} < 2$, then $0 < c_k < c_{k+1} < 2$. Thus $c_n > c_{n-1} > \cdots > c_1 > c_n$, a contradiction. Similarly, if $c_{k+1} > 2$

then $c_k > c_{k+1} > 2$. It follows that $c_n < c_{n-1} < \cdots < c_1 < c_n$, a contradiction. Hence, we must have $c_k = 2$ for all k.

Therefore, the only possible functions are $f_k(x) = 0$ for all x, k, and $f_k(x) = 2$ for all x, k.

Problem 4 Three straight lines k, l, and m are drawn through some fixed point P inside a triangle ABC such that:

(a) k meets the lines AB and AC in A_1 and A_2 ($A_1 \neq A_2$) respectively and $PA_1 = PA_2$;

(b) l meets the lines BC and BA in B_1 and B_2 ($B_1 \neq B_2$) respectively and $PB_1 = PB_2$;

(c) m meets the lines CA and CB in C_1 and C_2 ($C_1 \neq C_2$) respectively and $PC_1 = PC_2$.

Prove that the lines k, l, m are uniquely determined by the above conditions. Find the point P (and prove that there exists exactly one such point) for which the triangles AA_1A_2, BB_1B_2, and CC_1C_2 have the same area.

Solution. Let ℓ be the reflection of line AB about P. Because A_1 and A_2 are also mirror images of each other across P, A_2 must lie on ℓ. Thus A_2 must be the intersection of ℓ and line AC, and this intersection point is unique because lines AB and AC are not parallel. Therefore k must be the line passing through P and this point, and conversely this line satisfies condition (a). Similarly, lines l and m are also uniquely determined.

Now suppose that triangles AA_1A_2 and BB_1B_2 have the same area. Let Q be the midpoint of $\overline{AA_1}$. Because P is the midpoint of $\overline{A_1A_2}$, we have $[AQP] = \frac{1}{2}[AA_1P] = \frac{1}{4}[AA_1A_2]$. Similarly, let R be the midpoint of $\overline{BB_2}$. Then $[BRP] = \frac{1}{2}[BB_2P] = \frac{1}{4}[BB_1B_2]$. Therefore $[AQP] = [BRP]$, and because the heights from P in triangles AQP and BRP are equal, we must have $AQ = BR$.

Now because P and Q are the midpoints of $\overline{A_1A_2}$ and $\overline{A_1A}$, we have $PQ \parallel AA_2$ and hence $PQ \parallel AC$. This implies that Q lies between A and B. Similarly, R lies between A and B as well. Because $AQ = BR$, Q and R are equidistant from the midpoint C' of $\overline{AB}$. Therefore the homothety about C' that maps A to Q also maps B to R. This homothety also maps line AC to QP because $AC \parallel QP$, and it also maps line BC to RP. Hence it maps C to P, so that C', P, C are collinear and P lies on the median from C in triangle ABC.

Similarly, P must lie on the medians from A and B in triangle ABC, so it must be the centroid G of triangle ABC. Conversely, if $P = G$ then k, l, m are parallel to lines BC, CA, AB respectively, and $[AA_1A_2] = [BB_1B_2] = [CC_1C_2] = \frac{4}{9}[ABC]$.

Problem 5 A sequence of integers $\{a_n\}_{n \geq 1}$ satisfies the following recursive relation

$$a_{n+1} = a_n^3 + 1999 \quad \text{for } n = 1, 2, \ldots.$$

Prove that there exists at most one n for which a_n is a perfect square.

Solution. Consider the possible values of (a_n, a_{n+1}) modulo 4:

a_n	0	1	2	3
a_{n+1}	3	0	3	2

No matter what a_1 is, the terms $a_3, a_4, \ldots$ are all 2 or 3 (mod 4). However, all perfect squares are 0 or 1 (mod 4), so at most two terms (a_1 and a_2) can be perfect squares. If a_1 and a_2 are both perfect squares, then writing $a_1 = a^2$, $a_2 = b^2$ we have $a^6 + 1999 = b^2$ or $1999 = b^2 - (a^3)^2 = (b + a^3)(b - a^3)$. Because 1999 is prime, $b - a^3 = 1$ and $b + a^3 = 1999$. Thus $a^3 = \frac{1999-1}{2} = 999$, which is impossible. Hence at most one term of the sequence is a perfect square.

Problem 6 Find all real numbers $x_0, x_1, x_2, \ldots, x_{1998} \geq 0$ which satisfy

$$x_{i+1}^2 + x_{i+1}x_i + x_i^4 = 1$$

for $i = 0, 1, \ldots, 1998$, where $x_{1999} = x_0$.

Solution. Let R be the positive real root of $x^4 + 2x^2 - 1 = 0$, found using the quadratic formula:

$$R^2 = \frac{-2 + \sqrt{8}}{2} = -1 + \sqrt{2} \implies R = \sqrt{-1 + \sqrt{2}}.$$

If $x_n, x_{n+1} \geq R$ then $1 = x_{n+1}^2 + x_{n+1}x_n + x_n^4 \geq R^2 + R^2 + R^4 = 1$, with equality when $x_n = x_{n+1} = R$. Similarly, if $x_n, x_{n+1} \leq R$ then we must have $x_n = x_{n+1} = R$. Hence either $x_n = x_{n+1} = R$, $x_n < R < x_{n+1}$, or $x_n > R > x_{n+1}$.

Now if $x_0 < R$ then $x_1 > R$, $x_2 < R$, $\ldots$, $x_0 = x_{1999} > R$, a contradiction. Similarly, we cannot have $x_0 > R$. Therefore $x_0 = R$ and the only solution is

$$x_0 = x_1 = \cdots = x_{1999} = R = \sqrt{-1 + \sqrt{2}}.$$

Problem 7 Find all pairs (x, y) of positive integers such that
$$x^{x+y} = y^{y-x}.$$

Solution. Let $\gcd(x, y) = c$, and write $a = \frac{x}{c}$, $b = \frac{y}{c}$. Then the equation becomes
$$(ac)^{c(a+b)} = (bc)^{c(b-a)}$$
$$(ac)^{a+b} = (bc)^{b-a}$$
$$a^{a+b} c^{2a} = b^{b-a}.$$

Thus $a^{a+b} \mid b^{b-a}$. Because $\gcd(a, b) = 1$, a must equal 1. Therefore
$$b^{b-1} = c^2$$
is a perfect square. This is true exactly when b is odd, or when b is a perfect square. If $b = 2n + 1$, then $c = (2n + 1)^n$; if $b = n^2$, then $c = n^{n^2-1}$. Therefore (x, y) equals either
$$((2n+1)^n, (2n+1)^{n+1}) \quad \text{or} \quad (n^{n^2-1}, n^{n^2+1})$$
for some positive integer n, and any such pair indeed satisfies the given equations.

Problem 8 Let ℓ be a given straight line and let the points P and Q lie on the same side of the line ℓ. The points M, N lie on the line ℓ and satisfy $PM \perp \ell$ and $QN \perp \ell$. The point S lies between the lines PM and QN such that $PM = PS$ and $QN = QS$. The perpendicular bisectors of $\overline{SM}$ and $\overline{SN}$ meet at R. Let T be the second intersection of the line RS and the circumcircle of triangle PQR. Prove that $RS = ST$.

Solution. All angles are directed modulo $180°$. Let T' be the reflection of R across S; we wish to prove that $PRQT'$ is cyclic.

Because $RM = RS = RN$, $\angle RMN = \angle MNR = x$. Note that $PMRS$ is a kite which is symmetric with respect to line PR. Hence $\angle PRM = \angle SRP = y$, $\angle MPR = \angle RPS = z$, and $\angle PSR = \angle RMP = 90° + x$. Similarly, $\angle QRS = \angle NRQ = u$, $\angle SQR = \angle RQN = v$, and $\angle RSQ = \angle QNR = 90° + x$. In triangle MNR, $2(x + y + u) = 180°$. In triangle PMR, $y + z + 90° + x = 180°$ so that $2(x + y + z) = 180°$. Hence $2u = 2z$.

Orient our diagram so that line MN is horizontal and so that P and Q are above line MN. From the given information, S is above line MN and between lines PM and QN. Also, because R is the circumcenter of triangle MSN, R lies below between lines PM and QN and below S. This information allows us to safely conclude that $u = z$, or $\angle SPR = \angle SRQ$.

Similarly, $y = v$ and $\angle PRS = \angle SQR$. Thus $\triangle PSR \sim \triangle RSQ$. Let A and B be the midpoints of $\overline{PR}$ and $\overline{QR}$, respectively. Then $\overline{SA}$ and $\overline{SB}$ are corresponding medians in similar triangles PRS and RQS. Hence $\triangle ASR \sim \triangle BSQ$. It follows that

$$\angle ASB = \angle ASR + \angle RSB = \angle BSQ + \angle RSB = \angle RSQ = 90° + x.$$

Thus $\angle ASB + \angle ARB = \angle ASB + \angle PRQ = 90° + x + y + z = 180°$ and $ASRB$ is cyclic. Because $PRQT'$ is the image of $ARBS$ under a homothety about R with ratio 2, it follows that $PQRT'$ is cyclic as well.

Problem 9 Consider the following one player game. On the plane, a finite set of selected lattice points and segments is called a *position* in this game if the following hold:

(i) the endpoints of each selected segment are lattice points;

(ii) each selected segment is parallel to a coordinate axis, or to the line $y = x$, or to the line $y = -x$;

(iii) each selected segment contains exactly 5 lattice points and all of them are selected;

(iv) any two selected segments have at most one common point.

A move in this game consists of selecting a new lattice point and a new segment such that the new set of selected lattice points and segments is a position. Prove or disprove that there exists an initial position such that the game has infinitely many moves.

Solution. There exists no position so that the game can last for infinitely many moves.

Given any segment, let its *extreme point* be its leftmost, upper-leftmost, highest, or upper-rightmost point (depending on whether the segment is parallel to $y = 0$, $y = -x$, $x = 0$, or $y = x$). Let any segment's other four lattice points form the segment's *mini-segment*. Observe that no two mini-segments pointing in the same direction can intersect.

Also, given any position, call a lattice point a *missing point* if it is the extreme point of a selected segment in one direction, but does not lie on any other selected segment pointing in the same direction. (Notice that during the game, a lattice point might switch between being missing and not-missing.)

Lemma. *Given an integer $N > 0$, if the game continues forever then at some point at least N missing points will exist at the same time.*

Proof. Suppose we have k lines that contain at least one selected segment. Some $\lceil \frac{k}{4} \rceil$ of them must point in the same direction. Each of these lines contains at least one different missing point: its leftmost, upper-leftmost, highest, or upper-rightmost extreme point. Therefore it is enough to show that k gets arbitrarily large.

For sake of contradiction, suppose that k is bounded. Then all the selected segments will lie on some finite number of lines. These lines have only a finite set S of t intersection points, so from some position onward no new selected point will be in S. At this point suppose we have s selected segments, and p selected points outside of S.

After n more moves there will be $s + n$ mini-segments, and $p + n$ selected points outside of S. Each mini-segment contains 4 points for a total count of $4(s + n)$. On the other hand, each of the t points lies on at most 4 mini-segments, and each of the $p + n$ other points lies on at most 1 mini-segment, yielding a total count of at most $4t + (p + n)$. Thus $4(s + n) \leq 4t + (p + n)$, but this is false for large enough n — a contradiction. ∎

Now suppose in our original position we have s selected segments and p selected points. From the lemma, eventually we will have more than $4(p - s)$ missing points. Suppose this happens after n moves, when we have $s + n$ selected segments and $p + n$ selected points. Then as in the lemma, we will have $s + n$ mini-segments each containing 4 points — for a total count of $4(s + n)$. On the other hand, each of the missing points lies on at most 3 mini-segments, and all the other selected points lie on at most 4 mini-segments each. Thus our total count is less than $3 \cdot 4(p - s) + 4 \cdot (p + n - 4(p - s)) = 4(s + n)$, a contradiction. This completes the proof.

3 Balkan Mathematical Olympiad

Problem 1 Given an acute-angled triangle ABC, let D be the midpoint of minor arc $\overparen{BC}$ of circumcircle ABC. Let E and F be the respective images of D under reflections about BC and the center of the circumcircle. Finally, let K be the midpoint of AE. Prove that:

(a) the circle passing through the midpoints of the sides of the triangle ABC also passes through K.

(b) the line passing through K and the midpoint of BC is perpendicular to AF.

Solution.

(a) Let M, B_1, and C_1 denote the midpoints of sides BC, CA, and AB, respectively. We have $\triangle ABC \sim \triangle MB_1C_1$ and $\angle C_1 M B_1 = \angle A$. Also, $BECD$ is a rhombus, with $\angle BEC = \angle CDB = 180° - \angle A$. The homothety centered at A with ratio $\frac{1}{2}$ maps triangle BEC to triangle $C_1 K B_1$. Thus, $\angle C_1 K B_1 + \angle C_1 M B_1 = \angle BEC + \angle A = 180°$, so $MC_1 K B_1$ is cyclic.

(b) Because $ED = 2EM$ and $EA = 2EK$, $MK \parallel AD$. $\overline{DF}$ is a diameter, so $AD \perp AF$. Hence also $MK \perp AF$.

Problem 2 Let $p > 2$ be a prime number such that $3 \mid (p-2)$. Let

$$\mathcal{S} = \{y^2 - x^3 - 1 \mid x \text{ and } y \text{ are integers}, 0 \leq x, y \leq p-1\}.$$

Prove that at most p elements of $\mathcal{S}$ are divisible by p.

Solution.

Lemma. *Given a prime p and a positive integer $k > 1$, if k and $p-1$ are relatively prime then $x^k \equiv y^k \Rightarrow x \equiv y \pmod{p}$ for all x, y.*

Proof: If $y \equiv 0 \pmod{p}$ the claim is obvious. Otherwise, note that $x^k \equiv y^k \Longrightarrow (xy^{-1})^k \equiv 1 \pmod{p}$, so it suffices to prove that $a^k \equiv 1 \Longrightarrow a \equiv 1 \pmod{p}$.

Because $\gcd(p-1, k) = 1$, there exist integers b and c such that $b(p-1) + ck = 1$. Thus, $a^k \equiv 1 \Longrightarrow a^c \equiv 1 \Longrightarrow a^{1-b(p-1)} \equiv 1 \pmod{p}$. If $a = 0$ this is impossible. Otherwise, by Fermat's Little Theorem, $(a^{-b})^{p-1} \equiv 1 \pmod{p}$ so that $a \equiv 1 \pmod{p}$, as desired.

Alternatively, again note that clearly $a \not\equiv 1 \pmod{p}$. Then let d be the order of a, the smallest positive integer such that $a^d \equiv 1 \pmod{p}$; we have $d \mid k$. Take the set $\{1, a, a^2, \ldots, a^{d-1}\}$. If it does not contain all of $1, 2, \ldots, p-1$ then pick some other element b and consider the set $\{b, ba, ba^2, \ldots, ba^{d-1}\}$. These two sets are disjoint, because otherwise $ba^i \equiv a^j \Rightarrow b \equiv a^{j-i} \pmod{p}$, a contradiction. Continuing similarly, we can partition $\{1, 2, \ldots, p-1\}$ into d-element subsets, and hence $d \mid p-1$. However, $d \mid k$ and $\gcd(k, p-1) = 1$, implying that $d = 1$. Therefore $a \equiv a^d \equiv 1 \pmod{p}$, as desired. ∎

Because $3 \mid p-2$, $\gcd(3, p-1) = 1$. Then from the claim, it follows that the set of elements $\{1^3, 2^3, \ldots, p^3\}$ equals $\{1, 2, \ldots, p\}$ modulo p. Hence, for each y with $0 \leq y \leq p-1$, there is exactly one x between 0 and $p-1$ such that $x^3 \equiv y^2 - 1 \pmod{p}$: that is, such that $p \mid y^2 - x^3 - 1$. Therefore S contains at most p elements divisible by p, as desired.

Problem 3 Let ABC be an acute triangle, and let M, N, and P be the feet of the perpendiculars from the centroid to the three sides. Prove that

$$\frac{4}{27} < \frac{[MNP]}{[ABC]} \leq \frac{1}{4}.$$

Solution. We begin by proving that

$$\frac{9[MNP]}{[ABC]} = \sin^2 A + \sin^2 B + \sin^2 C.$$

Let G be the centroid of triangle ABC, and let M, N, and P be on sides BC, AC, and AB, respectively. Also let $AB = c$, $BC = a$, $CA = b$, and $K = [ABC]$.

We have

$$[ABG] = \frac{K}{3} = \frac{1}{2} c \cdot GP \Longrightarrow GP = \frac{2K}{3c}.$$

Similarly, $GN = \frac{2K}{3b}$, so

$$[PGN] = \frac{1}{2} GP \cdot GN \sin A = \frac{2K^2 \sin A}{9bc} = \frac{K^2 a^2}{9Rabc}.$$

Summing this formula with the analogous ones for $[NGM]$ and $[MGP]$ yields

$$[MNP] = \frac{K^2(a^2 + b^2 + c^2)}{9Rabc}.$$

Dividing this by $[ABC] = K$ and then substituting $K = \frac{abc}{4R}$, $a = 2R \sin A$, $b = 2R \sin B$, and $c = 2R \sin C$ on the right yields

$$\frac{[MNP]}{[ABC]} = \frac{1}{9} \left(\sin^2 A + \sin^2 B + \sin^2 C \right),$$

as desired.

Hence the problem reduces to proving $\frac{4}{3} < \sin^2 A + \sin^2 B + \sin^2 C \leq \frac{9}{4}$.
Assume without loss of generality that $A \geq B \geq C$.

To prove the right inequality, first note that $A < 90° \Rightarrow B > 45°$. The function $\sin^2 x$ is concave on $[45°, 90°]$. We can thus apply Jensen's Inequality to give

$$\sin^2 A + \sin^2 B \leq 2\sin^2\left(\frac{A}{2} + \frac{B}{2}\right) = 2\cos^2\left(\frac{C}{2}\right).$$

Hence it suffices to prove

$$2\cos^2\left(\frac{C}{2}\right) + \sin^2 C \leq \frac{9}{4} \iff 1 + \cos C + 1 - \cos^2 C$$

$$\leq \frac{9}{4} \iff -(\cos C + \frac{1}{2})^2 \leq 0,$$

which is true.

For the left inequality, note that $\sin^2 x$ is an increasing function on $[0, 90°]$. We have $B \geq \frac{180° - A}{2}$, so

$$\sin^2 A + \sin^2 B + \sin^2 C > \sin^2 A + \cos^2\left(\frac{A}{2}\right)$$

$$= -\cos^2 A + \frac{1}{2}\cos A + \frac{3}{2}.$$

By assumption A is the largest angle, implying that $90° > A \geq 60°$ and $\frac{1}{2} \geq \cos A > 0$. Therefore

$$-\cos^2 A + \frac{1}{2}\cos A + \frac{3}{2} \geq \frac{3}{2} > \frac{4}{3},$$

as desired.

Problem 4 Let $\{x_n\}_{n\geq 0}$ be a nondecreasing sequence of nonnegative integers such that for every $k \geq 0$ the number of terms of the sequence which are less than or equal to k is finite; let this number be y_k. Prove that for all positive integers m and n,

$$\sum_{i=0}^{n} x_i + \sum_{j=0}^{m} y_j \geq (n+1)(m+1).$$

Solution. Under the given construction, $y_s \leq t$ if and only if $x_t > s$. This condition is equivalent to the condition that $y_s > t$ if and only if $x_t \leq s$. Thus the sequences $\{x_i\}$ and $\{y_j\}$ are dual, meaning that applying the given algorithm to $\{y_j\}$ will restore the original $\{x_i\}$.

To find $\sum_{i=0}^{n} x_i$, note that among $x_0, x_1, \ldots, x_n$ there are exactly y_0 terms equal to 0, $y_1 - y_0$ terms equal to 1, ..., and $y_{x_{n-1}} - y_{x_{n-2}}$ terms equal to x_{n-1}, while the remaining $n + 1 - x_{n-1}$ terms equal x_n. Hence,

$$\sum_{i=0}^{n} x_i = 0 \cdot (y_0) + 1 \cdot (y_1 - y_0) + \cdots$$

$$+ (x_n - 1) \cdot (y_{x_n - 1} - y_{x_n - 2}) + x_n \cdot (n + 1 - y_{x_n - 1})$$

$$= -y_0 - y_1 - \cdots - y_{x_n - 1} + (n+1)x_n.$$

First suppose that $x_n - 1 \geq m$, and write $x_n - 1 = m + k$ for $k \geq 0$. Because $x_n > m + k$, from our initial observations we have $y_{m+k} \leq n$. Then $n + 1 \geq y_{m+k} \geq y_{m+k-1} \geq \cdots \geq y_m$, so

$$\sum_{i=0}^{n} x_i + \sum_{j=0}^{m} y_j = (n+1)x_n - \left(\sum_{j=0}^{x_n - 1} y_j - \sum_{i=0}^{m} y_i \right)$$

$$= (n+1)x_n - \sum_{i=m+1}^{m+k} y_i$$

$$\geq (n+1)(m+k+1) - k \cdot (n+1)$$

$$= (n+1)(m+1),$$

as desired.

Next suppose that $x_n - 1 < m$. Then

$$x_n \leq m \Longrightarrow y_m > n \Longrightarrow y_m - 1 \geq n.$$

Because $\{x_i\}$ and $\{y_j\}$ are dual, we may therefore apply the same argument with the roles of the two sequences reversed. This completes the proof.

4 Czech and Slovak Match

Problem 1 For arbitrary positive numbers a, b, c, prove the inequality

$$\frac{a}{b+2c} + \frac{b}{c+2a} + \frac{c}{a+2b} \geq 1.$$

First Solution. Set $x = b+2c$, $y = c+2a$, $z = a+2b$. Then $a = \frac{1}{9}(4y+z-2x)$, $b = \frac{1}{9}(4z+x-2y)$, $c = \frac{1}{9}(4x+y-2z)$, so the desired inequality becomes

$$\frac{4y+z-2x}{9x} + \frac{4z+x-2y}{9y} + \frac{4x+y-2z}{9z} \geq 1,$$

which is equivalent to

$$\left(\frac{x}{y}+\frac{y}{x}\right) + \left(\frac{y}{z}+\frac{z}{y}\right) + \left(\frac{z}{x}+\frac{x}{z}\right) + 3\cdot\left(\frac{y}{x}+\frac{z}{y}+\frac{x}{z}\right) \geq 15.$$

This inequality is true because by the AM-GM inequality, the quantities in parentheses are at least 2, 2, 2, and 3, respectively. Alternatively, it is true by the AM-GM inequality on all 15 fractions of the form $\frac{x}{y}$ on the left-hand side (where $3\cdot\left(\frac{y}{x}+\frac{x}{z}+\frac{z}{y}\right)$ contributes nine such fractions).

Second Solution. By the Cauchy-Schwarz inequality

$$(u_1 v_1 + u_2 v_2 + u_3 v_3)^2 \leq (u_1^2+u_2^2+u_3^2)(v_1^2+v_2^2+v_3^2),$$

the quantity $(a+b+c)^2$ is at most

$$\left(\frac{a}{b+2c} + \frac{b}{c+2a} + \frac{c}{a+2b}\right)[a(b+2c)+b(c+2a)+c(a+2b)].$$

On the other hand, from the inequality $(a-b)^2 + (b-c)^2 + (c-a)^2 \geq 0$ we have $a(b+2c)+b(c+2a)+c(a+2b) \leq (a+b+c)^2$. Combining these gives

$$a(b+2c)+b(c+2a)+c(a+2b) \leq \left(\frac{a}{b+2c} + \frac{b}{c+2a} + \frac{c}{a+2b}\right)$$
$$[a(b+2c)+b(c+2a)+c(a+2b)],$$

which yields our desired inequality upon division by the (positive) expression $a(b+2c)+b(c+2a)+c(a+2b)$.

Problem 2 Let ABC be a nonisosceles acute triangle with altitudes $\overline{AD}$, $\overline{BE}$, and $\overline{CF}$. Let ℓ be the line through D parallel to line EF.

Let $P = \overleftrightarrow{BC} \cap \overleftrightarrow{EF}$, $Q = \ell \cap \overleftrightarrow{AC}$, and $R = \ell \cap \overleftrightarrow{AB}$. Prove that the circumcircle of triangle PQR passes through the midpoint of $\overline{BC}$.

Solution. Let M be the midpoint of $\overline{BC}$. Without loss of generality we may assume that $AB > AC$. Then the order of the points in question on line BC is B, M, D, C, P. Because $\angle BEC = \angle BFC = 90°$, $BCEF$ is cyclic so that $\angle QCB = 180° - \angle BCE = \angle EFB = \angle QRB$. Thus $RCQB$ is cyclic as well and

$$DB \cdot DC = DQ \cdot DR.$$

Quadrilateral $MRPQ$ is cyclic if and only if $DM \cdot DP = DQ \cdot DR$, so it remains to prove that $DB \cdot DC = DP \cdot DM$. Because the points B, C, E, F are concyclic, we have $PB \cdot PC = PE \cdot PF$. The circumcircle of triangle DEF (the so-called nine-point circle of triangle ABC) also passes through the midpoints of the sides of triangle ABC, which implies that $PE \cdot PF = PD \cdot PM$. Comparing both equalities shows that $PB \cdot PC = PD \cdot PM$. Denoting $MB = MC = u$, $MD = d$, $MP = p$, the last equality reads $(p+u)(p-u) = (p-d)p \iff u^2 = dp \iff (u+d)(u-d) = (p-d)d \iff DB \cdot DC = DP \cdot DM$. This completes the proof.

Problem 3 Find all positive integers k for which there exists a ten-element set M of positive numbers such that there are exactly k different triangles whose side lengths are three (not necessarily distinct) elements of M. (Triangles are considered different if they are not congruent.)

Solution. Given any 10-element set M of positive integers, there are exactly $\binom{12}{3}$ triples x, y, z ($x \leq y \leq z$) chosen from the numbers in M. Hence, $k \leq \binom{12}{3} = 220$. On the other hand, for each of the $\binom{11}{2}$ pairs x, y from M (with $x \leq y$) we can form the triangle with side lengths x, y, y. Hence, $k \geq \binom{11}{2} = 55$. Then applying the following lemma, the possible values of k are then $55, 56, \ldots, 220$.

Lemma. *Suppose we have some positive integers n and k, with*

$$\binom{n+1}{2} \leq k \leq \binom{n+2}{3}.$$

Then there exists an n-element set M of positive numbers, such that there are exactly k triangles whose side lengths are three (not necessarily distinct) elements of M.

Furthermore, there exists an n-element set S_n with numbers $x_1 < x_2 < \cdots < x_n$ such that $x_n < 2x_1$ and such that all the $\binom{n+1}{2}$ pairwise sums

$$x_i + x_j \quad (1 \leq i \leq j \leq n)$$

are distinct. Exactly $\binom{n+2}{3}$ triangles can be formed from the elements of S_n.

Proof: We prove the claim by induction on n; it clearly holds for $n = 1$. Now suppose the claim is true for n and that $\binom{n+2}{2} \leq k \leq \binom{n+3}{3}$.

If $\binom{n+2}{2} \leq k \leq \binom{n+2}{3} + (n+1)$, then first find the n-element set $M' = \{x_1, x_2, \ldots, x_n\}$ from which exactly $k - (n+1)$ triangles can be formed. Choose $x_{n+1} > 2 \cdot \max\{M'\}$, and write $M = M' \cup \{x_{n+1}\}$. Then exactly k triangles can be formed from the elements of M: $k-(n+1)$ triangles from $\{x_1, x_2, \ldots, x_n\}$, and an additional $n+1$ triangles with side lengths x_i, x_{n+1}, x_{n+1} for $i = 1, 2, \ldots, n+1$.

Otherwise $k = \binom{n+2}{3} + n + 1 + q$, where $q \in \{1, 2, \ldots, \binom{n+1}{2}\}$. To the set S_n described in the lemma, add an element x_{n+1} which is greater than x_n but smaller than precisely q of the $\binom{n+1}{2}$ original pairwise sums from S_n. This gives a set M from which exactly k triangles can be formed: $\binom{n+2}{3}$ triangles from $\{x_1, x_2, \ldots, x_n\}$, $n+1$ additional triangles with side lengths x_i, x_{n+1}, x_{n+1}, and exactly q more triangles with side lengths x_i, x_j, x_{n+1} (where $i, j \neq n+1$).

None of the numbers $x_i + x_{n+1}$ equals any of the original pairwise sums. Thus we can construct S_{n+1}, completing the proof of the lemma. ∎

Problem 4 Find all positive integers k for which the following assertion holds: if $F(x)$ is a polynomial with integer coefficients which satisfies $0 \leq F(c) \leq k$ for all $c \in \{0, 1, \ldots, k+1\}$, then

$$F(0) = F(1) = \cdots = F(k+1).$$

Solution. The claim is false for $k < 4$ — we have the counterexamples $F(x) = x(2-x)$ for $k = 1$, $F(x) = x(3-x)$ for $k = 2$, and $F(x) = x(4-x)(x-2)^2$ for $k = 3$.

Now suppose $k \geq 4$ is fixed and that $F(x)$ has the described property. First of all $F(k+1) - F(0) = 0$, because it is a multiple of $k+1$ whose absolute value is at most k. Hence $F(x) - F(0) = x(x - k - 1)G(x)$, where $G(x)$ is another polynomial with integer coefficients. Then we have

$$k \geq |F(c) - F(0)| = c(k + 1 - c)|G(c)|$$

for each $c = 1, 2, \ldots, k$. If $2 \le c \le k-1$ (such numbers c exist because $k \ge 4$) then
$$c(k+1-c) = 2(k-1) + (c-2)(k-1-c) \ge 2(k-1) > k,$$
which in view of our first inequality means that $|G(c)| < 1 \Longrightarrow G(c) = 0$. Thus $2, 3, \ldots, k-1$ are roots of the polynomial $G(x)$, so
$$F(x) - F(0) = x(x-2)(x-3)\cdots(x-k+1)(x-k-1)H(x),$$
where $H(x)$ is again a polynomial with integer coefficients. It remains to explain why $H(1) = H(k) = 0$. This is easy: both values $c=1$ and $c=k$ satisfy $k \ge |F(c) - F(0)| = (k-2)! \cdot k \cdot |H(c)|$. Because $(k-2)! > 1$, we must have $H(1) = H(k) = 0$.

Problem 5 Find all functions $f : (1, \infty) \to \mathbb{R}$ such that
$$f(x) - f(y) = (y-x)f(xy)$$
for all $x, y > 1$.

Solution. For every $t > 1$ we use the equation in turn for $(x, y) = (t, 2)$, $(t, 4)$ and $(2t, 2)$:
$$f(t) - f(2) = (2-t)f(2t),$$
$$f(t) - f(4) = (4-t)f(4t),$$
$$f(2t) - f(2) = (2-2t)f(4t).$$
We eliminate $f(t)$ by subtracting the second equation from the first, and then substitute for $f(2t)$ from the third. This yields the equality
$$f(4) + (t-3)f(2) = t(2t-5)f(4t)$$
for any $t > 1$. Taking $t = \frac{5}{2}$ we get $f(4) = \frac{1}{2}f(2)$, and feeding this back gives
$$\left(t - \frac{5}{2}\right) f(2) = t(2t-5)f(4t).$$
It follows that for any $t > 1$, $t \ne \frac{5}{2}$,
$$f(4t) = \frac{f(2)}{2t},$$
so by the middle of the three equations in the beginning of this solution we obtain
$$f(t) = f(4) + (4-t)f(4t) = \frac{1}{2}f(2) + \frac{(4-t)f(2)}{2t} = \frac{2f(2)}{t}.$$

This formula for $f(t)$ holds even for $t = \frac{5}{2}$, as can now be checked directly by applying the original equation to $x = \frac{5}{2}$ and $y = 2$ (and using $f(5) = \frac{2f(2)}{5}$). Setting $c = 2f(2)$, we must have

$$f(x) = \frac{c}{x}$$

for all x. This function has the required property for any choice of the real constant c.

Problem 6 Show that for any positive integer $n \geq 3$, the least common multiple of the numbers $1, 2, \ldots, n$ is greater than 2^{n-1}.

Solution. For any $n \geq 3$ we have

$$2^{n-1} = \sum_{k=0}^{n-1} \binom{n-1}{k} < \sum_{k=0}^{n-1} \binom{n-1}{\lfloor \frac{n-1}{2} \rfloor} = n \cdot \binom{n-1}{\lfloor \frac{n-1}{2} \rfloor}.$$

Hence it suffices to show that $n \cdot \binom{n-1}{\lfloor \frac{n-1}{2} \rfloor}$ divides $\text{lcm}(1, 2, \ldots, n)$. Using an argument involving prime factorizations, we will prove the more general assertion that for each $k < n$, $\text{lcm}(n, n-1, \ldots, n-k)$ is divisible by $n \cdot \binom{n-1}{k}$.

Let k and n be fixed natural numbers with $k < n$, and let $p \leq n$ be an arbitrary prime. Let p^α be the highest power of p which divides $\text{lcm}(n, n-1, \ldots, n-k)$, where $p^\alpha \mid n-\ell$ for some ℓ. Then for each $i \leq \alpha$, we know that $p^i \mid n-\ell$. Thus exactly $\lfloor \frac{\ell}{p^i} \rfloor$ of $\{n-\ell+1, n-\ell+2, \ldots, n\}$ and exactly $\lfloor \frac{k-\ell}{p^i} \rfloor$ of $\{n-\ell-1, n-\ell-2, \ldots, n-k\}$ are multiples of p^i, so p^i divides $\lfloor \frac{\ell}{p^i} \rfloor + \lfloor \frac{k-\ell}{p^i} \rfloor \leq \lfloor \frac{k}{p^i} \rfloor$ of the remaining k numbers — that is, at most the number of multiples of p^i between 1 and k. It follows that p divides $n \cdot \binom{n-1}{k} = \frac{n(n-1)\cdots(n-\ell+1)(n-\ell-1)\cdots(n-k)}{k!} \cdot (n-\ell)$ at most α times, so that indeed $n \cdot \binom{n-1}{k} \mid \text{lcm}(n, n-1, \ldots, n-k)$.

5 Hungary-Israel Binational Mathematical Competition

Individual Round

Problem 1 Let S be the set of all partitions $(a_1, a_2, \ldots, a_k)$ of the number 2000, where $1 \leq a_1 \leq a_2 \leq \cdots \leq a_k$ and $a_1 + a_2 + \cdots + a_k = 2000$. Compute the smallest value that $k + a_k$ attains over all such partitions.

Solution. By the AM-GM inequality,
$$k + a_k \geq 2\sqrt{ka_k} \geq 2(\sqrt{2000}) > 89.$$
Because $k+a_k$ is an integer, it must be at least 90. 90 is indeed attainable, as $k + a_k = 90$ for the partition $(\underbrace{40, 40, \cdots, 40}_{50})$.

Problem 2 Prove or disprove the following claim: For any positive integer k, there exists a positive integer $n > 1$ such that the binomial coefficient $\binom{n}{i}$ is divisible by k for any $1 \leq i \leq n - 1$.

First Solution. The statement is false. To prove this, take $k = 4$ and assume by contradiction that there exists a positive integer $n > 1$ for which $\binom{n}{i}$ is divisible by 4 for every $1 \leq i \leq n - 1$. Then
$$0 \equiv \sum_{i=1}^{n-1} \binom{n}{i} = 2^n - 2 \equiv -2 \pmod{4},$$
a contradiction.

Second Solution. The claim is obviously true for $k = 1$; we prove that the set of positive integers $k > 1$ for which the claim holds is exactly the set of primes. First suppose that k is prime. Express n in base k, writing $n = n_0 + n_1 k + \cdots + n_m k^m$ where $0 \leq n_0, n_1, \ldots, n_m \leq k - 1$ and $n_m \neq 0$. Also suppose we have $1 \leq i \leq n - 1$, and write $i = i_0 + i_1 k + \cdots + i_m k^m$ where $0 \leq i_0, i_1, \ldots, i_m \leq k - 1$ (although perhaps $i_m = 0$). Lucas's Theorem states that
$$\binom{n}{i} \equiv \prod_{j=0}^{m} \binom{n_j}{i_j} \pmod{k}.$$

If $n = k^m$, then $n_0 = n_1 = \cdots = n_{m-1} = 0$. Also, $1 \le i \le n-1$ so that $i_m = 0$ but some other $i_{j'}$ is nonzero. Then $\binom{n_{j'}}{i_{j'}} = 0$, and indeed $\binom{n}{i} \equiv 0 \pmod{k}$ for all $1 \le i \le n-1$.

Suppose instead that $n \ne k^m$. If $n_m > 1$, then letting $i = k^m < n$ we have $\binom{n}{i} \equiv (n_m)(1)(1) \cdots (1) \equiv n_m \not\equiv 0 \pmod{k}$. Otherwise, some other $n_{j'} \ne 0$. Then setting $i = n_{j'} k^{j'} < n$ we have $\binom{n}{i} \equiv (1)(1) \cdots (1) \equiv 1 \not\equiv 0 \pmod{k}$. Therefore the claim holds for prime k exactly when $n = k^m$.

Now, suppose the claim holds for some $k > 1$ with the number n. If some prime p divides k, the claim must also hold for p with the number n. Thus n must equal a prime power p^m where $m \ge 1$. Then $k = p^r$ for some $r \ge 1$ as well, because if two primes p and q divided k then n would equal a perfect power of both p and q, which is impossible.

Choose $i = p^{m-1}$. Kummer's Theorem states that $p^t \mid \binom{n}{i}$ if and only if t is less than or equal to the number of carries in the addition $(n-i)+i$ in base p. There is only one such carry, between the p^{m-1} and p^m places:

$$
\begin{array}{rccccccc}
 & 1 & & & & & & \\
 & & 1 & 0 & 0 & \cdots & 0 & \\
+ & & p-1 & 0 & 0 & \cdots & 0 & \\
\hline
 & 1 & & 0 & 0 & 0 & \cdots & 0
\end{array}
$$

Thus, we must have $r \le 1$ and k must be prime, as claimed.

(Alternatively, for $n = p^m$ and $i = p^{m-1}$ we have

$$\binom{n}{i} = \prod_{j=0}^{p^{m-1}-1} \frac{p^m - j}{p^{m-1} - j}.$$

When $j = 0$ then $\frac{p^m - j}{p^{m-1} - j} = p$. Otherwise, $0 < j < p^{m-1}$ so that if $p^t < p^{m-1}$ is the highest power of p dividing j, then it is also the highest power of p dividing both $p^m - j$ and $p^{m-1} - j$. Therefore $\frac{p^m - j}{p^{m-1} - j}$ contributes one factor of p to $\binom{n}{i}$ when $j = 0$ and zero factors of p when $j > 0$. Thus p^2 does not divide *binomni*, and hence again $r \le 1$.)

Problem 3 Let ABC be a non-equilateral triangle with its incircle touching $\overline{BC}, \overline{CA}$, and $\overline{AB}$ at A_1, B_1, and C_1, respectively, and let H_1 be the orthocenter of triangle $A_1B_1C_1$. Prove that H_1 is on the line passing through the incenter and circumcenter of triangle ABC.

First Solution. Let ω_1, I, ω_2, and O be the incircle, incenter, circumcircle, and circumcenter of triangle ABC, respectively. Because $I \ne O$, line IO is well defined.

Let T be the center of the homothety with positive ratio that sends ω_1 to ω_2 and hence I to O. Also let A_2, B_2, C_2 be the midpoints of the arcs BC, CA, AB of ω_2 not containing A, B, C, respectively. Because rays IA_1 and OA_2 point in the same direction, T must send A_1 to A_2 and similarly B_1 to B_2 and C_1 to C_2.

Also, because the measures of arcs AC_2 and A_2B_2 add up to 180°, we know that $\overline{AA_2} \perp \overline{C_2B_2}$. Similarly, $\overline{BB_2} \perp \overline{C_2A_2}$ and $\overline{CC_2} \perp \overline{A_2B_2}$. Because lines AA_2, BB_2, CC_2 intersect at I, I is the orthocenter of triangle $A_2B_2C_2$. Hence I is the image of H_1 under the defined homothety. Therefore, T, H_1, I are collinear. From before, T, I, O are collinear. It follows that H_1, I, O are collinear, as desired.

Second Solution. Define $\omega_1, I, \omega_2,$ and O as before. Let ω_3 be the nine-point circle of triangle $A_1B_1C_1$, and let S be its center. Because I is the circumcenter of triangle $A_1B_1C_1$, S is the midpoint of $\overline{IH_1}$ and I, H_1, S are collinear.

Now invert the figure with respect to ω_1. The midpoints of $\overline{A_1B_1}$, $\overline{B_1C_1}, \overline{C_1A_1}$ are mapped to C, A, B, and thus ω_3 is mapped to ω_2. Thus I, O, S are collinear. Hence I, H_1, O, S are collinear, as desired.

Problem 4 Given a set X, define

$$X' = \{s - t \mid s, t \in X, s \neq t\}.$$

Let $S = \{1, 2, \ldots, 2000\}$. Consider two sets $A, B \subseteq S$, such that $|A||B| \geq 3999$. Prove that $A' \cap B' \neq \emptyset$.

Solution. Consider all $|A| \cdot |B| \geq 3999$ sums, not necessarily distinct, of the form $a + b$ where $a \in A, b \in B$. If both 2 and 4000 are of this form, then both A and B contain 1 and 2000 so that $2000 - 1 \in A' \cap B'$. Otherwise, each sum $a + b$ takes on one of at most 3998 values either between 2 and 3999, or between 3 and 4000. Thus by the pigeonhole principle, two of our $|A| \cdot |B|$ sums $a_1 + b_1$ and $a_2 + b_2$ are equal with $a_1, a_2 \in A, b_1, b_2 \in B$, and $(a_1, b_1) \neq (a_2, b_2)$. Then $a_1 \neq a_2$ (because otherwise we would have $b_1 = b_2$ and $(a_1, b_1) = (a_2, b_2)$), and therefore $A' \cap B'$ is nonempty because it contains $a_1 - a_2 = b_2 - b_1$.

Problem 5 Given an integer d, let

$$S = \{m^2 + dn^2 \mid m, n \in \mathbb{Z}\}.$$

Let $p, q \in S$ be such that p is a prime and $r = \frac{q}{p}$ is an integer. Prove that $r \in S$.

Solution. Note that
$$(x^2 + dy^2)(u^2 + dv^2) = (xu \pm dyv)^2 + d(xv \mp yu)^2.$$
Write $q = a^2 + db^2$ and $p = x^2 + dy^2$ for integers a, b, x, y. Reversing the above construction yields the desired result. Indeed, solving for u and v after setting $a = xu + dyv, b = xv - yu$ and $a = xu - dyv, b = xv + yu$ gives

$$u_1 = \frac{ax - dby}{p}, \quad v_1 = \frac{ay + bx}{p},$$

$$u_2 = \frac{ax + dby}{p}, \quad v_2 = \frac{ay - bx}{p}.$$

Note that
$$(ay + bx)(ay - bx) = (a^2 + db^2)y^2 - (x^2 + dy^2)b^2$$
$$\equiv 0 \pmod{p}.$$

Hence p divides one of $ay + bx, ay - bx$ so that one of v_1, v_2 is an integer. Without loss of generality, assume that v_1 is an integer. Because $r = u_1^2 + dv_1^2$ is an integer and u_1 is rational, u_1 is an integer as well and $r \in S$, as desired.

Problem 6 Let k and ℓ be two given positive integers, and let a_{ij}, $1 \le i \le k$ and $1 \le j \le \ell$, be $k\ell$ given positive numbers. Prove that if $q \ge p > 0$ then

$$\left(\sum_{j=1}^{\ell} \left(\sum_{i=1}^{k} a_{ij}^p \right)^{\frac{q}{p}} \right)^{\frac{1}{q}} \le \left(\sum_{i=1}^{k} \left(\sum_{j=1}^{\ell} a_{ij}^q \right)^{\frac{p}{q}} \right)^{\frac{1}{p}}.$$

First Solution. Define $b_j = \sum_{i=1}^{k} a_{ij}^p$ for $j = 1, 2, \ldots, \ell$, and denote the left-hand side of the required inequality by L and its right-hand side by R. Then

$$L^q = \sum_{j=1}^{\ell} b_j^{\frac{q}{p}}$$

$$= \sum_{j=1}^{\ell} \left(b_j^{\frac{q-p}{p}} \left(\sum_{i=1}^{k} a_{ij}^p \right) \right)$$

$$= \sum_{i=1}^{k} \left(\sum_{j=1}^{\ell} b_j^{\frac{q-p}{p}} a_{ij}^p \right).$$

Using Hölder's inequality it follows that

$$L^q \leq \sum_{i=1}^{k} \left[\left(\sum_{j=1}^{\ell} \left(b_j^{\frac{q-p}{p}}\right)^{\frac{q}{q-p}} \right)^{\frac{q-p}{q}} \left(\sum_{j=1}^{\ell} (a_{ij}^p)^{\frac{q}{p}} \right)^{\frac{p}{q}} \right]$$

$$= \sum_{i=1}^{k} \left[\left(\sum_{j=1}^{\ell} b_j^{\frac{q}{p}} \right)^{\frac{q-p}{q}} \left(\sum_{j=1}^{\ell} a_{ij}^q \right)^{\frac{p}{q}} \right]$$

$$= \left(\sum_{j=1}^{\ell} b_j^{\frac{q}{p}} \right)^{\frac{q-p}{q}} \cdot \left[\sum_{i=1}^{k} \left(\sum_{j=1}^{\ell} a_{ij}^q \right)^{\frac{p}{q}} \right] = 7L^{q-p} R^p.$$

The inequality $L \leq R$ follows by dividing both sides of $L^q \leq L^{q-p} R^p$ by L^{q-p} and taking the p-th root.

Second Solution. Let $r = \frac{q}{p}$, $c_{ij} = a_{ij}^p$. Then $r \geq 1$, and the given inequality is equivalent to the following inequality:

$$\left(\sum_{j=1}^{\ell} \left(\sum_{i=1}^{k} c_{ij} \right)^r \right)^{\frac{1}{r}} \leq \sum_{i=1}^{k} \left(\sum_{j=1}^{\ell} c_{ij}^r \right)^{\frac{1}{r}}$$

We shall prove this inequality by induction on k. For $k = 1$, we have equality. For $k = 2$, the inequality becomes Minkowski's inequality.

Suppose that $k \geq 3$ and the inequality holds for $k - 1$. Then by the induction assumption for $k - 1$ we have

$$\sum_{i=1}^{k-1} \left(\sum_{j=1}^{\ell} c_{ij}^r \right)^{\frac{1}{r}} + \left(\sum_{j=1}^{\ell} c_{kj}^r \right)^{\frac{1}{r}} \geq \left(\sum_{j=1}^{\ell} \left(\sum_{i=1}^{k-1} c_{ij} \right)^r \right)^{\frac{1}{r}} + \left(\sum_{j=1}^{\ell} c_{kj}^r \right)^{\frac{1}{r}}$$

Using Minkowski's inequality with $\tilde{c}_{1j} = \sum_{i=1}^{k-1} c_{ij}$, $\tilde{c}_{2j} = c_{kj}$, we have

$$\left(\sum_{j=1}^{\ell} \left(\sum_{i=1}^{k-1} c_{ij} \right)^r \right)^{\frac{1}{r}} + \left(\sum_{j=1}^{\ell} c_{kj}^r \right)^{\frac{1}{r}} \geq \left(\sum_{j=1}^{\ell} \left(\sum_{i=1}^{k} c_{ij} \right)^r \right)^{\frac{1}{r}},$$

completing the inductive step.

Team Round

Problem 1 Let ABC be a triangle and let P_1 be a point inside triangle ABC.

(a) Prove that the lines obtained by reflecting lines P_1A, P_1B, P_1C through the angle bisectors of $\angle A$, $\angle B$, $\angle C$, respectively, meet at a common point P_2.

(b) Let A_1, B_1, C_1 be the feet of the perpendiculars from P_1 to sides BC, CA and AB, respectively. Let A_2, B_2, C_2 be the feet of the perpendiculars from P_2 to sides BC, CA and AB, respectively. Prove that these six points A_1, B_1, C_1, A_2, B_2, C_2 lie on a circle.

Solution. (a) By the trigonometric form of Ceva's Theorem, we have
$$\frac{\sin \angle ABP_1 \sin \angle BCP_1 \sin \angle CAP_1}{\sin \angle P_1BC \sin \angle P_1CA \sin \angle P_1AB} = 1.$$
Now suppose that the given reflections of lines P_1A, P_1B, P_1C meet sides BC, CA, AB at points D, E, F, respectively. Then $\angle ABE = \angle P_1BC$, $\angle EBC = \angle ABP_1$, and so on. Thus
$$\frac{\sin \angle EBC \sin \angle FCA \sin \angle DAB}{\sin \angle BCF \sin \angle CAD \sin \angle ABE} = 1$$
as well. Therefore, again by the trigonometric form of Ceva's Theorem, the three new lines also concur.

(b) Note that $P_1A_1CB_1$ and $P_2A_2CB_2$ are both cyclic quadrilaterals because they each have two right angles opposite each other. Because $\angle P_1CB_1 = \angle A_2CP_2$, we have $\angle B_1P_1C = \angle CP_2A_2$, and, by the previous statement, that implies $\angle B_1A_1C = \angle CB_2A_2$, whence A_1, A_2, B_1, B_2 are concyclic. For similar reasons, A_1, A_2, C_1, C_2 are concyclic. Then all six points $A_1, B_1, C_1, A_2, B_2, C_2$ must be concyclic, or else the radical axes of circles $A_1A_2B_1B_2$, $B_1B_2C_1C_2$, $C_1C_2A_1A_2$ would not concur, contradicting the radical axis theorem.

Problem 2 An ant is walking inside the region bounded by the curve whose equation is $x^2 + y^2 + xy = 6$. Its path is formed by straight segments parallel to the coordinate axes. It starts at an arbitrary point on the curve and takes off inside the region. When reaching the boundary, it turns by $90°$ and continues its walk inside the region. When arriving at a point on the boundary which it has already visited, or where it cannot continue its walk according to the given rule, the ant stops. Prove that, sooner or later, and regardless of the starting point, the ant will stop.

First Solution. If the ant moves from (a, b) to (c, b), then a and c are the roots to $f(t) = t^2 + bt + b^2 - 6$. Thus $c = -a - b$. Similarly, if the ant moves from (a, b) in a direction parallel to the y-axis, it meets the curve at $(a, -a - b)$.

Let (a, b) be the starting point of the ant, and assume that the ant starts walking in a direction parallel to the x-axis; the case when it starts walking parallel to the y-axis is analogous. If after five moves the ant is still walking, then it will return to its original position after six moves:

$$(a, b) \to (-a - b, b) \to (-a - b, a) \to (b, a)$$
$$\to (b, -a - b) \to (a, -a - b) \to (a, b).$$

Therefore, the ant stops moving after at most six steps.

Second Solution. Rotate the curve by $45°$, where (x, y) is on our new curve C_1 when $\frac{1}{\sqrt{2}}(x - y, x + y)$ is on the original curve. The equation of the image of the curve under the rotation is

$$3x^2 + y^2 = 12.$$

Hence the curve is an ellipse, while the new directions of the ant's motion are inclined by $\pm 45°$ with respect to the x-axis. Next apply an affine transformation so that (x, y) is on our new curve C_2 when $(x, \sqrt{3}y)$ is on C_1. The ant's curve then becomes

$$x^2 + y^2 = 4,$$

a circle with radius 2, and the directions of the ant's paths are now inclined by $\pm 30°$ with respect to the x-axis.

Thus if the ant moves from P_1 to P_2 to P_3, then $\angle P_1 P_2 P_3$ will either equal $0°$, $60°$, or $120°$. Hence, as long as the ant continues moving, every two moves it travels to the other end of an arc measuring $0°$, $120°$, or $240°$ along the circle. Therefore after at most six moves, he must return to a position he visited earlier.

Problem 3

(a) In the plane, we are given a circle ω with unknown center, and an arbitrary point P. Is it possible to construct, using only a straightedge, the line through P and the center of the circle?

(b) In the plane, we are given a circle ω with unknown center, and a point Q on the circle. Construct the tangent to ω at the point Q, using only a straightedge.

(c) In the plane, we are given two circles ω_1 and ω_2 with unknown centers. Construct, with a straightedge only, the line through their centers when:

　(i) the two circles intersect;

(ii) the two circles touch each other, and their point of contact T is marked;

Solution. (a) It is not possible. First we prove that given a circle with unknown center O, it is impossible to construct its center using only a straightedge. Assume by contradiction that this construction *is* possible. Perform a projective transformation on the figure, taking O to O' and ω to another circle ω'. The drawn lines remain lines, and thus they still yield the point O'. On the other hand, those lines compose a construction which gives the center of ω'. However, O is not mapped to the center of ω', a contradiction.

Then the described construction must also be impossible. Otherwise, given a circle ω with unknown center O, we could mark a point P_1 and draw the line P_1O, and then mark another point P_2 not on line P_1O and draw the line P_2O. The intersection of these lines would yield O, which is impossible from above.

(b) Given a point A not on ω, suppose line ℓ_1 passes through A and hits ω at B_1 and C_1, and suppose another line ℓ_2 also passes through A and hits ω at B_2 and C_2. The line connecting $B_1B_2 \cap C_1C_2$ and $B_1C_2 \cap C_1B_2$ is the pole of A with respect to ω. Conversely, given any line ℓ we can pick two points on it that are not on ω, and construct the poles of these two points. If ℓ passes through the center of ω then these two poles are parallel; otherwise, they intersect at the polar of ℓ.

Now given ω and Q, mark another point R on the circle and then construct the polar T of line QR. (If the polar is the point at infinity, pick another point for R.) Then line TQ is the pole of Q — and this pole is tangent to ω at Q. Hence line QT is our desired line.

(c) (i) Draw the lines tangent to ω_1 at P and Q. They intersect at a point X_1, which by symmetry must lie on our desired line. Next draw the lines tangent to ω_2 at P and Q. They also intersect at a point X_2 that lies on our desired line. Thus line X_1X_2 passes through both circles' centers, as desired.

(ii) **Lemma.** *Suppose we have a trapezoid $ABCD$ with $AB \parallel CD$. Suppose that lines AD and BC intersect at M and that lines AC and BD intersect at N. Then line MN passes through the midpoints of $\overline{AB}$ and $\overline{CD}$.*

Proof. Without loss of generality suppose that M is on the same side of line CD as A and B. Let line MN intersect $\overline{CD}$ at P. By Ceva's theorem in triangle MDC applied to the cevians passing through N, we

have
$$\frac{DA}{AM} \cdot \frac{MB}{BC} \cdot \frac{CP}{PD} = 1.$$

Triangles MAB and MDC are similar, so $\frac{MA}{AN} = \frac{MB}{BC}$. Hence $\frac{CP}{PD} = 1$, as desired. ∎

Now construct a line through T intersecting ω_1 at A and ω_2 at C. Construct a different line through T hitting ω_1 at B and ω_2 at D. Under the homothety about T that maps ω_1 to ω_2, segment AB gets mapped to segment CD. Thus, $AB \parallel CD$. If lines AD and BC are parallel, pick different A and C; otherwise, using the construction in the lemma we can find the midpoint F_1 of $\overline{CD}$. Next, from the construction described in part (b), we can find the polar F_2 of line CD. Then line F_1F_2 passes through the center O_2 of ω_2.

Similarly, we can find a different line G_1G_2 passing through O_2. Hence O_2 is the intersection of lines F_1F_2 and G_1G_2. We can likewise find the center O_1 of ω_1; then drawing the line O_1O_2, we are done.

6 Iberoamerican Mathematical Olympiad

Problem 1 Find all positive integers n less than 1000 such that n^2 is equal to the cube of the sum of n's digits.

Solution. In order for n^2 to be a cube, n must be a cube itself. Because $n < 1000$ we must have $n = 1^3, 2^3, \ldots,$ or 9^3. Quick checks show that $n = 1$ and $n = 27$ work while $n = 8, 64,$ and 125 don't. As for $n \geq 6^3 = 216$, we have $n^2 \geq 6^6 > 27^3$. However, the sum of n's digits is at most $9 + 9 + 9 = 27$, implying that no $n \geq 6^3$ has the desired property. Thus $n = 1, 27$ are the only answers.

Problem 2 Given two circles ω_1 and ω_2, we say that ω_1 bisects ω_2 if they intersect and their common chord is a diameter of ω_2. (If ω_1 and ω_2 are identical, we still say that they bisect each other.) Consider two non-concentric fixed circles ω_1 and ω_2.

(a) Show that there are infinitely many circles ω that bisect both ω_1 and ω_2.

(b) Find the locus of the center of ω.

Solution. Suppose we have any circle ω with center O and radius r. Then we show that given any point P, there is a unique circle centered at P bisecting ω, AND that the radius of this circle is $\sqrt{r^2 + OP^2}$. If $O = P$ the claim is obvious. Otherwise, let $\overline{AB}$ be the diameter perpendicular to OP, so that $PA = PB = \sqrt{r^2 + OP^2}$.

Because $PA = PB$, there is a circle centered at P and passing through both A and B. This circle indeed bisects ω. Conversely, if circle Γ centered at P bisects ω along diameter $A'B'$, then both O and P lie on the perpendicular bisector of $\overline{A'B'}$. Thus $A'B' \perp OP$, and we must have $\overline{AB} = \overline{A'B'}$ and hence indeed

$$PA' = PB' = PA = PB = \sqrt{r^2 + OP^2}.$$

Now back to the original problem. Set up a coordinate system where ω_1 is centered at the origin $O_1 = (0,0)$ with radius r_1, and ω_2 is centered at $O_2 = (a, 0)$ with radius r_2. Given any point $P = (x, y)$, the circle centered at P bisecting ω_1 is the *same* as the circle centered at P bisecting ω_2 if and only if $\sqrt{r_1^2 + O_1P^2} = \sqrt{r_2^2 + O_2P^2}$; that is, if and only if

$$r_1^2 + x^2 + y^2 = r_2^2 + (x-a)^2 + y^2$$
$$2ax = r_2^2 - r_1^2 + a^2.$$

Therefore given any point P along the line
$$x = \frac{r_2^2 - r_1^2 + a^2}{2a},$$
some circle ω centered at P bisects both ω_1 and ω_2. Because there are infinitely many such points, there are infinitely many such circles.

Conversely, from above if ω does bisect both circles then it must be centered at a point on the given line — which is a line perpendicular to the line passing through the centers of ω_1 and ω_2. In fact, recall that the radical axis of ω_1 and ω_2 is the line $2a(a-x) = r_2^2 - r_1^2 + a^2$. Therefore the desired locus is the radical axis of ω_1 and ω_2, reflected across the perpendicular bisector of the segment joining the centers of the circles.

Problem 3 Let $P_1, P_2, \ldots, P_n$ ($n \geq 2$) be n distinct collinear points. Circles with diameter P_iP_j ($1 \leq i < j \leq n$) are drawn and each circle is colored in one of k given colors. All points that belong to more than one circle are not colored. Such a configuration is called a (n,k)-cover. For any given k, find all n such that for any (n,k)-cover there exist two lines externally tangent to two circles of the same color.

Solution. Without loss of generality label the points so that $P_1, P_2, \ldots, P_n$ lie on the line in that order from left to right. If $n \leq k+1$ points, then color any circle P_iP_j ($1 \leq i < j \leq n$) with the i-th color. Then any two circles sharing the same i-th color are internally tangent at P_i, so there do not exist two lines externally tangent to them.

However, if $n \geq k+2$ then some two of the circles with diameters P_1P_2, P_2P_3, ..., $P_{k+1}P_{k+2}$ must have the same color. Then there *do* exist two lines externally tangent to them. Therefore the solution is $n \geq k+2$.

Problem 4 Let n be an integer greater than 10 such that each of its digits belongs to the set $S = \{1, 3, 7, 9\}$. Prove that n has some prime divisor greater than or equal to 11.

Solution. Note that any product of any two numbers from $\{1, 3, 7, 9\}$ taken modulo 20 is still in $\{1, 3, 7, 9\}$. Therefore *any* finite product of such numbers is still in this set. Specifically, any number of the form $3^j 7^k$ is congruent to $1, 3, 7,$ or $9 \pmod{20}$.

Now if all the digits of $n \geq 10$ are in S, then its tens digit is odd and we cannot have $n \equiv 1, 3, 7,$ or $9 \pmod{20}$. Thus, n cannot be of the form $3^j 7^k$. Nor can n be divisible by 2 or 5 (otherwise, its last digit would not be 1, 3, 7, or 9). Hence n must be divisible by some prime greater than or equal to 11, as desired.

Problem 5 Let ABC be an acute triangle with circumcircle ω centered at O. Let $\overline{AD}$, $\overline{BE}$, and $\overline{CF}$ be the altitudes of ABC. Let line EF meet ω at P and Q.

(a) Prove that $AO \perp PQ$.

(b) If M is the midpoint of $\overline{BC}$, prove that
$$AP^2 = 2AD \cdot OM.$$

Solution. Let H be the orthocenter of triangle ABC, so that $AEHF$, $BFHD$, $CDHE$ are cyclic. Then
$$\angle AFE = \angle AHE = \angle DHB = 90° - \angle HBD$$
$$= 90° - \angle EBC = \angle BCE = C,$$

while $\angle OAF = \angle OAB = 90° - C$. Therefore $AO \perp EF$ and thus $AO \perp PQ$, as desired.

Let R be the circumradius of triangle ABC. Draw diameter AA' perpendicular to $\overline{PQ}$, intersecting $\overline{PQ}$ at T. Then
$$AT = AF\cos(90° - C) = AF \sin C = AC \cos A \sin C$$
$$= 2R \cos A \sin B \sin C.$$

By symmetry, $PT = TQ$. Thus by Power of a Point, $PT^2 = PT \cdot TQ = AT \cdot TA' = AT(2R - AT)$. Thus
$$AP^2 = AT^2 + PT^2 = AT^2 + AT(2R - AT) = 2R \cdot AT$$
$$= 4R^2 \cos A \sin B \sin C.$$

On the other hand, $AD = AC \sin C = 2R \sin B \sin C$, while $OM = OC \sin \angle OCM = R \sin(90° - A) = R \cos A$. Thus
$$AP^2 = 4R^2 \cos A \sin B \sin C = 2AD \cdot OM,$$

as desired.

Problem 6 Let AB be a segment and C a point on its perpendicular bisector. Construct $C_1, C_2, \ldots, C_n, \ldots$ as follows: $C_1 = C$, and for $n \geq 1$, if C_n is not on $\overline{AB}$, then C_{n+1} is the circumcenter of triangle ABC_n. Find all points C such that the sequence $\{C_n\}_{n \geq 1}$ is well defined for all n and such that the sequence eventually becomes periodic.

Solution. All angles are directed modulo $180°$ unless otherwise indicated. Then C_n is uniquely determined by $\theta_n = \angle AC_nB$. Furthermore, we have $\theta_{n+1} = 2\theta_n$ for all n. For this to be eventually periodic we must have $\theta_{j+1} = \theta_{k+1}$ or $2^j \theta_1 = 2^k \theta_1$ for some j, k; that is, $180°$ must divide $(2^k - 2^j)\theta$ for some k and (not using directed angles) $180° \cdot r = (2^k - 2^j)\theta_1$ for some integers p, k.

Therefore $\theta_1 = 180° \cdot \frac{r}{2^k - 2^j}$ must be a rational multiple $\frac{p}{q} \cdot 180°$ with p, q relatively prime. Also, if $q = 2^n$ for some integer n then $\theta_{n+1} = 180° \cdot p = 180°$, which we cannot have. Thus q must also not be a power of two.

Conversely, suppose we have such an angle $\frac{p}{q} \cdot 180°$ where p, q are relatively prime and q is not a power of 2. First, the sequence of points is well-defined because $\frac{2^n p}{q}$ will always have a nontrivial odd divisor in its denominator. Thus it will never be an integer and $\theta_{n+1} = \frac{2^n p}{q} \cdot 180°$ will never equal $180°$.

Next write $q = 2^j t$ for odd t, and let $k = \phi(t) + j$. Because $t \mid 2^{\phi(t)} - 1$ we have

$$\theta_{k+1} = 2^k \cdot \theta_1 = 2^{\phi(t)+j} \cdot \frac{p}{q} \cdot 180° = 2^{\phi(t)} \cdot \frac{p}{t} \cdot 180° \equiv \frac{p}{t} \cdot 180°,$$

while

$$\theta_{j+1} \equiv 2^j \frac{p}{q} \cdot 180° = \frac{p}{t} \cdot 180°.$$

Thus $\theta_{j+1} \equiv \theta_{k+1}$, so the sequence is indeed periodic.

Therefore the set of valid points C is all points C such that $\angle ACB$ (no longer directed) equals $\frac{p}{q} \cdot 180°$ for relatively prime positive integers p, q, where q is not a power of 2.

7 Olimpiada Matemática del Cono Sur

Problem 1 Find the smallest positive integer n such that the 73 fractions
$$\frac{19}{n+21}, \frac{20}{n+22}, \frac{21}{n+23}, \ldots, \frac{91}{n+93}$$
are all irreducible.

Solution. Note that the difference between the numerator and denominator of each fraction is $n + 2$. Then $n + 2$ must be relatively prime to each of the integers from 19 to 91. Because this list contains a multiple of each prime p less than or equal to 91, $n + 2$ must only have prime factors greater than 91. The smallest such number is 97, so $n = 95$.

Problem 2 Let ABC be a triangle with $\angle A = 90°$. Construct point P on $\overline{BC}$ such that if Q is the foot of the perpendicular from P to $\overline{AC}$ then $PQ^2 = PB \cdot PC$.

Solution. Draw D on ray AB such that $AB = BD$, and draw E on ray AC such that $AC = CE$. Next draw F on $\overline{DE}$ such that $\angle FBD = \angle BCA$. Finally, draw the line through F perpendicular to $\overline{AC}$. We claim that this line intersects $\overline{BC}$ and $\overline{AC}$ at our desired points P and Q.

Because $BD \parallel FQ$, we have $\angle BFQ = \angle FBD = \angle BCA = \angle BCQ$. Hence $BFCQ$ is cyclic. Furthermore, because $BD = BA$ and $DA \parallel FQ$, we have $PF = PQ$. Thus by Power of a Point, we have $PB \cdot PC = PF \cdot PQ = PQ^2$, as desired.

Problem 3 There are 1999 balls in a row. Each ball is colored either red or blue. Underneath each ball we write a number equal to the sum of the number of red balls to its right and blue balls to its left. Exactly three numbers each appear on an odd number of balls; determine these three numbers.

Solution. Call of a coloring of $4n - 1$ *good* if exactly 3 numbers each appear on an odd number of balls. We claim that these three numbers will then be $2n - 2$, $2n - 1$, and $2n$. Let $b_1, b_2, \ldots$ represent the ball colors (either B or R), and let $x_1, x_2, \ldots$ represent the numbers under the respective balls. Then $x_{k+1} - x_k = 1$ if $b_k = b_{k+1} = B$; $x_{k+1} - x_k = -1$ if $b_k = b_{k+1} = R$; and $x_k = x_{k+1}$ if $b_k \neq b_{k+1}$. Now, when $n = 1$, the only good colorings are RRR and BBB, which both satisfy the claim.

For the sake of contradiction, let $n_0 > 1$ be the least n for which the claim no longer holds. Now, let a *couple* be a pair of adjacent, different-colored balls. For our coloring of n_0 balls, one of the following cases is true:

(i) *There exist two or more disjoint couples in the coloring.* Removing the two couples decreases all the other x_i by exactly 2, while the numbers originally on the four removed balls are removed in pairs. Thus we have constructed a good coloring of $4(n_0 - 1) - 1$ balls for which the claim does not hold, a contradiction.

(ii) *The balls are colored $RR \cdots RBRR \cdots R$ or $BB \cdots BRBB \cdots B$.* Then $\{x_i\}$ is a nondecreasing series, with equality only under the balls BRB or RBR. Then $(4n_0 - 1) - 2$ numbers appear an odd number of times, but this cannot equal 3.

(iii) *There exists exactly one couple in the coloring.* Suppose, without loss of generality, that we have a string of m blue balls followed by a string of n red balls. It is trivial to check that m and n cannot equal 1. Then by removing the final two blue balls and the first two red balls, we construct an impossible coloring of $4(n_0 - 1) - 1$ balls as in case (i).

(iv) *All the balls are of the same color.* We then have $4n_0 - 1 > 3$ distinct numbers, a contradiction.

Thus, all good colorings satisfy our claim. Specifically, for $n_0 = 500$, we find that the three numbers must be 998, 999, and 1000.

Problem 4 Let A be a number with six digits, three of which are colored and are equal to $1, 2, 4$. Prove that it is always possible to obtain a multiple of 7 by doing one of the following:

(1) eliminate the three colored numbers;

(2) write the digits of A in a different order.

Solution. Note that modulo 7, the six numbers 421, 142, 241, 214, 124, 412 are congruent to 1, 2, 3, 4, 5, 6, respectively. Let the other digits besides 1, 2, and 4 be x, y, and z, appearing in that order from left to right. If the 3-digit number xyz is divisible by 7, we may eliminate the three colored numbers. If not, the 6-digit number $xyz000$ is also not divisible by 7, and we may add the appropriate permutation abc of 124 to $xyz000$ to make $xyzabc$ divisible by 7.

Problem 5 Consider a square of side length 1. Let S be a set of finitely many points on the sides of the square. Prove that there is a vertex of the square such that the arithmetic mean of the squares of the distances from the vertex to all the points in S is no less than $\frac{3}{4}$.

Solution. Let the four vertices of the square be $V_1, V_2, V_3,$ and V_4, and let the set of points be $\{P_1, P_2, \ldots, P_n\}$. For a given P_k, we may assume without loss of generality that P_k lies on side V_1V_2. Writing $x = P_kV_1$, we have

$$\sum_{i=1}^{4} P_k V_i^2 = x^2 + (1-x)^2 + (1+x^2) + (1+(1-x)^2) = 4\left(x - \frac{1}{2}\right)^2 + 3 \geq 3.$$

Hence $\sum_{i=1}^{4} \sum_{j=1}^{n} P_j V_i^2 \geq 3n$, or

$$\sum_{i=1}^{4} \left(\frac{1}{n} \sum_{j=1}^{n} P_j V_i^2\right) \geq 3.$$

Thus the average of $\frac{1}{n}\sum_{j=1}^{n} P_j V_i^2$ (for $i = 1, 2, 3, 4$) is at least $\frac{3}{4}$. Hence if we select the V_i for which $\frac{1}{n}\sum_{j=1}^{n} P_j V_i^2$ is maximized, we are guaranteed it will be at least $\frac{3}{4}$.

Problem 6 An ant crosses a circular disc of radius r and it advances in a straight line, but sometimes it stops. Whenever it stops, it turns 60°, each time in the opposite direction. (If the last time it turned 60° clockwise, this time it will turn 60° counterclockwise, and vice versa.) Find the maximum length of the ant's path.

Solution. Suppose the ant begins its path at P_0, stops at $P_1, P_2, \ldots, P_{n-1}$ and ends at P_n. Note that all the segments $P_{2i}P_{2i+1}$ are parallel to each other and that all the segments $P_{2i+1}P_{2i+2}$ are parallel to each other. We may then translate all the segments so as to form two segments P_0Q and QP_n where $\angle P_0QP_n = 120°$. Then $P_0P_n \leq 2r$, and the length of the initial path is equal to $P_0Q + QP_n$. Let $P_0P_n = c$, $P_0Q = a$, and $QP_n = b$. Then

$$(2r)^2 \geq c^2 = a^2 + b^2 + ab = (a+b)^2 - ab \geq (a+b)^2 - \frac{1}{4}(a+b)^2,$$

so $\frac{4}{\sqrt{3}}r \geq a+b$ with equality if and only if $a = b$. The maximum is therefore $\frac{4}{\sqrt{3}}r$. This maximum can be attained with the path where $\overline{P_0P_2}$ is a diameter of the circle, and $P_0P_1 = P_1P_2 = \frac{2}{\sqrt{3}}r$.

8 St. Petersburg City Mathematical Olympiad (Russia)

Problem 9.1 Let $x_0 > x_1 > \cdots > x_n$ be real numbers. Prove that

$$x_0 + \frac{1}{x_0 - x_1} + \frac{1}{x_1 - x_2} + \cdots + \frac{1}{x_{n-1} - x_n} \geq x_n + 2n.$$

Solution. For $i = 0, 1, \ldots, n-1$, we have $x_i - x_{i+1} > 0$. Thus, by the AM-GM inequality,

$$(x_i - x_{i+1}) + \frac{1}{(x_i - x_{i+1})} \geq 2.$$

Adding up these inequalities for $i = 0, 1, \ldots, n-1$ gives the desired result.

Problem 9.2 Let $f(x) = x^2 + ax + b$ be a quadratic trinomial with integral coefficients and $|b| \leq 800$. It is known also that $f(120)$ is prime. Prove that $f(x) = 0$ has no integer roots.

Solution. Suppose, by way of contradiction, that $f(x)$ had an integer root r_1. Writing $f(x) = (x - r_1)(x - r_2)$, we see that its other root must be $r_2 = -a - r_1$, also an integer.

Because $f(120) = (120 - r_1)(120 - r_2)$ is prime, one of $|120 - r_1|$ and $|120 - r_2|$ equals 1, and the other equals some prime p.

Without loss of generality, assume that $|120 - r_1| = 1$. Then $r_1 = 119$ or 121, and $|r_1| \geq 119$. Also, $|120 - r_2|$ is a prime, but the numbers $114, 115, \ldots, 126$ are all composite: $119 = 7 \cdot 17$, and all the other numbers are clearly divisible by 2, 3, 5, or 11. Therefore $|r_2| \geq 7$, and $|b| = |r_1 r_2| \geq |119 \cdot 7| > 800$, a contradiction.

Problem 9.3 The vertices of a regular n-gon ($n \geq 3$) are labelled with distinct integers from $\{1, 2, \ldots, n\}$. For any three vertices A, B, C with $AB = AC$, the number at A is either larger than the numbers at B and C, or less than both of them. Find all possible values of n.

Solution. Suppose that $n = 2^s t$, where $t \geq 3$ is odd. Look at the regular t-gon $P_1 \cdots P_t$ formed by every 2^s-th point. In this t-gon, some vertex A has the second-smallest number, and some vertex B has the smallest number. Because t is odd, there must be a third vertex C with

---

[1] Problems are numbered as they appeared in the contests. Problems that appeared more than once in the contests are only printed once in this book.

$AB = AC$. Then the number at C must also be smaller than A's number — a contradiction.

Alternatively, if P_1 has a bigger number than P_t and P_2, then P_2 has a smaller number than P_3, P_3 has a bigger number than P_4, and so on around the circle — so that P_t has a bigger number than P_1, a contradiction.

Now we prove by induction on $s \geq 0$ that we can satisfy the conditions for $n = 2^s$. For $s = 1$, label the single point 1. For the inductive step, note that if we can label a regular 2^s-gon with the numbers $a_1, \ldots, a_{2^s}$ in that order, then we can label a regular 2^{s+1}-gon with the numbers $a_1, a_1 + 2^s, a_2, a_2 + 2^s, \ldots, a_{2^s}, a_{2^s} + 2^s$, as desired.

(Alternatively, when $n = 2^s$ one could label the vertices as follows. For $i = 1, 2, \ldots, 2^s$, reverse each digit of the s-bit binary expansion of $i - 1$ and then add 1 to the result. Label the i-th vertex with this number.)

Problem 9.4 Points A_1, B_1, C_1 are chosen on the sides BC, CA, AB of an isosceles triangle ABC ($AB = BC$). It is known that $\angle BC_1A_1 = \angle CA_1B_1 = \angle A$. Let BB_1 and CC_1 meet at P. Prove that AB_1PC_1 is cyclic.

Solution. All angles are directed modulo $180°$.

Let the circumcircles of triangles AB_1C_1 and A_1B_1C intersect at P', so that $AB_1P'C_1$ and $CB_1P'A_1$ are cyclic. Then

$$\angle A_1P'C_1 = (180° - \angle C_1P'B_1) + (180° - \angle B_1P'A_1)$$
$$= \angle B_1AC_1 + \angle A_1CB_1 = \angle CAB + \angle BCA$$
$$= 180° - \angle ABC$$
$$= 180° - \angle C_1BA_1,$$

so $BA_1P'C_1$ is cyclic as well.

Now, $\angle BC_1A_1 = \angle A = \angle C$ implies that $\triangle BC_1A_1 \sim \triangle BCA$, so $BC_1 \cdot BA = BA_1 \cdot BC$. Thus B has equal power with respect to the circumcircles of triangles AB_1C_1 and A_1B_1C, so it lies on their radical axis B_1P'.

Similarly, $\angle CA_1B_1 = \angle A$ implies that $\triangle ABC \sim \triangle A_1B_1C$, so $CA \cdot CB_1 = CB \cdot CA_1$. Thus C has equal power with respect to the circumcircles of triangles B_1AC_1 and A_1BC_1, so it lies on their radical axis C_1P'.

Then P' lies on both line CC_1 and line BB_1, so it must *equal P*. Therefore AB_1PC_1 is indeed cyclic, as desired.

Problem 9.5 Find the set of possible values of the expression
$$f(x,y,z) = \left\{ \frac{xyz}{xy+yz+zx} \right\},$$
for positive integers x, y, z. Here $\{x\} = x - \lfloor x \rfloor$ is the fractional part of x.

Solution. Clearly $f(x, y, z)$ must be a nonnegative rational number below 1. We claim all such numbers are in the range of f. Suppose we have nonnegative integers p and q with $p < q$, and let $x_1 = y_1 = 2(q-1)$ and $z_1 = 1$. Then
$$f(x_1, y_1, z_1) = \left\{ \frac{4(q-1)^2}{4(q-1)^2 + 4(q-1)} \right\} = \frac{q-1}{q}.$$
Writing $X = \frac{xyz}{xy+yz+zx}$, notice that for any nonzero integer k we have
$$f(kx, ky, kz) = \{kX\} = \{k\lfloor X \rfloor + k\{X\}\} = \{k \cdot f(x,y,z)\}.$$
Then $f(p(q-1) \cdot x_1, p(q-1) \cdot y_1, p(q-1) \cdot z_1) = \frac{p}{q}$, so every nonnegative rational $\frac{p}{q} < 1$ is indeed in the range of f.

Problem 9.6 Let $\overline{AL}$ be the angle bisector of triangle ABC. Parallel lines ℓ_1 and ℓ_2 equidistant from A are drawn through B and C respectively. Points M and N are chosen on ℓ_1 and ℓ_2 respectively such that lines AB and AC meet lines LM and LN at the midpoints of $\overline{LM}$ and $\overline{LN}$ respectively. Prove that $LM = LN$.

Solution. Let A, B, C also represent the angles at those points in triangle ABC.

Let line ℓ pass through A perpendicular to AL. Next, draw M' and N' on ℓ so that $\angle ALM' = \angle ALN' = \frac{A}{2}$ (with M' and N' on the same sides of line AL as B and C, respectively). Finally, let ℓ hit ℓ_1 at Q.

We claim that M' lies on ℓ_1. Orient the figure so that ℓ_1 and ℓ_2 are vertical, and let $x = \angle QBA$. Note that $AM' = AL \tan \angle ALM' = AL \tan \frac{A}{2}$, so the horizontal distance between A and M' is
$$AM' \sin(180° - \angle AQB) = AL \tan \frac{A}{2} \cdot \sin(\angle QBA + \angle BAQ)$$
$$= AL \tan \frac{A}{2} \cdot \sin\left(x + 90° - \frac{A}{2}\right)$$
$$= AL \tan \frac{A}{2} \cdot \cos\left(\frac{A}{2} - x\right).$$

On the other hand, the horizontal distance between A and ℓ_1 is $AB \sin x$. Thus we need only prove that

$$AB \sin x = AL \tan \frac{A}{2} \cdot \cos \left(\frac{A}{2} - x \right).$$

By the law of sines on triangle ABL, we know that $AL \sin \left(\frac{A}{2} + B \right) = AB \sin B$. Hence we must prove

$$\sin \left(\frac{A}{2} + B \right) \cdot \sin x = \sin B \cdot \tan \frac{A}{2} \cdot \cos \left(\frac{A}{2} - x \right)$$

$$\iff \sin \left(\frac{A}{2} + B \right) \cdot \cos \frac{A}{2} \cdot \sin x$$

$$= \sin B \cdot \sin \frac{A}{2} \cdot \cos \left(\frac{A}{2} - x \right)$$

$$\iff (\sin(A + B) + \sin B) \cdot \sin x$$

$$= \sin B \cdot (\sin(A - x) + \sin x) \iff \sin C \cdot \sin x$$

$$= \sin B \cdot \sin(A - x) \iff AB \sin x$$

$$= AC \sin(A - x).$$

$AB \sin x$ is the distance between A and ℓ_1, and $AC \sin(A - x)$ is the distance between A and ℓ_2 — and these distances are equal. Therefore, M' indeed lies on ℓ_1.

Now let lines AB and LM' intersect at P. Because $\angle LAP = \angle PLA = \frac{A}{2}$, $\overline{AP}$ is the median to the hypotenuse of right triangle $M'AL$. Thus line AB hits the midpoint of $\overline{LM'}$, so $M = M'$.

Similarly, $N = N'$. Therefore, $\angle LMN = 90° - \frac{A}{2} = \angle LNM$ and $LM = LN$, as desired.

Problem 9.7 A *corner* is the figure resulted by removing 1 unit square from a 2×2 square. Prove that the number of ways to cut a 998×999 rectangle into corners, where two corners *can* form a 2×3 rectangle, does not exceed the number of ways to cut a 1998×2000 rectangle into corners, so that *no* two form a 2×3 rectangle.

Solution. Take any tiling of a 998×999 rectangle with corners, and add a 2×999 block underneath also tiled with corners:

Next, enlarge this 1000 × 999 board to twice its size, and replace each large corner by four normal-sized corners as follows:

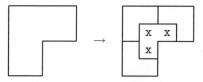

For each corner in the tiling of the 1000 × 999 board, none of the four new corners can be half of a 2 × 3 rectangle. Also, the "central" corners (like the one marked with x's above) have the same orientations as the original corners tiling the 1000 × 999 rectangle.

Thus, different tilings of a 998 × 999 rectangle turn into different tilings of a 2000 × 1998 board where no two corners form a 2 × 3 rectangle — which implies the desired result.

Problem 9.8 A convex n–gon ($n > 3$) is divided into triangles by non-intersecting diagonals. Prove that one can mark $n-1$ segments among these diagonals and sides of the polygon so that no set of marked segments forms a closed polygon and no vertex belongs to exactly two segments.

Solution. After we mark some segments, let the *degree* $d(V)$ of a vertex V be the number of marked segments it is on. Also, we say we *mark up* a triangulated n-gon with respect to side AB if we mark $n-1$ segments, no marked segments form a closed polygon, and none of the $n-2$ vertices besides A and B have degree 2.

Lemma. *Given any triangulated convex n-gon ($n \geq 3$) and any two adjacent vertices A and B, we can mark up the n-gon with respect to side AB in three ways (i), (ii), (iii), each satisfying the corresponding condition from the following list:*

(i) $\overline{AB}$ *is marked, and* $d(A) \geq 2$.

(ii) $\overline{AB}$ *is marked, and* $d(A) \neq 2$.

(iii) $(d(A), d(B)) \neq (1, 1)$ *or* $(2, 2)$.

Proof For $n = 3$, given a triangle ABC we can mark sides AB and (i) AC, (ii) BC, and (iii) AC. Now suppose that $n \geq 4$ and that the claims are true for all smaller n; we prove that they are true for n as well.

$\overline{AB}$ must be part of some triangle ABC of drawn segments. Then either (a) $\overline{AC}$ is a side of the polygon but $\overline{BC}$ is not; (b) $\overline{BC}$ is a side but $\overline{AC}$ is not; or (c) neither $\overline{AC}$ nor $\overline{BC}$ is a side.

(a) Let P be the $(n-1)$-gon formed by the vertices not including A.
(i) Apply (ii) to P so that $d(C) \neq 2$ and $\overline{BC}$ is marked; then unmark $\overline{BC}$, and mark $\overline{AC}$ and $\overline{AB}$.
(ii) Apply (ii) to P so that $d(C) \neq 2$ and $\overline{BC}$ is marked; then mark $\overline{AB}$.
(iii) Apply (i) to P so that $d(B) \geq 2$ and $\overline{BC}$ is marked. If $d(C) \neq 2$, then mark $\overline{AB}$; otherwise mark $\overline{AC}$.

(b) Let P be the $(n-1)$-gon formed by the vertices besides B.
(i) Apply (ii) to P so that $d(C) \neq 2$ and $\overline{AC}$ is marked; then mark $\overline{AB}$.
(ii) Apply (ii) to P so that $d(C) \neq 2$ and $\overline{AC}$ is marked. If $d(A) = 2$, then mark $\overline{AB}$; otherwise unmark $\overline{AC}$ and mark $\overline{AB}$ and $\overline{BC}$.
(iii) Repeat the construction in (i).

(c) Let P be the polygon formed by A, C, and the vertices in between (not on the same side of line AC as B); and let Q be the polygon formed by B, C, and the vertices in between (not on the same side of line BC as A).

(i) Apply (i) to P and Q so that $d(C) \geq 2+2 = 4$ and $\overline{AC}$, $\overline{BC}$ are marked; then unmark $\overline{BC}$ and mark $\overline{AB}$.
(ii) Apply (i) to P and Q so that $d(C) \geq 2+2 = 4$ and $\overline{AC}$, $\overline{BC}$ are marked. If $d(A) = 2$, unmark $\overline{BC}$ and mark $\overline{AB}$; otherwise, unmark $\overline{AC}$ and mark $\overline{AB}$.
(iii) Repeat the construction in (ii). ∎

Now to the main result. Because there are $n \geq 4$ sides but only $n-2$ triangles, some triangle contains two adjacent sides XY and YZ. Let P be the $(n-1)$-gon formed by the vertices not including Y, and apply (iii) to P and vertices X, Z.

Then if $d(X)$ or $d(Z)$ equals 2, mark $\overline{XY}$ or $\overline{YZ}$, respectively. Otherwise, if $d(X)$ or $d(Z)$ equals 1, mark $\overline{YZ}$ or $\overline{XY}$, respectively. Otherwise, both $d(X)$ and $d(Z)$ are at least 3, and we can mark either $\overline{XY}$ or $\overline{YZ}$ to finish off.

Problem 10.1 The sequence $\{x_n\}$ of positive integers is formed by the following rule: $x_1 = 10^{999} + 1$, and for every $n \geq 2$, the number x_n is obtained from the number $11x_{n-1}$ by rubbing out its first digit. Is the sequence bounded?

Solution. If x_{n-1} has k digits, then $x_{n-1} < 10^k$ so that $11x_{n-1} < 11 \cdot 10^k = 1100\ldots 0$. Thus if $11x_{n-1}$ has $k+2$ digits, its first two digits are 1 and 0. Rubbing out its first digit leaves x_n with at most k digits. Otherwise, $11x_{n-1}$ has at most $k+1$ digits, and rubbing out its first digit still leaves x_n with at most k digits. Therefore the number of digits in each x_n is bounded, and the x_n are bounded as well.

Problem 10.2 Prove that any positive integer less than $n!$ can be represented as a sum of no more than n positive integer divisors of $n!$.

Solution. Fix n, and write $a_k = \frac{n!}{k!}$ for each $k = 1, 2, \ldots, n$. Suppose we have some number m with $a_k \leq m < a_{k-1}$ where $2 \leq k \leq n$. Then consider the number $d = a_k \left\lfloor \frac{m}{a_k} \right\rfloor$. We have $0 \leq m-d < a_k$; furthermore, because $s = \left\lfloor \frac{m}{a_k} \right\rfloor < \frac{a_{k-1}}{a_k} = k$, we know that $\frac{n!}{d} = \frac{k!}{s}$ is an integer. Thus from m we can subtract d, a factor of $n!$, to obtain a number less than a_k.

Then if we start with any positive integer $m < n! = a_1$, then by subtracting at most one factor of $n!$ from m we can obtain an integer less than a_2; by subtracting at most one more factor of $n!$ we can obtain an integer less than a_3; and so on, so that we can represent m as the sum of at most $n-1$ positive integer divisors of $n!$.

Problem 10.5 How many 10-digit numbers divisible by 66667 are there whose decimal representation contains only the digits 3, 4, 5, and 6?

Solution. Suppose that $66667n$ had 10 digits, all of which were 3, 4, 5, and 6. Then

$$3333333333 \leq 66667n \leq 6666666666 \quad \Rightarrow \quad 50000 \leq n \leq 99999.$$

Now consider the following cases:

(i) $n \equiv 0 \pmod{3}$. Then

$$66667n = \frac{2}{3}n \cdot 10^5 + \frac{1}{3}n,$$

the five digits of $2 \cdot \frac{n}{3}$ followed by the five digits of $\frac{n}{3}$. These digits are all 3, 4, 5, or 6 if and only if $\frac{n}{3} = 33333$ and $n = 99999$.

(ii) $n \equiv 1 \pmod{3}$. Then

$$66667n = \frac{2}{3}(n-1) \cdot 10^5 + \frac{1}{3}(n+2) + 66666,$$

the five digits of $\frac{2}{3}(n-1)$ followed by the five digits of $\frac{1}{3}(n+2) + 66666$. Because $\frac{1}{3}(n+2) + 66666$ must be between 66667 and 99999, its digits cannot all be 3, 4, 5, or 6. Hence there are no satisfactory $n \equiv 1 \pmod{3}$.

(iii) $n \equiv 2 \pmod 3$. Let $a = \frac{1}{3}(n-2)$. Then

$$66667n = \left(\frac{2}{3}(n-2) + 1\right) \cdot 10^5 + \frac{1}{3}(n-2) + 33334,$$

the five digits of $x = 2a+1$ followed by the five digits of $y = a+33334$. The units digits in x and y are between 3 and 6 if and only if the units digit in a is 1 or 2. In this case the other digits in x and y are all between 3 and 6 if and only if the other digits in a are 2 or 3. Thus there are thirty-two satisfactory a — we can choose each of its five digits from two options — and each a corresponds to a satisfactory $n = 3a+2$.

Therefore there is exactly one satisfactory $n \equiv 0 \pmod 3$, and thirty-two satisfactory $n \equiv 2 \pmod 3$ — making a total of thirty-three values of n and thirty-three ten-digit numbers.

Problem 10.6 The numbers $1, 2, \ldots, 100$ are arranged in the squares of a 10×10 table in the following way: the numbers $1, \ldots, 10$ are in the bottom row in increasing order, numbers $11, \ldots, 20$ are in the next row in increasing order, and so on. One can choose any number and two of its neighbors in two opposite directions (horizontal, vertical, or diagonal). Then either the number is increased by 2 and its neighbors are decreased by 1, or the number is decreased by 2 and its neighbors are increased by 1. After several such operations the table again contains all the numbers $1, 2, \ldots, 100$. Prove that they are in the original order.

Solution. Label the table entries $a_{11}, a_{12}, \ldots$, with a_{ij} in the ith row and jth column, where the bottom-left corner is in the first row and first column. Also, let $b_{ij} = 10(i-1) + j$ be the number originally in the ith row and jth column.

Observe that

$$P = \sum_{i,j=1}^{10} a_{ij} b_{ij}$$

is invariant — this is because every time entries a_{mn}, a_{pq}, a_{rs} are changed (with $m+r = 2p$ and $n+s = 2q$), P increases or decreases by $b_{mn} - 2b_{pq} + b_{rs}$, or

$$10\big((m-1) + (r-1) - 2(p-1)\big) + (n+s-2q) = 0.$$

(For example, if entries a_{35}, a_{46}, a_{57} are changed then P changes by $\pm(35 - 2 \cdot 46 + 57) = 0$.)

At first, $P = \sum_{i,j=1}^{10} b_{ij}b_{ij}$. At the end, the entries a_{ij} equal the b_{ij} in some order, and $P = \sum_{i,j=1}^{10} a_{ij}b_{ij}$. By the rearrangement inequality, this is at least $\sum_{i,j=1}^{10} b_{ij}b_{ij}$, with equality only when each $a_{ij} = b_{ij}$.

We know equality *does* occur because P is invariant. Therefore the a_{ij} do indeed equal the b_{ij} in the same order, and thus the entries $1, 2, \ldots, 100$ appear in their original order.

Problem 10.7 Quadrilateral $ABCD$ is inscribed in circle ω centered at O. The bisector of $\angle ABD$ meets $\overline{AD}$ and ω at points K and M respectively. The bisector of $\angle CBD$ meets $\overline{CD}$ and ω at points L and N respectively. Suppose that $KL \parallel MN$. Prove that the circumcircle of triangle MON goes through the midpoint of $\overline{BD}$.

Solution. All angles are directed modulo $180°$. Let P, Q, R be the midpoints of $\overline{DB}, \overline{DA}, \overline{DC}$ respectively. Points M and N are the midpoints of arcs AD and DC, respectively. Hence M, Q, O and N, R, O are collinear, so that $\angle MON = \angle QOR$. Furthermore, $\angle DQO = \angle DRO = 90°$ implies that $DQOR$ is cyclic, and $\angle QOR = 180° - \angle RDQ = 180° - \angle CDA = \angle ABC$.

In addition, $\angle QPR = \angle QPD + \angle DPR = \angle ABD + \angle DBC = \angle ABC$. Then to prove our claim, it suffices to show that $\triangle PQM \sim \triangle PRN$ (with the same orientation) because then

$$\angle MPN = \angle QPR - \angle QPM + \angle RPN = \angle QPR = \angle ABC = \angle MON$$

so that $MPON$ would be cyclic. (Alternatively, it is possible that triangles PQM and PRN are both degenerate.)

Now, let $a = AB, b = BC, c = CD, D = DA, e = AC$, and $f = BD$. Suppose line MN hits lines AD and CD at E and F, respectively. Then

$$\angle DEF = \frac{1}{2}\left(\widehat{DN} + \widehat{AM}\right)$$
$$= \frac{1}{2}\left(\widehat{NC} + \widehat{MD}\right)$$
$$= \angle EFD,$$

so $DE = DF$. Because $KL \parallel EF$, we have $DK = DL$. Furthermore, by the Angle Bisector Theorem on triangles ABD and CBD, $DK = d \cdot \frac{f}{a+f}$ and $DL = c \cdot \frac{f}{b+f}$, so that

$$d(b+f) = c(a+f)$$
$$(c-d)f = bd - ac. \qquad (1)$$

If $c = d$ then we must have $bd = ac$, so $a = b$. Then $\overline{BD}$ is a diameter of ω so $P = O$ and the claim is obvious. Otherwise $c - d, bd - ac \neq 0$, and $f = \frac{bd-ac}{c-d}$.

Now we prove that $\triangle PQM \sim \triangle PRN$. Observe that

$$\angle PQM = \angle PQD + \angle DQM = \angle BAD + 90°$$
$$= \angle BCD + 90° = \angle PRD + \angle DRN = \angle PRN,$$

and also note that $\angle PQM$ and $\angle PRN$ are both obtuse.

So, we need only prove that

$$\frac{PQ}{QM} = \frac{PR}{RN} \iff \frac{\frac{BA}{2}}{AQ\tan\angle MAQ}$$
$$= \frac{\frac{BC}{2}}{CR\tan\angle RCN} \iff \frac{\frac{BA}{2}}{\frac{AD}{2}\tan\angle MCD}$$
$$= \frac{\frac{BC}{2}}{\frac{CD}{2}\tan\angle DAN} \iff \frac{a}{d\tan\angle ACM} = \frac{b}{c\tan\angle NAC}. \quad (2)$$

Because lines CM, AN are angle bisectors of $\angle ACD$, $\angle DAC$, they intersect at the incenter I of triangle ACD. Also, let T be the point where the incircle of triangle ACD hits $\overline{AC}$. Then

$$\tan\angle ACM = \frac{IT}{TC} = \frac{IT}{\frac{1}{2}(e+c-d)}$$

and

$$\tan\angle NAC = \frac{IT}{TA} = \frac{IT}{\frac{1}{2}(e+d-c)},$$

so

$$\frac{\tan\angle ACM}{\tan\angle NAC} = \frac{e+d-c}{e+c-d}.$$

Thus (2) is equivalent to

$$ac(e+c-d) = bd(e+d-c) \iff e$$
$$= \frac{(ac+bd)(c-d)}{bd-ac}.$$

By Ptolemy's Theorem and (1), we have

$$e = \frac{ac+bd}{f} = \frac{(ac+bd)(c-d)}{bd-ac},$$

as desired. Therefore triangle PQM is similar to triangle PRN, $\angle MPN = \angle MON$, and $MPON$ is indeed cyclic.

Problem 11.1 There are 150 red, 150 blue, and 150 green balls flying under the big top in the circus. There are exactly 13 green balls inside every blue one, and exactly 5 blue balls and 19 green balls inside every red one. (A ball is considered to be "inside" another ball even if it is not immediately inside it; for example, if a green ball is inside a blue ball and that blue ball is inside a red ball, then the green ball is also inside the red ball.) Prove that some green ball is not contained in any of the other 449 balls.

Solution. Suppose, by way of contradiction, that every green ball were in some other ball. Notice that no red ball is inside a blue ball, because then the blue ball would contain at least 19 green balls.

Look at the red balls that are not inside any other red balls — suppose there are m of them, and throw them away along with all the balls they contain. Then we throw all the red balls and anything inside a red ball, including $19m$ green balls. Also, because no red ball is inside a blue ball, there are still exactly 13 green balls inside each of the remaining blue balls.

Now look at the blue balls left that are not inside any other blue balls — suppose there are n of them, and throw them away along with all the balls they contain. Then we throw away all the remaining blue balls and the $13n$ more green balls inside.

At this point, all the green balls are gone and hence we must have $150 = 19m + 13n$. Taking this equation modulo 13, we find that $7 \equiv 6m \Rightarrow m \equiv 12 \pmod{13}$. Then $m \geq 12$ and

$$150 = 19m + 13n \geq 19 \cdot 12 = 228,$$

a contradiction. Thus our original assumption was false, and some green ball is contained in no other ball.

Problem 11.3 Let $\{a_n\}$ be an arithmetic sequence of positive integers. For every n, let p_n be the largest prime divisor of a_n. Prove that the sequence $\{\frac{a_n}{p_n}\}$ is unbounded.

Solution. Let d be the common difference of the arithmetic sequence and N be an arbitrary number. Choose primes $p < q$ bigger than N and relatively prime to d. Then there is some term a_k divisible by pq. Because $p_k \geq q > p$, we have $p \mid \frac{a_k}{p_k}$ so that

$$\frac{a_k}{p_k} \geq p > N.$$

Thus for any N, there exists k such that $\frac{a_k}{p_k} > N$, so the sequence $\{\frac{a_n}{p_n}\}$ is unbounded.

Problem 11.4 All positive integers not exceeding 100 are written on both sides of 50 cards (each number is written exactly once). The cards are put on a table so that Vasya only knows the numbers on the top side of each card. Vasya can choose several cards, turn them upside down, and then find the sum of all 50 numbers now on top. What is the maximum sum Vasya can be sure to obtain or beat?

Solution. The answer is

$$2525 = \frac{1}{2}(1 + 2 + 3 + \cdots + 100).$$

Vasya can always obtain or beat this: if the 50 numbers on top add to this or more, he is done; otherwise, if they add to less, Vasya can flip all of them.

This might be the best Vasya can do. Suppose that the numbers on top are 26, 27, 28, ..., 75, with sum 2525, and that the numbers on the k cards Vasya flips over are the first k numbers in the sequence 1, 2, ..., 25, 76, 77, ..., 100. If he flips over 0 through 25 cards, his sum decreases, and if he flips over more, his sum is at most

$$1 + 2 + \cdots + 25 + 76 + 77 + \cdots + 100 = 2525.$$

Either way, Vasya cannot obtain a sum of more than 2525, as claimed.

Problem 11.5 Two players play the following game. They in turn write on a blackboard different divisors of 100! (except 1). A player loses if after his turn, the greatest common divisor of the all the numbers written becomes 1. Which of the players has a winning strategy?

Solution. The second player has a winning strategy. Notice that every prime $p < 100$ divides an *even* number of factors of 100!: the factors it divides can be split into disjoint pairs $(k, 97k)$ — or, if $p = 97$, into the pairs $(k, 89k)$. (Note that none of these factors is 1, because 1 is not divisible by p.)

If the first player writes down a prime p, the second player can write down any other number divisible by p; if the first player writes down a composite number, the second player can write down a prime p dividing that number. Either way, from now on the players can write down a new number $q \mid 100!$ without losing if and only if it is divisible by p. Because there are an even number of such q, the second player will write down the last acceptable number and the first player will lose.

Problem 11.7 A connected graph G has 500 vertices, each with degree 1, 2, or 3. We call a black-and-white coloring of these vertices *interesting* if more than half of the vertices are white but no two white vertices are connected. Prove that it is possible to choose several vertices of G so that in any interesting coloring, more than half of the chosen vertices are black.

Solution. We first give an algorithm to (temporarily) erase edges from G so that our graph consists of *chains* of vertices — sequences of vertices $V_1, \ldots, V_n$ where each V_i is adjacent to V_{i+1} — and possibly one leftover vertex.

First, as long as G still contains any cycle erase an edge from that cycle. Such erasures eventually make G a tree (a connected graph with no cycles). Look at the leaves (vertices with degree 1) that are not part of a chain yet. If all of them are adjacent to a degree-3 vertex, then we must have exactly four unchained vertices left with one central vertex adjacent to the other 3. Remove one of the edges, and we are done. Otherwise, one of the leaves is *not* adjacent to a degree-3 vertex. Travel along the graph from this vertex until we reach a degree-3 vertex V, and erase the edge going into V. This creates a new chain, while leaving all the unchained vertices in a smaller, still-connected tree. We can then repeat the algorithm on this tree until we are finished.

Besides our lone vertex, every vertex is in an *odd chain* (a chain with an odd number of vertices) or an *even chain* (a chain with an even number of vertices). For our chosen vertices, in each odd chain pick one vertex adjacent to one of the ends. Even with all the original edges back in place, for any interesting coloring observe that in any of our chains at most every other vertex can be white. Thus in any even chain, at most half the vertices are white. Furthermore, if a chosen vertex is white, then in its odd chain there is at least one more black vertex than white; if a chosen vertex is black, then there is at most one more white vertex than black.

Suppose there are b odd chains with a black chosen vertex, and w odd chains with a white chosen vertex. If there is a lone vertex, there are at most $1 + b - w$ more white vertices than black in our graph so that $1 + b - w > 0$. We know $1 + b - w$ must be even because $1 + b + w \equiv 500 \pmod{2}$. Then $1 + b - w \geq 2$ and $b - w \geq 1$, implying that we have more black chosen vertices than white.

Finally, if there is *no* lone vertex, then there are at most $b - w$ more white vertices than black in our graph. Thus $b - w > 0$ and we still have more black chosen vertices than white. Therefore either way, for every interesting coloring we have more black chosen vertices than white, as desired.

Problem 11.8 Three conjurers show a trick. They give a spectator a pack of cards with numbers $1, 2, \ldots, 2n + 1$ ($n > 6$). The spectator takes one card and arbitrarily distributes the rest evenly between the first and the second conjurers. Without communicating with each other, these conjurers study their cards, each chooses an ordered pair of their cards, and gives these pairs to the third conjurer. The third conjurer studies these four cards and announces which card is taken by the spectator. Explain how such a trick can be done.

Solution. We will have each of the first two conjurers use their ordered pairs to communicate the sums of their card values modulo $2n + 1$. With this information, the third conjurer can simply subtract these two sums from

$$1 + 2 + \cdots + (2n + 1) = (2n + 1)(n + 1) \equiv 0 \,(\text{mod } 2n + 1)$$

to determine the remaining card.

From now on, all entries of ordered pairs are taken modulo $2n + 1$. Also, let $(a, b)_k$ denote the ordered pair $(a + k, b + k)$ and let its *difference* be $b - a$ (taken modulo $2n + 1$ between 1 and $2n$).

Let

$$(0, 2n)_k, (0, 1)_k, (n, 2)_k, (n, 2n - 1)_k, (4, n + 1)_k, \text{ and } (2n - 3, n + 1)_k$$

all represent the sum k (mod $2n + 1$). These pairs' differences are $2n$, 1, $n+3$, $n-1$, $n-3$, $n+5$. Because $n > 5$, these differences are all distinct.

If n is odd then let $(1, 2n)_k, (2, 2n - 1)_k, \ldots, (n - 1, n + 2)_k$ also represent the sum k (mod $2n + 1$). These pairs' differences are all odd: $2n - 1, 2n - 3, \ldots, 3$. Furthermore, they are all different from the one odd difference, 1, that we found in the last paragraph.

Similarly, if n is even then let $(2n, 1)_k, (2n - 1, 2)_k, \ldots, (n + 2, n - 1)_k$ represent the sum k (mod $2n + 1$). These pairs' differences are all even: $2, 4, \ldots, 2n - 2$. They are all different from the one even difference, $2n$, that we found two paragraphs ago.

Note that if two of the assigned pairs $(a_1, b_1)_{k_1}$ and $(a_2, b_2)_{k_2}$ are equal, then their differences must be equal and we must have $b_1 - a_1 \equiv$

$b_2 - a_2 \pmod{2n+1}$. Because we found that the differences $b - a$ are distinct, we must have $(a_1, b_1) = (a_2, b_2)$ and therefore $k_1 = k_2$ as well. Thus any pair (a, b) is assigned to at most one sum, and our choices are well-defined.

Now, suppose that one of the first two conjurers has cards whose values sum to $k \pmod{2n+1}$. Assume by way of contradiction that he could not give any pair $(a, b)_k$ described above. Then, letting $S_k = \{s + k \mid s \in S\}$, he has at most one card from each of the three triples $\{0, 1, 2n\}_k$, $\{2, n, 2n - 1\}_k$, $\{4, n+1, 2n-3\}_k$, and he has at most one card from each of the $n - 4$ pairs $\{3, 2n - 2\}_k$, $\{5, 2n - 4\}_k$, $\{6, 2n - 5\}_k$, ..., $\{n - 1, n + 2\}_k$. However, these sets partition *all* of $\{0, 1, 2, \ldots, 2n\}$, so the magician must then have at most $3 + (n - 4) = n - 1$ cards — a contradiction. Thus our assumption was false, and both conjurers can indeed communicate the desired sums. This completes the proof.

3

2000 National Contests: Problems

1 Belarus

Problem 1 Let M be the intersection point of the diagonals AC and BD of a convex quadrilateral $ABCD$. The bisector of angle ACD hits ray BA at K. If $MA \cdot MC + MA \cdot CD = MB \cdot MD$, prove that $\angle BKC = \angle CDB$.

Problem 2 In an equilateral triangle of $\frac{n(n+1)}{2}$ pennies, with n pennies along each side of the triangle, all but one penny shows heads. A *move* consists of choosing two adjacent pennies with centers A and B and flipping every penny on line AB. Determine all initial arrangements — the value of n and the position of the coin initially showing tails — from which one can make all the coins show tails after finitely many moves.

Problem 3 We are given triangle ABC with $\angle C = 90°$. Let M be the midpoint of the hypotenuse $\overline{AB}$, H be the foot of the altitude $\overline{CH}$, and P be a point inside the triangle such that $AP = AC$. Prove that $\overline{PM}$ bisects angle BPH if and only if $\angle A = 60°$.

Problem 4 Does there exist a function $f : \mathbb{N} \to \mathbb{N}$ such that

$$f(f(n-1)) = f(n+1) - f(n)$$

for all $n \geq 2$?

Problem 5 In a convex polyhedron with m triangular faces, exactly four edges meet at each vertex. Find the minimum possible value of m.

Problem 6

(a) Prove that $\{n\sqrt{3}\} > \frac{1}{n\sqrt{3}}$ for every positive integer n, where $\{x\}$ denotes the fractional part of x.

(b) Does there exist a constant $c > 1$ such that $\{n\sqrt{3}\} > \frac{c}{n\sqrt{3}}$ for every positive integer n?

Problem 7 Let $M = \{1, 2, \ldots, 40\}$. Find the smallest positive integer n for which it is possible to partition M into n disjoint subsets such that whenever a, b, and c (not necessarily distinct) are in the same subset, $a \neq b + c$.

Problem 8 A positive integer is called *monotonic* if its digits in base 10, read from left to right, are in nondecreasing order. Prove that for each $n \in \mathbb{N}$, there exists an n-digit monotonic number which is a perfect square.

Problem 9 Given a pair $(\vec{a}, \vec{b})$ of vectors in the plane, a *move* consists of choosing a nonzero integer k and then changing $(\vec{a}, \vec{b})$ to either (i) $(\vec{a} + 2k\vec{b}, \vec{b})$ or (ii) $(\vec{a}, \vec{b} + 2k\vec{a})$. A *game* consists of applying a finite sequence of moves, alternating between moves of types (i) and (ii), to some initial pair of vectors.

(a) Is it possible to obtain the pair $((1, 0), (2, 1))$ during a game with initial pair $((1, 0), (0, 1))$, if the first move is of type (i)?

(b) Find all pairs $((a, b), (c, d))$ that can be obtained during a game with initial pair $((1, 0), (0, 1))$, where the first move can be of either type.

Problem 10 Prove that
$$\frac{a^3}{x} + \frac{b^3}{y} + \frac{c^3}{z} \geq \frac{(a+b+c)^3}{3(x+y+z)}$$
for all positive real numbers a, b, c, x, y, z.

Problem 11 Let P be the intersection point of the diagonals AC and BD of the convex quadrilateral $ABCD$ in which $AB = AC = BD$. Let O and I be the circumcenter and incenter of triangle ABP, respectively. Prove that if $O \neq I$, then lines OI and CD are perpendicular.

2 Bulgaria

Problem 1 A line ℓ is drawn through the orthocenter of acute triangle ABC. Prove that the reflections of ℓ across the sides of the triangle are concurrent.

Problem 2 There are 2000 white balls in a box. There are also unlimited supplies of white, green, and red balls, initially outside the box. During each turn, we can replace two balls in the box with one or two balls as follows: two whites with a green, two reds with a green, two greens with a white and red, a white and green with a red, or a green and red with a white.

(a) After finitely many of the above operations there are three balls left in the box. Prove that at least one of them is a green ball.

(b) Is it possible after finitely many operations to have only one ball left in the box?

Problem 3 The incircle of the isosceles triangle ABC touches the legs AC and BC at points M and N, respectively. A line t is drawn tangent to minor arc MN, intersecting $\overline{NC}$ and $\overline{MC}$ at points P and Q, respectively. Let T be the intersection point of lines AP and BQ.

(a) Prove that T lies on $\overline{MN}$;

(b) Prove that the sum of the areas of triangles ATQ and BTP is smallest when t is parallel to line AB.

Problem 4 We are given $n \geq 4$ points in the plane such that the distance between any two of them is an integer. Prove that at least $\frac{1}{6}$ of these distances are divisible by 3.

Problem 5 In triangle ABC, $\overline{CH}$ is an altitude, and cevians $\overline{CM}$ and $\overline{CN}$ bisect angles ACH and BCH, respectively. The circumcenter of triangle CMN coincides with the incenter of triangle ABC. Prove that $[ABC] = \frac{AN \cdot BM}{2}$.

Problem 6 Let $a_1, a_2, \ldots$ be a sequence such that $a_1 = 43$, $a_2 = 142$, and $a_{n+1} = 3a_n + a_{n-1}$ for all $n \geq 2$. Prove that

(a) a_n and a_{n+1} are relatively prime for all $n \geq 1$;

(b) for every natural number m, there exist infinitely many natural numbers n such that $a_n - 1$ and $a_{n+1} - 1$ are both divisible by m.

Problem 7 In convex quadrilateral $ABCD$, $\angle BCD = \angle CDA$. The bisector of angle ABC intersects $\overline{CD}$ at point E. Prove that $\angle AEB = 90°$ if and only if $AB = AD + BC$.

Problem 8 In the coordinate plane, a set of 2000 points $\{(x_1, y_1), (x_2, y_2), \ldots, (x_{2000}, y_{2000})\}$ is called *good* if $0 \leq x_i \leq 83, 0 \leq y_i \leq 1$ for $i = 1, 2, \ldots, 2000$ and $x_i \neq x_j$ when $i \neq j$. Find the largest positive integer n such that, for any good set, some unit square contains exactly n of the points in the set.

Problem 9 We are given the acute triangle ABC.

(a) Prove that there exist unique points A_1, B_1, and C_1 on $\overline{BC}$, $\overline{CA}$, and $\overline{AB}$, respectively, with the following property: If we project any two of the points onto the corresponding side, the midpoint of the projected segment is the third point.

(b) Prove that triangle $A_1 B_1 C_1$ is similar to the triangle formed by the medians of triangle ABC.

Problem 10 Let $p \geq 3$ be a prime number and $a_1, a_2, \ldots, a_{p-2}$ be a sequence of positive integers such that p does not divide either a_k or $a_k^k - 1$ for all $k = 1, 2, \ldots, p-2$. Prove that the product of some terms of the sequence is congruent to 2 modulo p.

Problem 11 Let D be the midpoint of base AB of the isosceles acute triangle ABC. Choose a point E on $\overline{AB}$, and let O be the circumcenter of triangle ACE. Prove that the line through D perpendicular to $\overline{DO}$, the line through E perpendicular to $\overline{BC}$, and the line through B parallel to $\overline{AC}$ are concurrent.

Problem 12 Let n be a positive integer. A *binary sequence* is a sequence of integers, all equal to 0 or 1. Let $\mathcal{A}$ be the set of all binary sequences with n terms, and let $\mathbf{0} \in \mathcal{A}$ be the sequence of all zeroes. The sequence $c = c_1, c_2, \ldots, c_n$ is called the sum $a + b$ of $a = a_1, a_2, \ldots, a_n$ and $b = b_1, b_2, \ldots, b_n$ if $c_i = 0$ when $a_i = b_i$ and $c_i = 1$ when $a_i \neq b_i$. Let $f : \mathcal{A} \to \mathcal{A}$ be a function with $f(\mathbf{0}) = \mathbf{0}$ such that whenever the sequences a and b differ in exactly k terms, the sequences $f(a)$ and $f(b)$ also differ in exactly k terms. Prove that if a, b, and c are sequences from $\mathcal{A}$ such that $a + b + c = \mathbf{0}$, then $f(a) + f(b) + f(c) = \mathbf{0}$.

3 Canada

Problem 1 Let $a_1, a_2, \ldots, a_{2000}$ be a sequence of integers each lying in the interval $[-1000, 1000]$. Suppose that $\sum_{i=1}^{2000} a_i = 1$. Show that the terms in some nonempty subsequence of $a_1, a_2, \ldots, a_{2000}$ sum to zero.

Problem 2 Let $ABCD$ be a quadrilateral with $\angle CBD = 2\angle ADB$, $\angle ABD = 2\angle CDB$, and $AB = CB$. Prove that $AD = CD$.

Problem 3 Suppose that the real numbers $a_1, a_2, \ldots, a_{100}$ satisfy

(i) $a_1 \geq a_2 \geq \cdots \geq a_{100} \geq 0$,

(ii) $a_1 + a_2 \leq 100$, and

(iii) $a_3 + a_4 + \cdots + a_{100} \leq 100$.

Determine the maximum possible value of $a_1^2 + a_2^2 + \cdots + a_{100}^2$, and find all possible sequences $a_1, a_2, \ldots, a_{100}$ for which this maximum is achieved.

4 China

Problem 1 In triangle ABC, $BC \leq CA \leq AB$. Let R and r be the circumradius and inradius, respectively, of triangle ABC. As a function of $\angle C$, determine whether $BC + CA - 2R - 2r$ is positive, negative, or zero.

Problem 2 Define the infinite sequence $a_1, a_2, \ldots$ recursively as follows: $a_1 = 0$, $a_2 = 1$, and
$$a_n = \frac{1}{2}na_{n-1} + \frac{1}{2}n(n-1)a_{n-2} + (-1)^n \left(1 - \frac{n}{2}\right)$$
for all $n \geq 3$. Find an explicit formula for
$$f_n = a_n + 2\binom{n}{1}a_{n-1} + 3\binom{n}{2}a_{n-2} + \cdots + n\binom{n}{n-1}a_1.$$

Problem 3 A table tennis club wishes to organize a doubles tournament, a series of matches where in each match one pair of players competes against a pair of two different players. Let a player's *match number* for a tournament be the number of matches he or she participates in. We are given a set $A = \{a_1, a_2, \ldots, a_k\}$ of distinct positive integers all divisible by 6. Find with proof the minimal number of players among whom we can schedule a doubles tournament such that

(i) each participant belongs to at most 2 pairs;

(ii) any two different pairs have at most 1 match against each other;

(iii) if two participants belong to the same pair, they never compete against each other; and

(iv) the set of the participants' match numbers is exactly A.

Problem 4 We are given an integer $n \geq 2$. For any ordered n-tuple of real numbers $A = (a_1, a_2, \ldots, a_n)$, let A's *domination score* be the number of values $k \in \{1, 2, \ldots, n\}$ such that $a_k > a_j$ for all $1 \leq j < k$. Consider all permutations $A = (a_1, a_2, \ldots, a_n)$ of $(1, 2, \ldots, n)$ with domination score 2. Find with proof the arithmetic mean of the first elements a_1 of these permutations.

Problem 5 Find all positive integers n such that there exist integers n_1, $n_2, \ldots, n_k > 3$ with
$$n = n_1 n_2 \cdots n_k = 2^{\frac{1}{2^k}(n_1-1)(n_2-1)\cdots(n_k-1)} - 1.$$

Problem 6 An exam paper consists of 5 multiple-choice questions, each with 4 different choices; 2000 students take the test, and each student chooses exactly one answer per question. Find the least possible value of n such that among any n of the students' answer sheets, there exist 4 of them among which any two have at most 3 common answers.

5 Czech and Slovak Republics

Problem 1 Show that

$$\sqrt[3]{\frac{a}{b}} + \sqrt[3]{\frac{b}{a}} \le \sqrt[3]{2(a+b)\left(\frac{1}{a}+\frac{1}{b}\right)}$$

for all positive real numbers a and b, and determine when equality occurs.

Problem 2 Find all convex quadrilaterals $ABCD$ for which there exists a point E inside the quadrilateral with the following property: Any line which passes through E and intersects sides $\overline{AB}$ and $\overline{CD}$ divides the quadrilateral $ABCD$ into two parts of equal area.

Problem 3 An isosceles triangle ABC is given with base $\overline{AB}$ and altitude $\overline{CD}$. Point P lies on $\overline{CD}$. Let E be the intersection of line AP with side BC, and let F be the intersection of line BP with side AC. Suppose that the incircles of triangle ABP and quadrilateral $PECF$ are congruent. Show that the incircles of the triangles ADP and BCP are also congruent.

Problem 4 In the plane are given 2000 congruent triangles of area 1, which are images of a single triangle under different translations. Each of these triangles contains the centroids of all the others. Show that the area of the union of these triangles is less than $\frac{22}{9}$.

6 Estonia

Problem 1 Five real numbers are given such that, no matter which three of them we choose, the difference between the sum of these three numbers and the sum of the remaining two numbers is positive. Prove that the product of all these 10 differences (corresponding to all the possible triples of chosen numbers) is less than or equal to the product of the squares of these five numbers.

Problem 2 Prove that it is not possible to divide any set of 18 consecutive positive integers into two disjoint sets A and B, such that the product of the elements in A equals the product of the elements in B.

Problem 3 Let M, N, and K be the points of tangency of the incircle of triangle ABC with the sides of the triangle, and let Q be the center of the circle drawn through the midpoints of $\overline{MN}$, $\overline{NK}$, and $\overline{KM}$. Prove that the incenter and circumcenter of triangle ABC are collinear with Q.

Problem 4 Find all functions $f : \mathbb{Z}^+ \to \mathbb{Z}^+$ such that
$$f(f(f(n))) + f(f(n)) + f(n) = 3n$$
for all $n \in \mathbb{Z}^+$.

Problem 5 In a triangle ABC we have $AC \neq BC$. Take a point X in the interior of this triangle and let $\alpha = \angle A$, $\beta = \angle B$, $\phi = \angle ACX$, and $\psi = \angle BCX$. Prove that
$$\frac{\sin\alpha \sin\beta}{\sin(\alpha - \beta)} = \frac{\sin\phi \sin\psi}{\sin(\phi - \psi)}$$
if and only if X lies on the median of triangle ABC drawn from the vertex C.

Problem 6 We call an infinite sequence of positive integers an F-*sequence* if every term of this sequence (starting from the third term) equals the sum of the two preceding terms. Is it possible to decompose the set of all positive integers into

(a) a finite;

(b) an infinite

number of F-sequences having no common members?

7 Hungary

Problem 1 Find all positive primes p for which there exist positive integers n, x, y such that $p^n = x^3 + y^3$.

Problem 2 Is there a polynomial f of degree 1999 with integer coefficients, such that $f(n), f(f(n)), f(f(f(n))), \ldots$ are pairwise relatively prime for any integer n?

Problem 3 The feet of the angle bisectors of triangle ABC are X, Y, and Z. The circumcircle of triangle XYZ cuts off three segments from lines AB, BC, and CA. Prove that two of these segments' lengths add up to the third segment's length.

Problem 4 Let k and t be relatively prime integers greater than 1. Starting from the permutation $(1, 2, \ldots, n)$ of the numbers $1, 2, \ldots, n$, we may swap two numbers if their difference is either k or t. Prove that we can get any permutation of $1, 2, \ldots, n$ with such steps if and only if $n \geq k + t - 1$.

Problem 5 For any positive integer k, let $e(k)$ denote the number of positive even divisors of k, and let $o(k)$ denote the number of positive odd divisors of k. For all $n \geq 1$, prove that $\sum_{k=1}^{n} e(k)$ and $\sum_{k=1}^{n} o(k)$ differ by at most n.

Problem 6 Given a triangle in the plane, show how to construct a point P inside the triangle which satisfies the following condition: if we drop perpendiculars from P to the sides of the triangle, the feet of the perpendiculars determine a triangle whose centroid is P.

Problem 7 Given a natural number k and more than 2^k different integers, prove that a set S of $k + 2$ of these numbers can be selected such that for any positive integer $m \leq k + 2$, all the m-element subsets of S have different sums of elements.

8 India

Problem 1 Let ABC be a nonequilateral triangle. Suppose there is an interior point P such that the three cevians through P all have the same length λ where $\lambda < \min\{AB, BC, CA\}$. Show that there is another interior point $P' \neq P$ such that the three cevians through P' also are of equal length.

Problem 2 Let m, n be positive integers such that $m \leq n^2/4$ and every prime divisor of m is less than or equal to n. Show that m divides $n!$.

Problem 3 Let G be a graph with $n \geq 4$ vertices and m edges. If $m > n(\sqrt{4n-3}+1)/4$ show that G has a 4-cycle.

Problem 4 Suppose $f : \mathbb{Q} \to \{0, 1\}$ is a function with the property that for $x, y \in \mathbb{Q}$, if $f(x) = f(y)$ then $f(x) = f((x+y)/2) = f(y)$. If $f(0) = 0$ and $f(1) = 1$ show that $f(q) = 1$ for all rational numbers q greater than or equal to 1.

9 Iran

Problem 1 Call two circles in three-dimensional space *pairwise tangent* at a point P if they both pass through P and the lines tangent to each circle at P coincide. Three circles not all lying in a plane are pairwise tangent at three distinct points. Prove that there exists a sphere which passes through the three circles.

Problem 2 We are given a sequence $c_1, c_2, \ldots$ of natural numbers. For any natural numbers m, n with $1 \leq m \leq \sum_{i=1}^{n} c_i$, we can choose natural numbers $a_1, a_2, \ldots, a_n$ such that

$$m = \sum_{i=1}^{n} \frac{c_i}{a_i}.$$

For each i, find the maximum value of c_i.

Problem 3 Circles C_1 and C_2 with centers O_1 and O_2, respectively, meet at points A and B. Lines $O_1 B$ and $O_2 B$ intersect C_2 and C_1 at F and E, respectively. The line parallel to EF through B meets C_1 and C_2 at M and N. Given that B lies between M and N, prove that $MN = AE + AF$.

Problem 4 Two triangles ABC and $A'B'C'$ lie in three-dimensional space. The sides of triangle ABC have lengths greater than or equal to a, and the sides of triangle $A'B'C'$ have lengths greater than or equal to a'. Prove that one can select one vertex from triangle ABC and one vertex from triangle $A'B'C'$ such that the distance between them is at least $\sqrt{\frac{a^2 + a'^2}{3}}$.

Problem 5 The function $f : \mathbb{N} \to \mathbb{N}$ is defined recursively with $f(1) = 1$ and

$$f(n+1) = \begin{cases} f(n) + 2 & \text{if } n = f(f(n) - n + 1) \\ f(n) + 1 & \text{otherwise.} \end{cases}$$

for all $n \geq 1$.

(a) Prove that $f(f(n) - n + 1) \in \{n, n+1\}$.

(b) Find an explicit formula for f.

Problem 6 Find all functions $f : \mathbb{N} \to \mathbb{N}$ such that

(i) $f(m) = 1$ if and only if $m = 1$;

(ii) if $d = \gcd(m, n)$, then $f(mn) = \frac{f(m) f(n)}{f(d)}$; and

(iii) for every $m \in \mathbb{N}$, we have $f^{2000}(m) = m$.

Problem 7 The n tennis players, $A_1, A_2, \ldots, A_n$, participate in a tournament. Any two players play against each other at most once, and $k \leq \frac{n(n-1)}{2}$ matches take place. No draws occur, and in each match the winner adds 1 point to his tournament score while the loser adds 0. For nonnegative integers $d_1, d_2, \ldots, d_n$, prove that it is possible for $A_1, A_2, \ldots, A_n$ to obtain the tournament scores $d_1, d_2, \ldots, d_n$, respectively, if and only if the following conditions are satisfied:

(i) $\sum_{i=1}^{n} d_i = k$.

(ii) For every subset $X \subseteq \{A_1, \ldots, A_n\}$, the number of matches taking place among the players in X is at most $\sum_{A_j \in X} d_j$.

Problem 8 Isosceles triangles $A_3 A_1 O_2$ and $A_1 A_2 O_3$ are constructed externally along the sides of a triangle $A_1 A_2 A_3$ with $O_2 A_3 = O_2 A_1$ and $O_3 A_1 = O_3 A_2$. Let O_1 be a point on the opposite side of line $A_2 A_3$ as A_1 with $\angle O_1 A_3 A_2 = \frac{1}{2} \angle A_1 O_3 A_2$ and $\angle O_1 A_2 A_3 = \frac{1}{2} \angle A_1 O_2 A_3$, and let T be the foot of the perpendicular from O_1 to $\overline{A_2 A_3}$. Prove that $A_1 O_1 \perp O_2 O_3$ and that $\frac{A_1 O_1}{O_2 O_3} = 2 \frac{O_1 T}{A_2 A_3}$.

Problem 9 Given a circle Γ, a line d is drawn not intersecting Γ. M, N are two points varying on line d such that the circle with diameter $\overline{MN}$ is externally tangent to Γ. Prove that there exists a point P in the plane such that for any such segment MN, $\angle MPN$ is constant.

Problem 10 Suppose that a, b, c are real numbers such that for any positive real numbers $x_1, x_2, \ldots, x_n$, we have

$$\left(\frac{\sum_{i=1}^n x_i}{n} \right)^a \cdot \left(\frac{\sum_{i=1}^n x_i^2}{n} \right)^b \cdot \left(\frac{\sum_{i=1}^n x_i^3}{n} \right)^c \geq 1.$$

Prove that the vector (a, b, c) has the form $p(-2, 1, 0) + q(1, -2, 1)$ for some nonnegative real numbers p and q.

10 Israel

Problem 1 Define $f(n) = n!$. Let
$$a = 0.f(1)f(2)f(3)\ldots.$$
In other words, to obtain the decimal representation of a write the decimal representations of $f(1), f(2), f(3), \ldots$ in a row. Is a rational?

Problem 2 ABC is a triangle whose vertices are lattice points. Two of its sides have lengths which belong to the set $\{\sqrt{17}, \sqrt{1999}, \sqrt{2000}\}$. What is the maximum possible area of triangle ABC?

Problem 3 The points A, B, C, D, E, F lie on a circle, and the lines AD, BE, CF concur. Let P, Q, R be the midpoints of $\overline{AD}, \overline{BE}, \overline{CF}$, respectively. Two chords AG, AH are drawn such that $AG \parallel BE$ and $AH \parallel CF$. Prove that triangles PQR and DGH are similar.

Problem 4 A square $ABCD$ is given. A *triangulation* of the square is a partition of the square into triangles such that any two triangles are either disjoint, share only a common vertex, or share only a common side. A *good triangulation* of the square is a triangulation in which all the triangles are acute.

(a) Give an example of a good triangulation of the square.

(b) What is the minimal number of triangles required for a good triangulation?

11 Italy

Problem 1 Let $ABCD$ be a convex quadrilateral, and write $\alpha = \angle DAB$; $\beta = \angle ADB$; $\gamma = \angle ACB$; $\delta = \angle DBC$; and $\epsilon = \angle DBA$. Assuming that $\alpha < 90°$, $\beta + \gamma = 90°$, and $\delta + 2\epsilon = 180°$, prove that
$$(DB + BC)^2 = AD^2 + AC^2.$$

Problem 2 Given a fixed integer $n > 1$, Alberto and Barbara play the following game, starting with the first step and then alternating between the second and third:

- Alberto chooses a positive integer.
- Barbara picks an integer greater than 1 which is a multiple or divisor of Alberto's number, possibly choosing Alberto's number itself.
- Alberto adds or subtracts 1 from Barbara's number.

Barbara wins if she succeeds in picking n by her fiftieth move. For which values of n does she have a winning strategy?

Problem 3 Let $p(x)$ be a polynomial with integer coefficients such that $p(0) = 0$ and $0 \leq p(1) \leq 10^7$, and such that there exist integers a, b satisfying $p(a) = 1999$ and $p(b) = 2001$. Determine the possible values of $p(1)$.

12 Japan

Problem 1 We *shuffle* a line of cards labelled $a_1, a_2, \ldots, a_{3n}$ from left to right by rearranging the cards into the new order

$$a_3, a_6, \ldots, a_{3n}, a_2, a_5, \ldots, a_{3n-1}, a_1, a_4, \cdots, a_{3n-2}.$$

For example, if six cards are labelled $1, 2, \ldots, 6$ from left to right, then shuffling them twice changes their order as follows:

$$1, 2, 3, 4, 5, 6 \longrightarrow 3, 6, 2, 5, 1, 4 \longrightarrow 2, 4, 6, 1, 3, 5.$$

Starting with 192 cards labelled $1, 2, \ldots, 192$ from left to right, is it possible to obtain the order $192, 191, \ldots, 1$ after a finite number of shuffles?

Problem 2 In the plane are given distinct points A, B, C, P, Q, no three of which are collinear. Prove that

$$AB + BC + CA + PQ < AP + AQ + BP + BQ + CP + CQ.$$

Problem 3 Given a natural number $n \geq 3$, prove that there exists a set A_n with the following two properties:

(i) A_n consists of n distinct natural numbers.

(ii) For any $a \in A_n$, the product of all the other elements in A_n has remainder 1 when divided by a.

Problem 4 We are given finitely many lines in the plane. Let an *intersection point* be a point where at least two of these lines meet, and let a *good intersection point* be a point where exactly two of these lines meet. Given that there are at least two intersection points, find the minimum number of good intersection points.

13 Korea

Problem 1 Show that given any prime p, there exist integers x, y, z, w satisfying $x^2 + y^2 + z^2 - wp = 0$ and $0 < w < p$.

Problem 2 Find all functions $f : \mathbb{R} \to \mathbb{R}$ satisfying
$$f(x^2 - y^2) = (x - y)\left(f(x) + f(y)\right)$$
for all $x, y \in \mathbb{R}$.

Problem 3 We are given a convex cyclic quadrilateral $ABCD$. Let P, Q, R, S be the intersections of the exterior angle bisectors of $\angle ABD$ and $\angle ADB$, $\angle DAB$ and $\angle DBA$, $\angle ACD$ and $\angle ADC$, $\angle DAC$ and $\angle DCA$, respectively. Show that the four points P, Q, R, S are concyclic.

Problem 4 Let p be a prime number such that $p \equiv 1 \pmod{4}$. Evaluate
$$\sum_{k=1}^{p-1} \left(\left\lfloor \frac{2k^2}{p} \right\rfloor - 2 \left\lfloor \frac{k^2}{p} \right\rfloor \right).$$

Problem 5 Consider the following L-shaped figures, each made of four unit squares:

Let m and n be integers greater than 1. Prove that an $m \times n$ rectangular region can be tiled with such figures if and only if mn is a multiple of 8.

Problem 6 The real numbers a, b, c, x, y, z satisfy $a \geq b \geq c > 0$ and $x \geq y \geq z > 0$. Prove that
$$\frac{a^2 x^2}{(by + cz)(bz + cy)} + \frac{b^2 y^2}{(cz + ax)(cx + az)} + \frac{c^2 z^2}{(ax + by)(ay + bx)}$$
is at least $\frac{3}{4}$.

14 Mongolia

Problem 1 Let rad$(1) = 1$, and for $k > 1$ let rad(k) equal the product of the prime divisors of k. A sequence of natural numbers $a_1, a_2, \ldots$ with arbitrary first term a_1 is defined recursively by the relation $a_{n+1} = a_n + \text{rad}(a_n)$. Show that for any positive integer N, the sequence $a_1, a_2, \ldots$ contains some N consecutive terms in arithmetic progression.

Problem 2 The circles $\omega_1, \omega_2, \omega_3$ in the plane are pairwise externally tangent to each other. Let P_1 be the point of tangency between circles ω_1 and ω_3, and let P_2 be the point of tangency between circles ω_2 and ω_3. A and B, both different from P_1 and P_2, are points on ω_3 such that $\overline{AB}$ is a diameter of ω_3. Line AP_1 intersects ω_1 again at X, line BP_2 intersects ω_2 again at Y, and lines AP_2 and BP_1 intersect at Z. Prove that X, Y, and Z are collinear.

Problem 3 A function $f : \mathbb{R} \to \mathbb{R}$ satisfies the following conditions:

(i) $|f(a) - f(b)| \leq |a - b|$ for any real numbers $a, b \in \mathbb{R}$.

(ii) $f(f(f(0))) = 0$.

Prove that $f(0) = 0$.

Problem 4 The bisectors of angles A, B, C of a triangle ABC intersect its sides at points A_1, B_1, C_1. Prove that if the quadrilateral $BA_1B_1C_1$ is cyclic, then
$$\frac{BC}{AC+AB} = \frac{AC}{AB+BC} - \frac{AB}{BC+AC}.$$

Problem 5 Which integers can be represented in the form $\frac{(x+y+z)^2}{xyz}$ where x, y, and z are positive integers?

Problem 6 In a country with n towns the cost of travel from the i-th town to the j-th town is x_{ij}. Suppose that the total cost of any route passing through each town exactly once and ending at its starting point does not depend on which route is chosen. Prove that there exist numbers $a_1, \ldots, a_n$ and $b_1, \ldots, b_n$ such that $x_{ij} = a_i + b_j$ for all integers i, j with $1 \leq i < j \leq n$.

15 Poland

Problem 1 Let $n \geq 2$ be a given integer. How many solutions does the system of equations
$$\begin{cases} x_1 + x_n^2 = 4x_n \\ x_2 + x_1^2 = 4x_1 \\ \vdots \\ x_n + x_{n-1}^2 = 4x_{n-1} \end{cases}$$
have in nonnegative real numbers $x_1, \ldots, x_n$?

Problem 2 The sides AC and BC of a triangle ABC have equal length. Let P be a point inside triangle ABC such that $\angle PAB = \angle PBC$ and let M be the midpoint of $\overline{AB}$. Prove that $\angle APM + \angle BPC = 180°$.

Problem 3 A sequence $p_1, p_2, \ldots$ of prime numbers satisfies the following condition: for $n \geq 3$, p_n is the greatest prime divisor of $p_{n-1} + p_{n-2} + 2000$. Prove that the sequence is bounded.

Problem 4 For an integer $n \geq 3$ consider a pyramid with vertex S and the regular n-gon $A_1 A_2 \ldots A_n$ as a base, such that all the angles between lateral edges and the base equal $60°$. Points $B_2, B_3, \ldots$ lie on $\overline{A_2 S}, \overline{A_3 S}, \ldots, \overline{A_n S}$, respectively, such that $A_1 B_2 + B_2 B_3 + \cdots + B_{n-1} B_n + B_n A_1 < 2 A_1 S$. For which n is this possible?

Problem 5 Given a natural number $n \geq 2$, find the smallest integer k with the following property: Every set consisting of k cells of an $n \times n$ table contains a nonempty subset S such that in every row and in every column of the table, there is an even number of cells belonging to S.

Problem 6 Let P be a polynomial of odd degree satisfying the identity
$$P(x^2 - 1) = P(x)^2 - 1.$$
Prove that $P(x) = x$ for all real x.

16 Romania

Problem 1 A function $f : \mathbb{R}^2 \to \mathbb{R}$ is called *olympic* if it has the following property: given $n \geq 3$ distinct points $A_1, A_2, \ldots, A_n \in \mathbb{R}^2$, if $f(A_1) = f(A_2) = \cdots = f(A_n)$ then the points $A_1, A_2, \ldots, A_n$ are the vertices of a convex polygon. Let $P \in \mathbb{C}[X]$ be a non-constant polynomial. Prove that the function $f : \mathbb{R}^2 \to \mathbb{R}$, defined by $f(x, y) = |P(x + iy)|$, is olympic if and only if all the roots of P are equal.

Problem 2 Let $n \geq 2$ be a positive integer. Find the number of functions $f : \{1, 2, \ldots, n\} \to \{1, 2, 3, 4, 5\}$ which have the following property: $|f(k+1) - f(k)| \geq 3$ for $k = 1, 2, \ldots, n-1$.

Problem 3 Let $n \geq 1$ be a positive integer and $x_1, x_2, \ldots, x_n$ be real numbers such that $|x_{k+1} - x_k| \leq 1$ for $k = 1, 2, \ldots, n-1$. Show that
$$\sum_{k=1}^{n} |x_k| - \left|\sum_{k=1}^{n} x_k\right| \leq \frac{n^2 - 1}{4}.$$

Problem 4 Let n, k be arbitrary positive integers. Show that there exist positive integers $a_1 > a_2 > a_3 > a_4 > a_5 > k$ such that
$$n = \pm \binom{a_1}{3} \pm \binom{a_2}{3} \pm \binom{a_3}{3} \pm \binom{a_4}{3} \pm \binom{a_5}{3},$$
where $\binom{a}{3} = \frac{a(a-1)(a-2)}{6}$.

Problem 5 Let $P_1 P_2 \cdots P_n$ be a convex polygon in the plane. Assume that for any pair of vertices P_i, P_j, there exists a vertex V of the polygon such that $\angle P_i V P_j = 60°$. Show that $n = 3$.

Problem 6 Show that there exist infinitely many 4-tuples of positive integers (x, y, z, t) such that the four numbers' greatest common divisor is 1 and such that
$$x^3 + y^3 + z^2 = t^4.$$

Problem 7 Given an odd integer a, find (as a function of a) the least positive integer n such that 2^{2000} is a divisor of $a^n - 1$.

Problem 8 Let ABC be an acute triangle and let M be the midpoint of segment BC. Consider the interior point N such that $\angle ABN = \angle BAM$ and $\angle ACN = \angle CAM$. Prove that $\angle BAN = \angle CAM$.

Problem 9 Find, with proof, whether there exists a sphere with interior S, a circle with interior C, and a function $f : S \to C$ such that the distance between $f(A)$ and $f(B)$ is greater than or equal to AB for all points A and B in S.

Problem 10 Let $n \geq 3$ be an odd integer and $m \geq n^2 - n + 1$ be an integer. The sequence of polygons $P_1, P_2, \ldots, P_m$ is defined as follows:

(i) P_1 is a regular polygon with n vertices.

(ii) For $k > 1$, P_k is the regular polygon whose vertices are the midpoints of the sides of P_{k-1}.

Find, with proof, the maximum number of colors which can be used such that for every coloring of the vertices of these polygons, one can find four vertices A, B, C, D which have the same color and form an isosceles trapezoid.

17 Russia

Problem 1 Sasha tries to determine some positive integer $X \leq 100$. He can choose any two positive integers M and N that are less than 100 and ask the question, "What is the greatest common divisor of the numbers $X + M$ and N?" Prove that Sasha can determine the value of X after 7 questions.

Problem 2 Let O be the center of the circumcircle ω of an acute-angled triangle ABC. The circle ω_1 with center K passes through the points A, I, C and intersects sides AB and BC at points M and N. Let L be the reflection of K across line MN. Prove that $BL \perp AC$.

Problem 3 There are several cities in a state and a set of roads, each road connecting two cities. It is known that at least 3 roads go out of every city. Prove that there exists a cyclic path (that is, a path where the last road ends where the first road begins) such that the number of roads in the path is not divisible by 3.

Problem 4 Let $x_1, x_2, \ldots, x_n$ be real numbers ($n \geq 2$), satisfying the conditions $-1 < x_1 < x_2 < \cdots < x_n < 1$ and

$$x_1^{13} + x_2^{13} + \cdots + x_n^{13} = x_1 + x_2 + \cdots + x_n.$$

Prove that

$$x_1^{13}y_1 + x_2^{13}y_2 + \cdots + x_n^{13}y_n < x_1 y_1 + x_2 y_2 + \cdots + x_n y_n$$

for any real numbers $y_1 < y_2 < \cdots < y_n$.

Problem 5 Let $\overline{AA_1}$ and $\overline{CC_1}$ be the altitudes of an acute-angled nonisosceles triangle ABC. The bisector of the acute angle between lines AA_1 and CC_1 intersects sides AB and BC at P and Q, respectively. Let H be the orthocenter of triangle ABC and let M be the midpoint of $\overline{AC}$; and let the bisector of $\angle ABC$ intersect $\overline{HM}$ at R. Prove that $PBQR$ is cyclic.

Problem 6 Five stones which appear identical all have different weights; Oleg knows the weight of each stone. Given any stone x, let $m(x)$ denote its weight. Dmitrii tries to determine the order of the weights of the stones. He is allowed to choose any three stones A, B, C and ask Oleg the question, "Is it true that $m(A) < m(B) < m(C)$?" Oleg then responds "yes" or "no." Can Dmitrii determine the order of the weights with at most nine questions?

Problem 7 Find all functions $f : \mathbb{R} \to \mathbb{R}$ that satisfy the inequality
$$f(x+y) + f(y+z) + f(z+x) \geq 3f(x+2y+3z)$$
for all $x, y, z \in \mathbb{R}$.

Problem 8 Prove that the set of all positive integers can be partitioned into 100 nonempty subsets such that if three positive integers a, b, c satisfy $a + 99b = c$, then two of them belong to the same subset.

Problem 9 Let $ABCDE$ be a convex pentagon on the coordinate plane. Each of its vertices are lattice points. The five diagonals of $ABCDE$ form a convex pentagon $A_1B_1C_1D_1E_1$ inside of $ABCDE$. Prove that this smaller pentagon contains a lattice point on its boundary or within its interior.

Problem 10 Let $a_1, a_2, \ldots, a_n$ be a sequence of nonnegative real numbers, not all zero. For $1 \leq k \leq n$, let
$$m_k = \max_{1 \leq i \leq k} \frac{a_{k-i+1} + a_{k-i+2} + \cdots + a_k}{i}.$$
Prove that for any $\alpha > 0$, the number of integers k which satisfy $m_k > \alpha$ is less than $\frac{a_1 + a_2 + \cdots + a_n}{\alpha}$.

Problem 11 Let $a_1, a_2, a_3, \ldots$ be a sequence with $a_1 = 1$ satisfying the recursion
$$a_{n+1} = \begin{cases} a_n - 2 & \text{if } a_n - 2 \notin \{a_1, a_2, \ldots, a_n\} \text{ and } a_n - 2 > 0 \\ a_n + 3 & \text{otherwise.} \end{cases}$$
Prove that for every integer $k > 1$, we have $a_n = k^2 = a_{n-1} + 3$ for some n.

Problem 12 There are black and white checkers on some squares of a $2n \times 2n$ board, with at most one checker on each square. First, we remove every black checker that is in the same column as any white checker. Next, we remove every white checker that is in the same row as any remaining black checker. Prove that for some color, at most n^2 checkers of this color remain.

Problem 13 Let E be a point on the median CD of triangle ABC. Let S_1 be the circle passing through E and tangent to line AB at A, intersecting side AC again at M; let S_2 be the circle passing through E and tangent to line AB at B, intersecting side BC again at N. Prove that the circumcircle of triangle CMN is tangent to circles S_1 and S_2.

Problem 14 One hundred positive integers, with no common divisor greater than one, are arranged in a circle. To any number, we can add the greatest common divisor of its neighboring numbers. Prove that using this operation, we can transform these numbers into a new set of pairwise coprime numbers.

Problem 15 M is a finite set of real numbers such that given three distinct elements from M, we can choose two of them whose sum also belongs to M. What is the largest number of elements that M can have?

Problem 16 A positive integer n is called *perfect* if the sum of all its positive divisors, excluding n itself, equals n. For example, 6 is perfect since 6 = 1 + 2 + 3. Prove that

(a) if a perfect number larger than 6 is divisible by 3, then it is also divisible by 9.

(b) if a perfect number larger than 28 is divisible by 7, then it is also divisible by 49.

Problem 17 Circles ω_1 and ω_2 are internally tangent at N. The chords BA and BC of ω_1 are tangent to ω_2 at K and M, respectively. Let Q and P be the midpoints of the arcs AB and BC not containing the point N. Let the circumcircles of triangles BQK and BPM intersect at B and B_1. Prove that BPB_1Q is a parallelogram.

Problem 18 There is a finite set of congruent square cards, placed on a rectangular table with their sides parallel to the sides of the table. Each card is colored in one of k colors. For any k cards of different colors, it is possible to pierce some two of them with a single pin. Prove that all the cards can be pierced by $2k - 2$ pins.

Problem 19 Prove the inequality

$$\sin^n(2x) + (\sin^n x - \cos^n x)^2 \leq 1.$$

Problem 20 The circle ω is inscribed in the quadrilateral $ABCD$, and O is the intersection point of the lines AB and CD. The circle ω_1 is tangent to side BC at K and is tangent to lines AB and CD at points lying outside $ABCD$; the circle ω_2 is tangent to side AD at L and is also tangent to lines AB and CD at points lying outside $ABCD$. If O, K, L are collinear, prove that the midpoint of side BC, the midpoint of side AD, and the center of ω are collinear.

Problem 21 Every cell of a 100×100 board is colored in one of 4 colors so that there are exactly 25 cells of each color in every column and in every row. Prove that one can choose two columns and two rows so that the four cells where they intersect are colored in four different colors.

Problem 22 The nonzero real numbers a, b satisfy the equation
$$a^2b^2(a^2b^2 + 4) = 2(a^6 + b^6).$$
Prove that a and b are not both rational.

Problem 23 Find the smallest odd integer n such that some n-gon (not necessarily convex) can be partitioned into parallelograms whose interiors do not overlap.

Problem 24 Two pirates divide their loot, consisting of two sacks of coins and one diamond. They decide to use the following rules. On each turn, one pirate chooses a sack and takes $2m$ coins from it, keeping m for himself and putting the rest into the other sack. The pirates alternate taking turns until no more moves are possible; the pirate who makes the last move takes the diamond. For what initial amounts of coins can the first pirate guarantee that he will obtain the diamond?

Problem 25 Do there exist coprime integers $a, b, c > 1$ such that $2^a + 1$ is divisible by b, $2^b + 1$ is divisible by c, and $2^c + 1$ is divisible by a?

Problem 26 $2n+1$ segments are marked on a line. Each of the segments intersects at least n other segments. Prove that one of these segments intersects all the other segments.

Problem 27 The circles S_1 and S_2 intersect at points M and N. Let A and D be points on S_1 and S_2 such that lines AM and AN intersect S_2 at B and C, lines DM and DN intersect S_1 at E and F, and the triples A, E, F and D, B, C lie on opposite sides of line MN. Prove that there is a fixed point O such that for any points A and D that satisfy the condition $AB = DE$, the quadrilateral $AFCD$ is cyclic with center O.

Problem 28 Let the set M consist of the 2000 numbers $10^1 + 1, 10^2 + 1, \ldots, 10^{2000} + 1$. Prove that at least 99% of the elements of M are not prime.

Problem 29 There are 2 counterfeit coins among 5 coins that look identical. Both counterfeit coins have the same weight and the other three real coins have the same weight. The five coins do not all weigh the same, but it is unknown whether the weight of each counterfeit coin is more or less than the weight of each real coin. Find the minimal number of weighings needed to find at least one real coin, and describe how to do so. (The balance scale reports the difference between the weights of the objects in two pans.)

Problem 30 Let $ABCD$ be a parallelogram with $\angle A = 60°$. Let O be the circumcenter of triangle ABD. Line AO intersects the external angle bisector of angle BCD at K. Find the value $\frac{AO}{OK}$.

Problem 31 Find the smallest integer n such that an $n \times n$ square can be partitioned into 40×40 and 49×49 squares, with both types of squares present in the partition.

Problem 32 Prove that there exist 10 distinct real numbers $a_1, a_2, \ldots, a_{10}$ such that the equation
$$(x - a_1)(x - a_2) \cdots (x - a_{10}) = (x + a_1)(x + a_2) \cdots (x + a_{10})$$
has exactly 5 different real roots.

Problem 33 We are given a cylindrical region in space, whose altitude is 1 and whose base has radius 1. Find the minimal number of balls of radius 1 needed to cover this region.

Problem 34 The sequence $a_1, a_2, \ldots, a_{2000}$ of real numbers satisfies the condition
$$a_1^3 + a_2^3 + \cdots + a_n^3 = (a_1 + a_2 + \cdots + a_n)^2$$
for all n, $1 \leq n \leq 2000$. Prove that every element of the sequence is an integer.

Problem 35 The bisectors $\overline{AD}$ and $\overline{CE}$ of a triangle ABC intersect at I. Let ℓ_1 be the reflection of line AB across line CE, and let ℓ_2 be the reflection of line BC across line AD. If lines ℓ_1 and ℓ_2 intersect at some point K, prove that $KI \perp AC$.

Problem 36 There are 2000 cities in a country, some pairs of which are connected by a direct airplane flight. For every city A the number of cities connected with A by direct flights equals $1, 2, 4, \ldots,$ or 1024. Let $S(A)$ be the number of routes from A to other cities (different from A) with at most one intermediate landing. Prove that the sum of $S(A)$ over all 2000 cities A cannot be equal to 10000.

Problem 37 A heap of balls consists of one thousand 10-gram balls and one thousand 9.9-gram balls. We wish to pick out two heaps of balls with equal numbers of balls in them but different total weights. What is the minimal number of weighings needed to do this?

Problem 38 Let D be a point on side AB of triangle ABC. The circumcircle of triangle BCD intersects line AC at C and M, and the circumcircle of triangle ACD intersects line BC at C and N. Let O be the center of the circumcircle of triangle CMN. Prove that $OD \perp AB$.

Problem 39 Every cell of a 200×200 table is colored black or white. The difference between the number of black and white cells is 404. Prove that some 2×2 square contains an odd number of white cells.

Problem 40 Is there a function $f : \mathbb{R} \to \mathbb{R}$ such that
$$|f(x+y) + \sin x + \sin y| < 2$$
for all $x, y \in \mathbb{R}$?

Problem 41 For any odd integer $a_0 > 5$, consider the sequence $a_0, a_1, a_2, \ldots,$ where
$$a_{n+1} = \begin{cases} a_n^2 - 5 & \text{if } a_n \text{ is odd} \\ \frac{a_n}{2} & \text{if } a_n \text{ is even} \end{cases}$$
for all $n \geq 0$. Prove that this sequence is not bounded.

Problem 42 Let $\ell_a, \ell_b, \ell_c,$ and ℓ_d be the external angle bisectors of angles $DAB, ABC, BCD,$ and CDA, respectively. The pairs of lines ℓ_a and ℓ_b, ℓ_b and ℓ_c, ℓ_c and ℓ_d, ℓ_d and ℓ_a intersect at points K, L, M, N, respectively. Suppose that the perpendiculars to line AB passing through K, to line BC passing through L, and to line CD passing through M are concurrent. Prove that $ABCD$ can be inscribed in a circle.

Problem 43 There are 2000 cities in a country, and each pair of cities is connected by either no roads or exactly one road. A *cyclic path* is a nonempty, connected path of roads such that each city is at the end of either 0 or 2 roads in the path. For every city, there at most N cyclic paths which both pass through this city and contain an odd number of roads. Prove that the country can be separated into $2N + 2$ republics such that any two cities from the same republic are not connected by a road.

Problem 44 Prove the inequality
$$\frac{1}{\sqrt{1+x^2}} + \frac{1}{\sqrt{1+y^2}} \le \frac{2}{\sqrt{1+xy}}$$
for $0 \le x, y \le 1$.

Problem 45 The incircle of triangle ABC touches side AC at K. A second circle S with the same center intersects all the sides of the triangle. Let E and F be the intersection points on $\overline{AB}$ and $\overline{BC}$ closer to B; let B_1 and B_2 be the intersection points on $\overline{AC}$ with B_1 closer to A. Finally, let P be the intersection point of segments B_2E and B_1F. Prove that points B, K, P are collinear.

Problem 46 Each of the numbers $1, 2, \ldots, N$ is colored black or white. We are allowed to simultaneously change the colors of any three numbers in arithmetic progression. For which numbers N can we always make all the numbers white?

18 Taiwan

Problem 1 In an acute triangle ABC, $AC > BC$ and M is the midpoint of $\overline{AB}$. Let altitudes $\overline{AP}$ and $\overline{BQ}$ meet at H, and let lines AB and PQ meet at R. Prove that the two lines RH and CM are perpendicular.

Problem 2 Let $\phi(k)$ denote the number of positive integers n satisfying $\gcd(n,k) = 1$ and $n \leq k$. Suppose that $\phi(5^m - 1) = 5^n - 1$ for some positive integers m, n. Prove that $\gcd(m,n) > 1$.

Problem 3 Let $A = \{1, 2, \ldots, n\}$, where n is a positive integer. A subset of A is *connected* if it is a nonempty set which consists of one element or of consecutive integers. Determine the greatest integer k for which A contains k distinct subsets $A_1, A_2, \ldots, A_k$ such that the intersection of any two distinct sets A_i and A_j is connected.

Problem 4 Let $f : \mathbb{N} \to \mathbb{N} \cup \{0\}$ be defined recursively by $f(1) = 0$ and

$$f(n) = \max_{1 \leq j \leq \lfloor \frac{n}{2} \rfloor} \{f(j) + f(n-j) + j\}$$

for all $n \geq 2$. Determine $f(2000)$.

19 Turkey

Problem 1 Find the number of ordered quadruples (x, y, z, w) of integers with $0 \leq x, y, z, w \leq 36$ such that
$$x^2 + y^2 \equiv z^3 + w^3 \pmod{37}.$$

Problem 2 Given a circle with center O, the two tangent lines from a point S outside the circle touch the circle at points P and Q. Line SO intersects the circle at A and B, with B closer to S. Let X be an interior point of minor arc PB, and let line OS intersect lines QX and PX at C and D, respectively. Prove that
$$\frac{1}{AC} + \frac{1}{AD} = \frac{2}{AB}.$$

Problem 3 For any two positive integers n and p, prove that there are exactly $(p+1)^{n+1} - p^{n+1}$ functions
$$f : \{1, 2, \ldots, n\} \to \{-p, -p+1, \ldots, p\}$$
such that $|f(i) - f(j)| \leq p$ for all $i, j \in \{1, 2, \ldots, n\}$.

Problem 4 In an acute triangle ABC with circumradius R, altitudes $\overline{AD}, \overline{BE}, \overline{CF}$ have lengths h_1, h_2, h_3, respectively. If t_1, t_2, t_3 are the lengths of the tangents from A, B, C, respectively, to the circumcircle of triangle DEF, prove that
$$\sum_{i=1}^{3} \left(\frac{t_i}{\sqrt{h_i}}\right)^2 \leq \frac{3}{2} R.$$

Problem 5

(a) Prove that for each positive integer n, the number of ordered pairs (x, y) of integers satisfying
$$x^2 - xy + y^2 = n$$
is finite and divisible by 6.

(b) Find all ordered pairs (x, y) of integers satisfying
$$x^2 - xy + y^2 = 727.$$

Problem 6 Given a triangle ABC, the internal and external bisectors of angle A intersect line BC at points D and E, respectively. Let F be the point (different from A) where line AC intersects the circle ω with diameter $\overline{DE}$. Finally, draw the tangent at A to the circumcircle of triangle ABF, and let it hit ω at A and G. Prove that $AF = AG$.

Problem 7 Show that it is possible to cut any triangular prism of infinite length with a plane such that the resulting intersection is an equilateral triangle.

Problem 8 Given a square $ABCD$, the points M, N, K, L are chosen on sides AB, BC, CD, DA, respectively, such that lines MN and LK are parallel and such that the distance between lines MN and LK equals AB. Show that the circumcircles of triangles ALM and NCK intersect each other, while those of triangles LDK and MBN do not.

Problem 9 Let $f : \mathbb{R} \to \mathbb{R}$ be a function such that
$$\left| f(x+y) - f(x) - f(y) \right| \leq 1$$
for all $x, y \in \mathbb{R}$. Show that there exists a function $g : \mathbb{R} \to \mathbb{R}$ with $|f(x) - g(x)| \leq 1$ for all $x \in \mathbb{R}$, and with $g(x+y) = g(x) + g(y)$ for all $x, y \in \mathbb{R}$.

20 United Kingdom

Problem 1 Two intersecting circles C_1 and C_2 have a common tangent which touches C_1 at P and C_2 at Q. The two circles intersect at M and N. Prove that the triangles MNP and MNQ have equal areas.

Problem 2 Given that x, y, z are positive real numbers satisfying $xyz = 32$, find the minimum value of
$$x^2 + 4xy + 4y^2 + 2z^2.$$

Problem 3

(a) Find a set A of ten positive integers such that no six distinct elements of A have a sum which is divisible by 6.

(b) Is it possible to find such a set if "ten" is replaced by "eleven"?

21 United States of America

Problem 1 Call a real-valued function f *very convex* if
$$\frac{f(x)+f(y)}{2} \geq f\left(\frac{x+y}{2}\right) + |x-y|$$
holds for all real numbers x and y. Prove that no very convex function exists.

Problem 2 Let S be the set of all triangles ABC for which
$$5\left(\frac{1}{AP}+\frac{1}{BQ}+\frac{1}{CR}\right) - \frac{3}{\min\{AP,BQ,CR\}} = \frac{6}{r},$$
where r is the inradius and P,Q,R are the points of tangency of the incircle with sides AB, BC, CA, respectively. Prove that all triangles in S are isosceles and similar to one another.

Problem 3 A game of solitaire is played with R red cards, W white cards, and B blue cards. A player plays all the cards one at a time. With each play he accumulates a penalty. If he plays a blue card, then he is charged a penalty which is the number of white cards still in his hand. If he plays a white card, then he is charged a penalty which is twice the number of red cards still in his hand. If he plays a red card, then he is charged a penalty which is three times the number of blue cards still in his hand. Find, as a function of R, W, and B, the minimal total penalty a player can amass and all the ways in which this minimum can be achieved.

Problem 4 Find the smallest positive integer n such that if n unit squares of a 1000×1000 unit-square board are colored, then there will exist three colored unit squares whose centers form a right triangle with legs parallel to the edges of the board.

Problem 5 Let $A_1A_2A_3$ be a triangle and let ω_1 be a circle in its plane passing through A_1 and A_2. Suppose there exist circles $\omega_2, \omega_3, \ldots, \omega_7$ such that for $k = 2, 3, \ldots, 7$, ω_k is externally tangent to ω_{k-1} and passes through A_k and A_{k+1}, where $A_{n+3} = A_n$ for all $n \geq 1$. Prove that $\omega_7 = \omega_1$.

Problem 6 Let $a_1, b_1, a_2, b_2, \ldots, a_n, b_n$ be nonnegative real numbers. Prove that
$$\sum_{i,j=1}^{n} \min\{a_ia_j, b_ib_j\} \leq \sum_{i,j=1}^{n} \min\{a_ib_j, a_jb_i\}.$$

22 Vietnam

Problem 1 Two circles ω_1 and ω_2 are given in the plane, with centers O_1 and O_2, respectively. Let M_1' and M_2' be two points on ω_1 and ω_2, respectively, such that the lines $O_1 M_1'$ and $O_2 M_2'$ intersect. Let M_1 and M_2 be points on ω_1 and ω_2, respectively, such that when measured clockwise the angles $\angle M_1' O M_1$ and $\angle M_2' O M_2$ are equal.

(a) Determine the locus of the midpoint of $\overline{M_1 M_2}$.

(b) Let P be the point of intersection of lines $O_1 M_1$ and $O_2 M_2$. The circumcircle of triangle $M_1 P M_2$ intersects the circumcircle of triangle $O_1 P O_2$ at P and another point Q. Prove that Q is fixed, independent of the locations of M_1 and M_2.

Problem 2 Suppose that all circumcircles of the four faces of a tetrahedron have congruent radii. Show that any two opposite edges of the tetrahedron are congruent.

Problem 3 Two circles C_1 and C_2 intersect at two points P and Q. The common tangent of C_1 and C_2 closer to P than to Q touches C_1 and C_2 at A and B, respectively. The tangent to C_1 at P intersects C_2 at E (distinct from P) and the tangent to C_2 at P intersects C_1 at F (distinct from P). Let H and K be two points on the rays AF and BE, respectively, such that $AH = AP$, $BK = BP$. Prove that the five points A, H, Q, K, B lie on the same circle.

Problem 4 Let a, b, c be pairwise coprime positive integers. An integer $n \geq 1$ is said to be *stubborn* if it cannot be written in the form
$$n = bcx + cay + abz$$
for any positive integers x, y, z. Determine, as a function of a, b, and c, the number of stubborn integers.

Problem 5 Let $\mathbb{R}^+$ denote the set of positive real numbers, and let $a, r > 1$ be real numbers. Suppose that $f : \mathbb{R}^+ \to \mathbb{R}$ is a function such that $(f(x))^2 \leq a x^r f\left(\frac{x}{a}\right)$ for all $x > 0$.

(a) If $f(x) < 2^{2000}$ for all $x < \frac{1}{2^{2000}}$, prove that $f(x) \leq x^r a^{1-r}$ for all $x > 0$.

(b) Construct such a function $f : \mathbb{R}^+ \to \mathbb{R}$ (not satisfying the condition given in (a)) such that $f(x) > x^r a^{1-r}$ for all $x > 0$.

4

2000 Regional Contests: Problems

1 Asian Pacific Mathematical Olympiad

Problem 1 Compute the sum

$$S = \sum_{i=0}^{101} \frac{x_i^3}{1 - 3x_i + 3x_i^2}$$

where $x_i = \frac{i}{101}$ for $i = 0, 1, \ldots, 101$.

Problem 2 We are given an arrangement of nine circular slots along three sides of a triangle: one slot at each corner, and two more along each side. Each of the numbers $1, 2, \ldots, 9$ is to be written into exactly one of these circles, so that

(i) the sums of the four numbers on each side of the triangle are equal;

(ii) the sums of the squares of the four numbers on each side of the triangle are equal.

Find all ways in which this can be done.

Problem 3 Let ABC be a triangle with median $\overline{AM}$ and angle bisector $\overline{AN}$. Draw the perpendicular to line NA through N, hitting lines MA and BA at Q and P, respectively. Also let O be the point where the perpendicular to line BA through P meets line AN. Prove that $QO \perp BC$.

Problem 4 Let n, k be positive integers with $n > k$. Prove that

$$\frac{1}{n+1} \cdot \frac{n^n}{k^k(n-k)^{n-k}} < \frac{n!}{k!(n-k)!} < \frac{n^n}{k^k(n-k)^{n-k}}.$$

Problem 5 Given a permutation $(a_0, a_1, \ldots, a_n)$ of the sequence $0, 1, \ldots, n$, a transposition of a_i with a_j is called *legal* if $a_i = 0$, $i > 0$, and $a_{i-1} + 1 = a_j$. The permutation $(a_0, a_1, \ldots, a_n)$ is called *regular* if after finitely many legal transpositions it becomes $(1, 2, \ldots, n, 0)$. For which numbers n is the permutation $(1, n, n-1, \ldots, 3, 2, 0)$ regular?

2 Austrian-Polish Mathematics Competition

Problem 1 Find all positive integers N whose only prime divisors are 2 and 5, such that the number $N + 25$ is a perfect square.

Problem 2 For which integers $n \geq 5$ is it possible to color the vertices of a regular n-gon using at most 6 colors such that any 5 consecutive vertices have different colors?

Problem 3 Let the *3-cross* be the solid made up of one central unit cube with six other unit cubes attached to its faces, such as the solid made of the seven unit cubes centered at $(0, 0, 0)$, $(\pm 1, 0, 0)$, $(0, \pm 1, 0)$, and $(0, 0, \pm 1)$. Prove or disprove that the space can be tiled with 3-crosses in such a way that no two of them share any interior points.

Problem 4 In the plane the triangle $A_0 B_0 C_0$ is given. Consider all triangles ABC satisfying the following conditions: (i) lines AB, BC, and CA pass through points C_0, A_0, and B_0, respectively; (ii) $\angle ABC = \angle A_0 B_0 C_0$, $\angle BCA = \angle B_0 C_0 A_0$, and $\angle CAB = \angle C_0 A_0 B_0$. Find the locus of the circumcenter of all such triangles ABC.

Problem 5 We are given a set of 27 distinct points in the plane, no three collinear. Four points from this set are vertices of a unit square; the other 23 points lie inside this square. Prove that there exist three distinct points X, Y, Z in this set such that $[XYZ] \leq \frac{1}{48}$.

Problem 6 For all real numbers $a, b, c \geq 0$ such that $a + b + c = 1$, prove that

$$2 \leq (1 - a^2)^2 + (1 - b^2)^2 + (1 - c^2)^2 \leq (1 + a)(1 + b)(1 + c)$$

and determine when equality occurs for each of the two inequalities.

3 Balkan Mathematical Olympiad

Problem 1 Let E be a point inside nonisosceles acute triangle ABC lying on median $\overline{AD}$, and drop perpendicular $\overline{EF}$ to line BC. Let M be an arbitrary point on segment EF, and let N and P be the orthogonal projections of M onto lines AC and AB, respectively. Prove that the angle bisectors of $\angle PMN$ and $\angle PEN$ are parallel.

Problem 2 Find the maximum number of $1 \times 10\sqrt{2}$ rectangles one can remove from a 50×90 rectangle by using cuts parallel to the edges of the original rectangle.

Problem 3 Call a positive integer r a *perfect power* if it is of the form $r = t^s$ for some integers s, t greater than 1. Show that for any positive integer n, there exists a set S of n distinct positive integers with the following property: given any nonempty subset T of S, the arithmetic mean of the elements in T is a perfect power.

4 Mediterranean Mathematical Competition

Problem 1 We are given n different positive numbers $a_1, a_2, \ldots, a_n$ and the set $\{\sigma_1, \sigma_2, \ldots, \sigma_n\}$, where each $\sigma_i \in \{-1, 1\}$. Prove that there exist a permutation $(b_1, b_2, \ldots, b_n)$ of $(a_1, a_2, \ldots, a_n)$ and a set $\{\beta_1, \beta_2, \ldots, \beta_n\}$ where each $\beta_i \in \{-1, 1\}$, such that the sign of $\sum_{j=1}^{i} \beta_j b_j$ equals the sign of σ_i for all $1 \leq i \leq n$.

Problem 2 In the convex quadrilateral $ABCD$, $AC = BD$. Outwards along its sides are constructed equilateral triangles WAB, XBC, YCD, ZDA with centroids S_1, S_2, S_3, S_4, respectively. Prove that $S_1S_3 \perp S_2S_4$ if and only if $AC = BD$.

Problem 3 P, Q, R, S are the midpoints of sides BC, CD, DA, AB, respectively, of convex quadrilateral $ABCD$. Prove that

$$4(AP^2 + BQ^2 + CR^2 + DS^2) \leq 5(AB^2 + BC^2 + CD^2 + DA^2).$$

5 St. Petersburg City Mathematical Olympiad (Russia)

Problem 1 Let $\overline{AA_1}, \overline{BB_1}, \overline{CC_1}$ be the altitudes of an acute triangle ABC. The points A_2 and C_2 on line A_1C_1 are such that line CC_1 bisects $\overline{A_2B_1}$ and line AA_1 bisects $\overline{C_2B_1}$. Lines A_2B_1 and AA_1 meet at K, and lines C_2B_1 and CC_1 meet at L. Prove that lines KL and AC are parallel.

Problem 2 One hundred points are chosen in the coordinate plane. Show that at most $2025 = 45^2$ rectangles with vertices among these points have sides parallel to the axes.

Problem 3

(a) Find all pairs of distinct positive integers a, b such that $(b^2 + a) \mid (a^2 + b)$ and $b^2 + a$ is a power of a prime.

(b) Let a and b be distinct positive integers greater than 1 such that $(b^2 + a - 1) \mid (a^2 + b - 1)$. Prove that $b^2 + a - 1$ has at least two distinct prime factors.

Problem 4 In a country of 2000 airports, there are initially no airline flights. Two airlines take turns introducing new roundtrip nonstop flights. (Between any two cities, only one nonstop flight can be introduced.) The transport authority would like to achieve the goal that if any airport is shut down, one can still travel between any two other airports, possibly with transfers. The airline that causes the goal to be achieved loses. Which airline wins with perfect play?

Problem 5 We are given several monic quadratic polynomials, all with the same discriminant. The sum of any two of the polynomials has distinct real roots. Show that the sum of all of the polynomials also has distinct real roots.

Problem 6 On an infinite checkerboard are placed 111 nonoverlapping *corners*, L-shaped figures made of 3 unit squares. The collection has the following property: for any corner, the 2×2 square containing it is entirely covered by the corners. Prove that one can remove between 1 and 110 of the corners so that the property will be preserved.

Problem 7 We are given distinct positive integers $a_1, a_2, \ldots, a_{20}$. The set of pairwise sums $\{a_i + a_j \mid 1 \leq i \leq j \leq 20\}$ contains 201 elements. What is the smallest possible number of elements in the set $\{|a_i - a_j| \mid 1 \le i < j \le 20\}$, the set of positive differences between the integers?

Problem 8 Let $ABCD$ be an isosceles trapezoid with bases AD and BC. An arbitrary circle tangent to lines AB and AC intersects $\overline{BC}$ at M and N. Let X and Y be the intersections closer to D of the incircle of triangle BCD with $\overline{DM}$ and $\overline{DN}$, respectively. Show that line XY is parallel to line AD.

Problem 9 In each square of a chessboard is written a positive real number such that the sum of the numbers in each row is 1. It is known that for any eight squares, no two in the same row or column, the product of the numbers in these squares is no greater than the product of the numbers on the main diagonal. Prove that the sum of the numbers on the main diagonal is at least 1.

Problem 10 Is it possible to draw finitely many segments in three-dimensional space such that any two segments either share an endpoint or do not intersect, any endpoint of a segment is the endpoint of exactly two other segments, and any closed polygon made from these segments has at least 30 sides?

Problem 11 What is the smallest number of weighings on a balance scale needed to identify the individual weights of a set of objects known to weigh $1, 3, 3^2, \ldots, 3^{26}$ in some order? (The balance scale reports the difference between the weights of the objects in two pans.)

Problem 12 The line ℓ is tangent to the circumcircle of acute triangle ABC at B. Let K be the projection of the orthocenter of ABC onto ℓ, and let L be the midpoint of side AC. Show that triangle BKL is isosceles.

Problem 13 Two balls of negligible size move within a vertical 1×1 square at the same constant speed. Each travels in a straight path except when it hits a wall, in which case it reflects off the wall so that its angle of incidence equals its angle of reflection. Show that a spider, moving at the same speed as the balls, can descend straight down on a string from the top edge of the square to the bottom so that while the spider is within in the square, neither the spider nor its string is touching one of the balls.

Problem 14 Let $n \geq 3$ be an integer. Prove that for positive numbers $x_1 \leq x_2 \leq \cdots \leq x_n$,
$$\frac{x_n x_1}{x_2} + \frac{x_1 x_2}{x_3} + \cdots + \frac{x_{n-1} x_n}{x_1} \geq x_1 + x_2 + \cdots + x_n.$$

Problem 15 In the plane is given a convex n-gon $\mathcal{P}$ with area less than 1. For each point X in the plane, let $F(X)$ denote the area of the union of all segments joining X to points of $\mathcal{P}$. Show that the set of points X such that $F(X) = 1$ is a convex polygon with at most $2n$ sides.

Problem 16 What is the smallest number of unit segments that can be erased from the interior of a 2000×3000 rectangular grid so that no smaller rectangle remains intact?

Problem 17 Let $\overline{AA_1}$ and $\overline{CC_1}$ be altitudes of acute triangle ABC. The line through the incenters of triangles AA_1C and AC_1C meets lines AB and BC at X and Y, respectively. Prove that $BX = BY$.

Problem 18 Does there exist a 30-digit number such that the number obtained by taking any five of its consecutive digits is divisible by 13?

Problem 19 Let $ABCD$ be a convex quadrilateral, and M and N the midpoints of $\overline{AD}$ and $\overline{BC}$, respectively. Suppose A, B, M, N lie on a circle such that $\overline{AB}$ is tangent to the circumcircle of triangle BMC. Prove that $\overline{AB}$ is also tangent to the circumcircle of triangle AND.

Problem 20 Let $n \geq 3$ be a positive integer. For all positive numbers $a_1, a_2, \ldots, a_n$, show that
$$\frac{a_1 + a_2}{2} \frac{a_2 + a_3}{2} \cdots \frac{a_n + a_1}{2} \leq \frac{a_1 + a_2 + a_3}{2\sqrt{2}} \cdots \frac{a_n + a_1 + a_2}{2\sqrt{2}}.$$

Problem 21 A connected graph is said to be 2-connected if after removing any single vertex, the graph remains connected. Prove that given any 2-connected graph in which the degree of every vertex is greater than 2, it is possible to remove a vertex (and all edges adjacent to that vertex) so that the remaining graph is still 2-connected.

Problem 22 The perpendicular bisectors of sides AB and BC of nonequilateral triangle ABC meet lines BC and AB at A_1 and C_1, respectively. Let the bisectors of angles A_1AC and C_1CA meet at B', and define C' and A' analogously. Prove that the points A', B', C' lie on a line passing through the circumcenter of triangle ABC.

Problem 23 Is it possible to select 102 17-element subsets of a 102-element set, such that the intersection of any two of the subsets has at most 3 elements?

Glossary

Abel summation For an integer $n > 0$ and reals $a_1, a_2, \ldots, a_n$ and $b_1, b_2, \ldots, b_n$,

$$\sum_{i=1}^{n} a_i b_i = b_n \sum_{i=1}^{n} a_i + \sum_{i=1}^{n-1} \left((b_i - b_{i+1}) \sum_{j=1}^{i} a_j \right).$$

Arithmetic mean-geometric mean (AM-GM) inequality If $a_1, a_2, \ldots, a_n$ are n nonnegative numbers, then their **arithmetic mean** is defined as $\frac{1}{n} \sum_{i=1}^{n} a_i$ and their **geometric mean** is defined as $(a_1 a_2 \cdots a_n)^{\frac{1}{n}}$. The arithmetic mean-geometric mean inequality states that

$$\frac{1}{n} \sum_{i=1}^{n} a_i \geq (a_1 a_2 \cdots a_n)^{\frac{1}{n}}$$

with equality if and only if $a_1 = a_2 = \cdots = a_n$. The inequality is a special case of the **power mean inequality**.

Arithmetic mean-harmonic mean (AM-HM) inequality If $a_1, a_2, \ldots, a_n$ are n positive numbers, then their **arithmetic mean** is defined as $\frac{1}{n} \sum_{i=1}^{n} a_i$ and their **harmonic mean** is defined as $\frac{1}{\frac{1}{n} \sum_{i=1}^{n} \frac{1}{a_i}}$. The arithmetic mean-geometric mean inequality states that

$$\frac{1}{n} \sum_{i=1}^{n} a_i \geq \frac{1}{\frac{1}{n} \sum_{i=1}^{n} \frac{1}{a_i}}$$

with equality if and only if $a_1 = a_2 = \cdots = a_n$. Like the arithmetic mean-geometric mean inequality, this inequality is a special case of the **power mean inequality**.

Bernoulli's inequality For $x > -1$ and $a > 1$,
$$(1+x)^a \geq 1 + ax,$$
with equality when $x = 0$.

Binomial coefficient
$$\binom{n}{k} = \frac{n!}{k!(n-k)!},$$
the coefficient of x^k in the expansion of $(x+1)^n$.

Binomial theorem
$$(x+y)^n = \sum_{k=0}^{n} \binom{n}{k} x^{n-k} y^k.$$

Brocard angle See **Brocard points**.

Brocard points Given a triangle ABC, there exists a unique point P such that $\angle ABP = \angle BCP = \angle CAP$ and a unique point Q such that $\angle BAQ = \angle CBQ = \angle ACQ$. The points P and Q are the Brocard points of triangle ABC. Moreover, $\angle ABP$ and $\angle BAQ$ are equal; their value ϕ is the Brocard angle of triangle ABC.

Cauchy–Schwarz inequality For any real numbers $a_1, a_2, \ldots, a_n$, and $b_1, b_2, \ldots, b_n$
$$\sum_{i=1}^{n} a_i^2 \cdot \sum_{i=1}^{n} b_i^2 \geq \left(\sum_{i=1}^{n} a_i b_i \right)^2,$$
with equality if and only if a_i and b_i are proportional, $i = 1, 2, \ldots, n$.

Centrally symmetric A geometric figure is centrally symmetric (centrosymmetric) about a point O if, whenever P is in the figure and O is the midpoint of a segment PQ, then Q is also in the figure.

Centroid of a triangle Point of intersection of the medians.

Centroid of a tetrahedron Point of the intersection of the segments connecting the midpoints of the opposite edges, which is the same as the point of intersection of the segments connecting each vertex with the centroid of the opposite face.

Glossary

Ceva's theorem and its trigonometric form Let AD, BE, CF be three cevians of triangle ABC. The following are equivalent:

(i) AD, BE, CF are concurrent;

(ii) $\dfrac{AF}{FB} \cdot \dfrac{BD}{DC} \cdot \dfrac{CE}{EA} = 1;$

(iii) $\dfrac{\sin \angle ABE}{\sin \angle EBC} \cdot \dfrac{\sin \angle BCF}{\sin \angle FCA} \cdot \dfrac{\sin \angle CAD}{\sin \angle DAB} = 1.$

Cevian A cevian of a triangle is any segment joining a vertex to a point on the opposite side.

Chinese remainder theorem Let k be a positive integer. Given integers $a_1, a_2, \ldots, a_k$ and pairwise relatively prime positive integers $n_1, n_2, \ldots, n_k$, there exists a unique integer a such that $0 \leq a < \prod_{i=1}^{k} n_i$ and $a \equiv a_i \pmod{n_i}$ for $i = 1, 2, \ldots, k$.

Circumcenter Center of the circumscribed circle or sphere.

Circumcircle Circumscribed circle.

Congruence For integers a, b, and n with $n \geq 1$, $a \equiv b \pmod{n}$ (or "a is congruent to b modulo n") means that $a - b$ is divisible by n.

Concave up (down) function A function $f(x)$ is concave up (down) on $[a, b]$ if $f(x)$ lies under the line connecting $(a_1, f(a_1))$ and $(b_1, f(b_1))$ for all
$$a \leq a_1 < x < b_1 \leq b.$$

Cyclic polygon Polygon that can be inscribed in a circle.

de Moivre's formula For any angle α and for any integer n,
$$(\cos \alpha + i \sin \alpha)^n = \cos n\alpha + i \sin n\alpha.$$

Desargues' theorem Two triangles have corresponding vertices joined by lines which are concurrent or parallel if and only if the intersections of corresponding sides are collinear.

Euler's formula Let O and I be the circumcenter and incenter, respectively, of a triangle with circumradius R and inradius r. Then
$$OI^2 = R^2 - 2rR.$$

Euler line The orthocenter, centroid and circumcenter of any triangle are collinear. The centroid divides the distance from the orthocenter to the circumcenter in the ratio of 2 : 1. The line on which these three points lie is called the Euler line of the triangle.

Euler's theorem Given relatively prime integers a and m with $m \geq 1$, $a^{\phi(m)} \equiv a \pmod{m}$, where $\phi(m)$ is the number of positive integers less than or equal to m and relatively prime to m.

Excircles or escribed circles Given a triangle ABC, there are four circles tangent to the lines AB, BC, CA. One is the inscribed circle, which lies in the interior of the triangle. One lies on the opposite side of line BC from A, and is called the excircle (escribed circle) opposite A, and similarly for the other two sides. The excenter opposite A is the center of the excircle opposite A; it lies on the internal angle bisector of A and the external angle bisectors of B and C.

Excenters See **excircles**.

Exradii The radii of the three excircles of a triangle.

Fermat number A number of the form 2^{2^n} for some positive integer n.

Fermat's little theorem If p is prime, then $a^p \equiv a \pmod{p}$.

Feuerbach circle The feet of the three altitudes of any triangle, the midpoints of the three sides, and the midpoints of segments from the three vertices to the orthocenter, all lie on the same circle, the Feuerbach circle or the **nine-point circle** of the triangle. Let R be the circumradius of the triangle. The nine-point circle of the triangle has radius $R/2$ and is centered at the midpoint of the segment joining the orthocenter and the circumcenter of the triangle.

Feuerbach's theorem The nine-point circle of a triangle is tangent to the incircle and to the three excircles of the triangle.

Fibonacci sequence The sequence $F_0, F_1, \ldots$ defined recursively by $F_0 = 0$, $F_1 = 1$, and $F_{n+2} = F_{n+1} + F_n$ for all $n \geq 0$.

Generating function If $a_0, a_1, a_2, \ldots$ is a sequence of numbers, then the generating function for the sequence is the infinite series

$$a_0 + a_1 x + a_2 x^2 + \cdots.$$

Glossary

If f is a function such that

$$f(x) = a_0 + a_1 x + a_2 x^2 + \cdots,$$

then we also refer to f as the generating function for the sequence.

Harmonic conjugates   Let A, C, B, D be four points on a line in that order. If the points C and D divide AB internally and externally in the same ratio, (i.e., $AC : CB = AD : DB$), then the points C and D are said to be harmonic conjugates of each other with respect to the points A and B, and AB is said to be **harmonically divided** by the points C and D. If C and D are harmonic with respect to A and B, then A and B are harmonic with respect to C and D.

Harmonic range   The four points A, B, C, D are referred to as a harmonic range, denoted by $(ABCD)$, if C and D are harmonic conjugates with respect to A and B.

Helly's theorem   If $n > d$ and $C_1, \ldots, C_n$ are convex subsets of $\mathbb{R}^d$, each $d + 1$ of which have nonempty intersection, then there is a point in common to all the sets.

Heron's formula   The area of a triangle with sides a, b, c is equal to

$$\sqrt{s(s-a)(s-b)(s-c)},$$

where $s = (a + b + c)/2$.

Homothety   A homothety (central similarity) is a transformation that fixes one point O (its center) and maps each point P to a point P' for which O, P, P' are collinear and the ratio $OP : OP' = k$ is constant (k can be either positive or negative), where k is called the **magnitude** of the homothety.

Homothetic triangles   Two triangles ABC and DEF are homothetic if they have parallel sides. Suppose that $AB \parallel DE$, $BC \parallel EF$, and $CA \parallel FD$. Then lines AD, BE, and CF concur at a point X, as given by a special case of Desargues' theorem. Furthermore, some homothety centered at X maps triangle ABC onto triangle DEF.

Incenter   Center of inscribed circle.

Incircle   Inscribed circle.

Inversion of center O and ratio r Given a point O in the plane and a real number $r > 0$, the inversion through O with radius r maps every point $P \neq O$ to the point P' on the ray $\overrightarrow{OP}$ such that $OP \cdot OP' = r^2$. We also refer to this map as inversion through ω, the circle with center O and radius r. Key properties of inversion are:

1. Lines through O invert to themselves (though the individual points on the line are not all fixed).
2. Lines not through O invert to circles through O and vice versa.
3. Circles not through O invert to other circles not through O.
4. A circle other than ω inverts to itself (as a whole, not point-by-point) if and only if it is orthogonal to ω, that is, it intersects ω and the tangents to the circle and to ω at either intersection point are perpendicular.

Jensen's inequality If f is concave up on an interval $[a, b]$ and $\lambda_1, \lambda_2, \ldots, \lambda_n$ are nonnegative numbers with sum equal to 1, then

$$\lambda_1 f(x_1) + \lambda_2 f(x_2) + \cdots + \lambda_n f(x_n) \geq f(\lambda_1 x_1 + \lambda_2 x_2 + \cdots + \lambda_n x_n)$$

for any $x_1, x_2, \ldots, x_n$ in the interval $[a, b]$. If the function is concave down, the inequality is reversed.

Kummer's Theorem Given nonnegative integers a and b and a prime p, $p^t \mid \binom{a+b}{a}$ if and only if t is less than or equal to the number of carries in the addition $a + b$ in base p.

Lattice point In the Cartesian plane, the lattice points are the points (x, y) for which x and y are both integers.

Law of sines In a triangle ABC with circumradius equal to R one has

$$\frac{\sin A}{BC} = \frac{\sin B}{AC} = \frac{\sin C}{AB} = 2R.$$

Lucas's theorem Let p be a prime; let a and b be two positive integers such that

$$a = a_k p^k + a_{k-1} p^{k-1} + \cdots + a_1 p + a_0,$$
$$b = b_k p^k + b_{k-1} p^{k-1} + \cdots + b_1 p + b_0,$$

where $0 \leq a_i, b_i < p$ are integers for $i = 0, 1, \ldots, k$. Then

$$\binom{a}{b} \equiv \binom{a_k}{b_k}\binom{a_{k-1}}{b_{k-1}} \cdots \binom{a_1}{b_1}\binom{a_0}{b_0} \pmod{p}.$$

Glossary

Matrix A matrix is a rectangular array of objects. A matrix A with m rows and n columns is an $m \times n$ matrix. The object in the i-th row and j-th column of matrix A is denoted $a_{i,j}$. If a matrix has the same number of rows as it has columns, then the matrix is called a square matrix. In a square $n \times n$ matrix A, the **main diagonal** consists of the elements $a_{1,1}, a_{2,2}, \ldots, a_{n,n}$.

Menelaus' theorem Given a triangle ABC, let F, G, H be points on lines BC, CA, AB, respectively. Then F, G, H are collinear if and only if, using directed lengths,

$$\frac{AH}{HB} \cdot \frac{BF}{FC} \cdot \frac{CG}{GA} = -1.$$

Minkowski's inequality Given a positive integer n, a real number $r \geq 1$, and positive reals $a_1, a_2, \ldots, a_n$ and $b_1, b_2, \ldots, b_n$, we have

$$\left(\sum_{i=1}^{n} (a_n + b_n)^r \right)^{1/r} \leq \left(\sum_{i=1}^{n} a_i^r \right)^{1/r} + \left(\sum_{i=1}^{n} b_i^r \right)^{1/r}.$$

Nine point circle See **Feuerbach circle**.

Orthocenter of a triangle Point of intersection of the altitudes.

Periodic function $f(x)$ is periodic with period $T > 0$ if

$$f(x + T) = f(x)$$

for all x.

Permutation Let S be a set. A permutation of S is a one-to-one function $\pi : S \to S$ that maps S onto S. If $S = \{x_1, x_2, \ldots, x_n\}$ is a finite set, then we may denote a permutation π of S by $\{y_1, y_2, \ldots, y_n\}$, where $y_k = \pi(x_k)$.

Pick's theorem Given a non self-intersecting polygon $\mathcal{P}$ in the coordinate plane whose vertices are at lattice points, let B denote the number of lattice points on its boundary and let I denote the number of lattice points in its interior. The area of $\mathcal{P}$ is given by the formula $I + \frac{1}{2}B - 1$.

Pigeonhole principle If n objects are distributed among $k < n$ boxes, some box contains at least two objects.

Pole-polar transformation Let C be a circle with center O and radius R. The pole-polar transformation with respect to C maps points different from O to lines, and lines that do not pass through O to points. If $P \neq O$ is a point then the **polar** of P is the line p' that is perpendicular to ray $\overrightarrow{OP}$ and satisfies
$$d(O,P)d(O,p') = R^2,$$
where $d(A,B)$ denote the distance between the objects A and B. If q is a line that does not pass through O, then the **pole** of q is the point Q' that has polar q. The pole-polar transformation with respect to the circle C is also called **reciprocation** in the circle C.

Polynomial in x of degree n Function of the form $f(x) = \sum_{k=0}^{n} a_k x^k$.

Power of a point theorem Given a fixed point P and a fixed circle ω, draw a line through P which intersects the circle at X and Y. The power of the point P with respect to ω is defined to be $PX \cdot PY$. The power of a point theorem states that this quantity is a constant; i.e., does not depend on which line was drawn. Note that it did not matter whether P was in, on or outside ω.

Power mean inequality Let $a_1, a_2, \ldots, a_n$ be any positive numbers for which $a_1 + a_2 + \cdots + a_n = 1$. For positive numbers $x_1, x_2, \ldots, x_n$ we define
$$M_{-\infty} = \min\{x_1, x_2, \ldots, x_k\},$$
$$M_{\infty} = \max\{x_1, x_2, \ldots, x_k\},$$
$$M_0 = x_1^{a_1} x_2^{a_2} \cdots x_n^{a_n},$$
$$M_t = (a_1 x_1^t + a_2 x_2^t + \cdots + a_k x_k^t)^{1/t},$$
where t is a non-zero real number. Then
$$M_{-\infty} \leq M_s \leq M_t \leq M_{\infty}$$
for $s \leq t$.

Ptolemy's theorem In a convex cyclic quadrilateral $ABCD$,
$$AC \cdot BD = AB \cdot CD + AD \cdot BC.$$

Pythagorean triple A set (a,b,c) of three numbers is a Pythagorean triple if there is a right-angled triangle with sides of lengths a, b, c. If c

Glossary

is the length of the hypotenuse, this is equivalent to the assertion:

$$c^2 = a^2 + b^2.$$

If a, b, c are integers, the triple is **primitive** if the greatest common divisor of a, b, c is 1. All primitive Pythagorean triples (a, b, c) are given parametrically by

$$a = 2uv, \quad b = u^2 - v^2, \quad c = u^2 + v^2,$$

where $u > v$ are relatively prime integers, not both odd.

Radical axis Let ω_1 and ω_2 be two non-concentric circles. The locus of all points of equal power with respect to these circles is called the radical axis of ω_1 and ω_2. Let $\omega_1, \omega_2, \omega_3$ be three circles whose centers are not collinear. There is exactly one point whose powers with respect to the three circles are all equal. This point is called the **radical center** of $\omega_1, \omega_2, \omega_3$.

Root of an equation Solution to the equation.

Root of unity Solution to the equation $z^n - 1 = 0$.

Root Mean Square-Arithmetic Mean Inequality For positive numbers $x_1, x_2, \ldots, x_n$,

$$\sqrt{\frac{x_1^2 + x_2^2 + \cdots + x_k^2}{n}} \geq \frac{x_1 + x_2 + \cdots + x_k}{n}.$$

Simson line For any point P on the circumcircle of $\triangle ABC$, the feet of the perpendiculars from P to the sides of $\triangle ABC$ all lie on a line called the Simson line of P with respect to $\triangle ABC$.

Stewart's theorem In a triangle ABC with cevian $\overline{AD}$, write $a = BC$, $b = CA$, $c = AB$, $m = BD$, $n = DC$, and $d = AD$. Then

$$d^2 a + man = c^2 n + b^2 m.$$

This formula can be used to express the lengths of the altitudes and angle bisectors of a triangle in terms of its side lengths.

Thue-Morse sequence The sequence $t_0, t_1, \ldots$, defined by $t_0 = 0$ and the recursive relations $t_{2k} = t_k$, $t_{2k+1} = 1 - t_{2k}$ for $k \geq 1$. The binary representation of n contains an odd number of 1's if and only if t_n is odd.

Triangular number A number of the form $n(n+1)/2$, where n is some positive integer.

Trigonometric identities

$$\sin^2 x + \cos^2 x = 1,$$
$$1 + \cot^2 x = \csc^2 x,$$
$$\tan^2 x + 1 = \sec^2 x;$$

addition and subtraction formulas:

$$\sin(a \pm b) = \sin a \cos b \pm \cos a \sin b,$$
$$\cos(a \pm b) = \cos a \cos b \mp \sin a \sin b,$$
$$\tan(a \pm b) = \frac{\tan a \pm \tan b}{1 \mp \tan a \tan b};$$

double-angle formulas:

$$\sin 2a = 2 \sin a \cos a$$
$$= \frac{2 \tan a}{1 + \tan^2 a},$$
$$\cos 2a = 2 \cos^2 a - 1 = 1 - 2 \sin^2 a$$
$$= \frac{1 - \tan^2 a}{1 + \tan^2 a},$$
$$\tan 2a = \frac{2 \tan a}{1 - \tan^2 a};$$

triple-angle formulas:

$$\sin 3a = 3 \sin a - 4 \sin^3 a,$$
$$\cos 3a = 4 \cos^3 a - 3 \cos a,$$
$$\tan 3a = \frac{3 \tan a - \tan^3 a}{1 - 3 \tan^2 a};$$

half-angle formulas:

$$\sin^2 \frac{a}{2} = \frac{1 - \cos a}{2},$$
$$\cos^2 \frac{a}{2} = \frac{1 + \cos a}{2};$$

sum-to-product formulas:

$$\sin a + \sin b = 2 \sin \frac{a+b}{2} \cos \frac{a-b}{2},$$

Glossary

$$\cos a + \cos b = 2\cos\frac{a+b}{2}\cos\frac{a-b}{2},$$

$$\tan a + \tan b = \frac{\sin(a+b)}{\cos a \cos b};$$

difference-to-product formulas:

$$\sin a - \sin b = 2\sin\frac{a-b}{2}\cos\frac{a+b}{2},$$

$$\cos a - \cos b = -2\sin\frac{a-b}{2}\sin\frac{a+b}{2},$$

$$\tan a - \tan b = \frac{\sin(a-b)}{\cos a \cos b};$$

product-to-sum formulas:

$$2\sin a \cos b = \sin(a+b) + \sin(a-b),$$

$$2\cos a \cos b = \cos(a+b) + \cos(a-b),$$

$$2\sin a \sin b = -\cos(a+b) + \cos(a-b).$$

Van der Warden's theorem Given positive integers n and k, there exists a positive integer N such that the following property holds: among any N consecutive integers, each colored in one of n colors, there exist k in arithmetic progression which share the same color.

Classification of Problems

Algebra

Belarus	99-10.1, 11.1; 99-S-6
Bulgaria	99-R4-1
Canada	99-1
China	99-2; 00-2
Czech and Slovak	99-6
Hungary	99-8, 11, 17
Iran	99-R2-1
Ireland	99-1, 6
Italy	00-3
Japan	99-3, 4
Korea	00-4
Mongolia	00-6
Poland	99-4
Romania	99-7.1, 7.2, 10.1; 99-S-3, 6, 9; 00-1, 4
Russia	99-R4-8.1, 8.2, 8.4, 10.5, 11.1, 11.8; 99-R5-9.1, 10.2, 10.5; 00-11, 32, 41
Slovenia	99-1
Ukraine	99-1

---

[1] Entries are written in the form "year-round-problem" or "year-problem," with the following abbreviations:
- R   round
- S   selection test
- I   individual round
- T   team round

(Algebra, continued)

United Kingdom	99-1, 4
Vietnam	99-1, 3
Asian Pacific	99-1; 00-1, 2
Austrian-Polish	99-3, 6
Czech-Slovak Match	99-4
Hungary-Israel	99-T-2
Iberoamerican	99-1, 6
St. Petersburg	99-9.2, 10.1, 10.2, 10.5

Combinatorics

Belarus	99-10.3, 11.2; 99-S-10, 12; 00-2, 9
Brazil	99-2, 4
Bulgaria	99-R3-3; 99-R4-4; 00-2, 00-12
Canada	99-4
China	99-3, 6; 00-3, 6
Czech and Slovak	99-4
France	99-4
Hong Kong	99-2
Hungary	99-7; 00-4
Iran	99-R1-6; 99-R3-4; 00-7
Ireland	99-4
Italy	99-4, 5; 00-2
Japan	99-1; 00-1
Korea	99-5; 00-5
Poland	99-2; 00-5
Romania	99-S-11, 12; 00-10
Russia	99-R4-8.7, 8.8, 9.1, 9.4, 9.5, 10.7, 11.3, 11.6; 99-R5-9.4, 9.5, 9.8, 10.1, 10.4, 10.8; 00-6, 12, 21, 24, 29, 31, 37, 39
Taiwan	99-2, 5
Turkey	99-3, 5; 00-3
United States	99-1, 5; 00-3, 4
Asian Pacific	00-5
Austrian-Polish	99-1; 00-2
Czech-Slovak Match	99-3
Iberoamerican	99-3

(Combinatorics, continued)

Olimpiada del Cono Sur	99-3
St. Petersburg	99-9.3, 9.7, 9.8; 00-2, 6, 11, 16

Combinatorial Geometry

Bulgaria	99-R3-6; 00-4, 8
Czech and Slovak	00-4
Hungary	99-18
Iran	99-R2-5; 99-R3-3
Israel	00-4
Japan	00-4
Russia	99-R4-10.3, 11.4; 99-R5-11.6; 00-9, 18, 23, 26, 33
Turkey	99-9
Asian Pacific	99-5
Austrian-Polish	99-9; 00-3, 5
Balkan	00-2

Combinatorial Number Theory

Belarus	99-11.3
Estonia	00-6
Hungary	99-4
Iran	99-R1-2
Ireland	99-5
Russia	99-R4-10.8; 00-14
Slovenia	99-4
United Kingdom	00-3
St. Petersburg	99-11.5, 11.8

Combinatorics and Sets

Belarus	00-7
Canada	00-1
China	00-4
Hungary	00-7
Iran	99-R3-1, 5
Italy	99-10
Romania	00-2
Russia	00-8, 15

(Combinatorics, continued)

Taiwan	00-3
Hungary-Israel	99-I-4
St. Petersburg	00-7, 00-23

Graph Theory

Belarus	00-5
India	00-3
Russia	00-36, 43
St. Petersburg	99-11.7; 00-4, 10, 21

Functional Equations

Belarus	99-S-1, 9; 00-4
Estonia	00-4
Hungary	99-4
India	00-4
Iran	99-R2-3; 00-5, 6
Ireland	99-7
Italy	99-9
Korea	99-2, 4; 00-2
Mongolia	00-3
Poland	00-6
Romania	00-9
Russia	99-R5-1.2; 00-7, 40
Slovenia	99-2
Taiwan	00-4
Turkey	99-7; 00-9
Vietnam	99-6, 10; 00-5
Czech-Slovak Match	99-5

Geometry

Belarus	99-10.4, 10.5, 11.4, 11.7; 99-S-5, 7, 8; 00-1, 3, 11

Classification of Problems

(Geometry, continued)

Brazil	99-1, 5
Bulgaria	99-R3-2, 5; 99-R4-5; 00-1, 3, 5, 7, 9, 11
Canada	99-2; 00-2
China	99-1
Czech and Slovak	99-2, 3, 5; 00-2, 3
Estonia	00-3, 5
France	99-5
Hungary	99-1, 2, 3, 9, 10; 00-3, 6
India	00-1
Iran	99-R1-3, 4; 99-R2-4; 99-R3-2; 00-1, 3, 8, 9
Ireland	99-3, 10
Israel	00-2, 00-3
Italy	99-1, 3, 8; 00-1
Korea	99-1; 00-3
Mongolia	00-2, 4
Poland	99-1, 6; 00-2, 4
Romania	99-7.3, 7.4, 8.3, 8.4, 9.1, 9.3; 99-S-2, 10; 00-5, 8
Russia	99-R4-8.3, 8.6, 9.2, 9.8, 10.2, 10.6; 99-R5-9.3, 9.7, 10.3, 11.3, 11.5, 11.7; 00-2, 5, 13, 17, 20, 27, 30, 35, 38, 42
Slovenia	99-3
Taiwan	99-4; 00-1
Turkey	99-1, 4, 6; 00-2, 6, 7, 8
United Kingdom	99-2, 7; 00-1
United States	99-6; 00-2, 5
Vietnam	99-2, 5, 8, 11; 00-1, 2, 3
Asian Pacific	99-3; 00-3
Austrian-Polish	99-4, 8; 00-4
Balkan	99-1; 00-1
Hungary-Israel	99-I-3; 99-T-1, 3
Iberoamerican	99-2, 5
Mediterranean	00-2, 3
Olimpiada del Cono Sur	99-2
St. Petersburg	99-9.4, 9.6, 10.7; 00-1, 8, 12, 13, 15, 17, 22

Inequalities

Belarus	99-10.2, 10.7, 11.5; 99-S-4; 00-6, 10
Brazil	99-3
Canada	99-5; 00-3
China	99-5
Czech and Slovak	00-1
Estonia	00-1
France	99-3
Hungary	99-5, 12, 14; 00-5
Iran	99-R1-1; 00-10
Ireland	99-8
Korea	99-6; 00-6
Poland	99-5
Romania	99-8.2, 9.2, 9.4, 10.3, 10.4; 99-S-4, 5, 7; 00-3
Russia	99-R4-9.3, 11.5; 99-R5-9.6, 10.7; 00-4, 10, 19
Turkey	99-2, 8
Ukraine	99-2
United Kingdom	00-2
United States	99-4; 00-1, 6
Vietnam	99-4, 7, 9
Asian Pacific	99-2; 00-4
Austrian-Polish	99-2; 00-6
Balkan	99-4
Czech-Slovak Match	99-1
Hungary-Israel	99-I-1, 6
St. Petersburg	99-9.1, 10.6, 11.4; 00-5, 9, 14, 20

Geometric Inequalities

China	00-1
Hungary	99-6, 15
Iran	99-R2-2; 00-4
Japan	99-5; 00-2
Romania	99-10.2; S-8, 13
Turkey	00-4
United States	99-2
Balkan	99-3
Olimpiada del Cono Sur	99-5, 6

Number Theory

Belarus	99-11.6; 99-S-2, 3, 11; 00-8
Bulgaria	99-R3-1, 4; 99-R4-2, 3, 6; 00-6, 10
Canada	99-3
China	99-4; 00-5
Czech and Slovak	99-1
Estonia	00-2
Hungary	99-3, 4, 13, 16; 00-1, 2
India	00-2
Iran	99-R1-5; 00-2
Ireland	99-2, 9
Israel	00-1
Italy	99-2, 6, 7
Japan	99-2; 00-3
Korea	99-3; 00-1
Mongolia	00-1, 5
Poland	99-3; 00-3
Romania	99-8.1; 99-S-1; 00-6, 7
Russia	99-R4-8.5, 9.7; 99-R5-11.1; 00-1, 16, 22, 25, 28
Taiwan	99-1, 3; 00-2
Turkey	00-1, 5
Ukraine	99-3
United States	99-3
Vietnam	99-12; 00-4
Asian Pacific	99-4
Austrian-Polish	99-5, 7; 00-1
Balkan	99-2; 00-3
Czech-Slovak Match	99-6
Hungary-Israel	99-I-2, 5
Iberoamerican	99-4
Olimpiada del Cono Sur	99-1, 4
St. Petersburg	99-9.5, 11.1, 11.3; 00-3, 18